Multi-Carrier Systems & Solutions 2009

Lecture Notes in Electrical Engineering

Volume 41

For other volumes published in this series, go to
http://www.springer.com/series/7818

Simon Plass · Armin Dammann · Stefan Kaiser
Khaled Fazel
Editors

Multi-Carrier Systems & Solutions 2009

Proceedings from the
7th International Workshop on
Multi-Carrier Systems & Solutions,
May 2009, Herrsching, Germany

Editors
Dr. Simon Plass
Xcitec GmbH
Einsteinring 22
85609 Aschheim b. München
Germany

Dr. Armin Dammann
Deutsches Zentrum für
Luft-And Raumfahrt (DLR)
Inst. Kommunikation und Navigation
Oberpfaffenhofen
82234 Weßling
Germany

Dr. Stefan Kaiser
DOCOMO Communications Laboratories
Europe GmbH
Landsbergerstr. 312
80687 München
Germany

Dr. Khaled Fazel
Rohde & Schwarz GmbH & Co. KG
Geschäftsbereich
Funkkommunikationssysteme
Hemminger Str. 41
D-70499 Weilimdorf
Germany

ISSN 1876-1100
ISBN 978-90-481-2529-6
DOI 10.1007/978-90-481-2530-2
Springer Dordrecht Heidelberg London New York

e-ISSN 1876-1119
e-ISBN 978-90-481-2530-2

Library of Congress Control Number: 2009926000

© Springer Science+Business Media B.V. 2009
No part of this work may be reproduced, stored in a retrieval system, or transmitted in any form or by any means, electronic, mechanical, photocopying, microfilming, recording or otherwise, without written permission from the Publisher, with the exception of any material supplied specifically for the purpose of being entered and executed on a computer system, for exclusive use by the purchaser of the work.

Cover design: eStudio Calamar S.L.

Printed on acid-free paper

Springer is part of Springer Science+Business Media (www.springer.com)

Preface

Since the principle of multi-carrier code division multiple access (MC-CDMA) was simultaneously proposed by Khaled Fazel et al. and Nathan Yee et al. at the IEEE International Symposium on Personal, Indoor and Mobile Radio Communications (PIMRC) in the year 1993, multi-carrier spread spectrum has rapidly become one of the most wide spread independent research topics on the field of mobile radio communications. Therefore, the International Workshop on Multi-Carrier Spread Spectrum (MC-SS) was initiated in the year 1997 and renamed to the International Workshop on Multi-Carrier Systems & Solutions in 2009.

The wireless standards 3GPP Long Term Evolution (LTE), WiMAX, IEEE 802.11a/n, and DxB have in common that they apply the spectrally efficient multi-carrier modulation in order to achieve very high rate data transmission. This trend is expected to continue in the future with the development of the next generation of mobile wireless communications under the IMT-Advanced standardization activities. Additional measures like advanced coding, spreading, and MIMO are combined with multi-carrier transmission to further enhance the efficiency of future systems.

The material summarized in this volume was selected for the seventh International Workshop on Multi-Carrier Systems & Solutions (MC-SS 2009) held in Herrsching, Germany from 05–06 May 2009. The workshop was organized by the Institute of Communications and Navigation of the German Aerospace Center (DLR) in Oberpfaffenhofen, Germany.

We would like to thank all those who have contributed to the success of the conference. First, the authors of paper submissions, this year reaching a total of 62 submissions from 25 countries with 36 accepted papers and two demonstration papers. Second, the industry partners for their research and industry exhibitions, and the invited paper authors and session chairs. Third, the members of the Technical Program Committee and all reviewers, whose expertise and work have contributed to configure a very attractive conference program. Finally, we would like to thank all sponsors for their logistic support.

May 2009

Simon Plass
Armin Dammann
Stefan Kaiser
Khaled Fazel

Program Committee

F. Adachi (Japan)
G. Bauch (Germany)
G. P. Fettweis (Germany)
B. Fleury (Denmark)
J.-F. Helard (France)
L. Hanzo (U.K.)
K. D. Kammeyer (Germany)
A. Klein (Germany)
W. A. Krzymien (Canada)
J. Lindner (Germany)
M. Moeneclaey (Belgium)
W. Mohr (Germany)

M. Morelli (Italy)
M. Nakagawa (Japan)
H. Ochiai (Japan)
S. Pfletschinger (Spain)
L. Ping (China)
M. Renfors (Finland)
H. Rohling (Germany)
S. Sand (Germany)
H. Sari (France)
M. Sawahashi (Japan)
M. Sternad (Sweden)
I. Viering (Germany)

Additional Reviewers

Koichi	Adachi	Yves	Lostanlen
Philipp	Albrecht	Tarcisio	Maciel
Gunther	Auer	Markus	Mayrock
Giacomo	Bacci	Christian	Mensing
Leonardo	Baltar	Marco	Moretti
Bernd	Baumgartner	David	Mottier
Maik	Bevermeier	Siva	Muruganathan
Andre	Bourdoux	Srikanth	Pagadarai
John	Cosmas	Christian	Pietsch
Ivan	Cosovic	Bernhard	Raaf
Antonio Alberto	D'Amico	Simone	Redana
Florence	Danilo-Lemoine	Patrick	Robertson
Robert	Fischer	Yukitoshi	Sanada
Matthieu	Gautier	Wolfgang	Sauer-Greff
Ingmar	Groh	Andreas	Saul
Thomas	Hindelang	Malte	Schellmann
Daesik	Hong	Martin	Senst
Koji	Ishibashi	Tommy	Svensson
Koji	Ishii	Poramate	Tarasak
Tommi	Jamsa	Jun	Tong
Katsutoshi	Kusume	Hao	Wang
Fotis	Lazarakis	Matthias	Wetz
Kwang Bok	Lee	Doris	Yacoub
Sungeun	Lee	Du	Yang
Geert	Leus	Rong	Zhang
Alexander	Linduska	Xin	Zhang
Gianluigi	Liva		

Contents

Part I General Issues

Multicarrier Signals: A Natural Enabler for Cognitive Positioning Systems 3
Marco Luise and Francesca Zanier

Linearisation of Transmitter and Receiver Nonlinearities in Optical OFDM Transmission 15
Henning Paul and Karl-Dirk Kammeyer

Multi-user OFDM Based on Braided Convolutional Codes 25
Michael Lentmaier, Marcos B.S. Tavares, Gerhard P. Fettweis, and Kamil Sh. Zigangirov

Interference Aware Subcarrier and Power Allocation in OFDMA-Based Cognitive Radio Networks 35
Sami M. Almalfouh and Gordon L. Stüber

An Efficient ICA Based Approach to Multiuser Detection in MIMO OFDM Systems 47
Mahdi Khosravy, Mohammad Reza Alsharif, and Katsumi Yamashita

Windowing in the Receiver for OFDM Systems in High-Mobility Scenarios 57
Eva Peiker, Werner G. Teich, and Jürgen Lindner

Part II Multiple-Input and Multiple-Output (MIMO)

MIMO–OFDM with Doppler Compensating Antennas in Rapidly Fading Channels 69
Peter Klenner, Lars Reichardt, Karl-Dirk Kammeyer, and Thomas Zwick

Resource Allocation Algorithms for Minimum Rates Scheduling in MIMO–OFDM Systems 79
Johannes Georg Klotz, Frederic Knabe, and Carolin Huppert

Transmit Antenna and Code Selection for Mimo Systems with Linear Receivers 89
Ismael Gutierrez, Joan Lluis Pijoan, Faouzi Bader, and Alain Mourad

Space-Time-Frequency Diversity in the Next Generation of Terrestrial Digital Video Broadcasting 101
Armin Dammann

Part III Channel Estimation & Characterization

Robust 2-D Channel Estimation for Staggered Pilot Grids in Multi-Carrier Systems: LTE Downlink as an Example 113
Muhammad Danish Nisar, Wolfgang Utschick, and Josef Forster

Block-IFDMA – Iterative Channel Estimation Versus Estimation with Interpolation Filters 123
Anja Sohl and Anja Klein

A Study on Channel Estimation Using EM Algorithm for Mobile WiMAX Systems 133
Masahiro Fujii, Yasufumi Kawahara, Yu Watanabe, and Makoto Itami

A Technique for Correcting Residual Frequency Offset in OFDM Systems 143
Young Mo Chung

A Joint Channel and Carrier Frequency Offset Estimation Based on Spread Pilot for Future Broadcasting Systems 153
Oudomsack Pierre Pasquero, Youssef Nasser, Matthieu Crussière, and Jean-François Hélard

Channel Characteristics of Different Floors for Joint Communications and Positioning 163
Wei Wang, Thomas Jost, and Armin Dammann

Part IV Long Term Evolution (LTE)

Uplink Power Control Performance in UTRAN LTE Networks 175
Robert Müllner, Carsten F. Ball, Kolio Ivanov, Johann Lienhart, and Peter Hric

Improving UMTS LTE Performance by UEP in High Order Modulation 185
Helge Lüders, Andreas Minwegen, and Peter Vary

Performance Evaluation of a Low-Complexity LTE Base Station Receiver 195
Xiaoyin Zhao and Andreas Ibing

Part V Peak-to-Average Power Ratio (PAPR)

On DT-CWT Based OFDM: PAPR Analysis 207
Mohamed Hussien Mohamed Nerma, Nidal S. Kamel, and Varun Jeoti Jagadish

SOCP Approach for PAPR Reduction Using Tone Reservation for the Future DVB-T/H Standards 219
Irène Mahafeno, Jean-François Hélard, and Yves Louet

Reduced-PAPR Code Allocation Strategy for MC-CDMA Transmissions 227
Filippo Giannetti, Vincenzo Lottici, Ivan Stupia, and Nunzio Aldo D'Andrea

Part VI Adaptive Transmission

Adaptive Multiuser OFDMA Systems with High Priority Users in the Presence of Imperfect CQI 239
Alexander Kühne and Anja Klein

Adaptive BICM-OFDM Systems 249
Carsten Bockelmann, Dirk Wübben, and Karl-Dirk Kammeyer

Power Allocation with Interference Constraint in Multicarrier Based Cognitive Radio Systems 259
Musbah Shaat and Faouzi Bader

Limited-Feedback Multiuser MIMO-OFDM Downlink with Spatial Multiplexing and Per-Chunk/Per-Antenna User Scheduling 269
Mohsen Eslami and Witold A. Krzymień

Part VII Performance Evaluation

Simple Series Form Formula of BER Performance of M-ary QAM/OFDM Signals Over Nonlinear Fading Channels 281
Yuichiro Goto, Akihiro Yamakita, and Fumiaki Maehara

A Novel Exponential Link Error Prediction Method for OFDM Systems 291
Ivan Stupia, Filippo Giannetti, Vincenzo Lottici, and Luc Vandendorpe

WIMAX Performance in the Airport Environment 301
Paola Pulini and Snjezana Gligorevic

Throughput Enhancement Through Femto-Cell Deployment 311
Zubin Bharucha, Harald Haas, Ivan Ćosović, and Gunther Auer

Part VIII Modulation & Demodulation

Phase Rotation/MC-CDMA for Uplink Transmission 323
Koichi Adachi and Masao Nakagawa

Hierarchical Modulation in DVB-T/H Mobile TV Transmission 333
Tomáš Kratochvíl

Efficient Compensation of Frequency Selective TX and RX IQ Imbalances in OFDM Systems 343
Deepaknath Tandur and Marc Moonen

Part IX Spectrum & Interference

Dynamic Cross-Layer Spectrum Allocation for Multi-Band High-Rate UWB Systems 355
Ayman Khalil, Matthieu Crussière, and Jean-François Hélard

Egress Reduction for OFDM Via Transmit Windowing – Framework and Comparison 365
Thomas Magesacher

Interference Mitigation for the Future Aeronautical Communication System in the L-Band 375
Sinja Brandes and Michael Schnell

Part X Demonstration

Ranging and Communications with Impulse Radio Ultrawideband 387
Álvaro Álvarez, David Blanco, Lorena de Celis, and Amparo Herrera

Generic SDR Platform Used for Multi-Carrier Aided Localization 397
Igor Arambasic, Javier Casajus, and Ivana Raos

Part I
General Issues

Multicarrier Signals: A Natural Enabler for Cognitive Positioning Systems

Marco Luise and Francesca Zanier

Abstract It is known that localization of a wireless terminal is based on time-delay estimation of a radio ranging signal. In this paper, we just tackle the issue of time delay estimation and we concentrate on the ultimate limits in the accuracy of such function. Starting from a simple frequency-domain formulation of the Cramér-Rao bound for time delay estimation, we show how a multicarrier signal can be formatted to obtain maximum estimation accuracy. Capitalizing on this, we also demonstrate that the inherent flexibility about power and frequency allocation possessed by such signal is expedient to the achievement of a cognitive localization system that adapts to the changing situation of availability and interference of a wideband radio channel

1 Introduction

Cognitive Radio (CR) is a paradigm for wireless communication in which either a network or a wireless node changes its transmission and/or reception parameters (signal format and bandwidth, frequency band etc.) to communicate efficiently avoiding interference with licensed or unlicensed users. Most readers of this contribution are already familiar with this notion, whilst not so many might have heard about the "twin" concept of *Cognitive Positioning* (CP). Modern wireless networks more and more expect availability of location information about the wireless terminals, driven by requirements coming from applications, or just for better network resources allocation. Thus, signal-intrinsic capability for accurate localization is a goal of 4G as well as Beyond-4G (B4G) networks. Cognitive systems strive for optimum spectrum efficiency by allocating capacity as requested in different, possibly disjoint frequency bands. Such approach is naturally enabled, as the title of this paper suggests, by the adoption of flexible MultiCarrier (MC) technologies, in all of its flavors: traditional OFDM, Filter-Bank Multicarrier Modulation (FBMCM), and

---

M. Luise (✉) and F. Zanier
University of Pisa, Dipartimento di Ingegneria dell'Informazione,
Via G. Caruso 16, 56122 PISA, Italy
e-mail: {marco.luise,francesca.zanier}@iet.unipi.it

possibly non-orthogonal formats with full time/frequency resource allocation. In the following, we will see that a multicarrier signal format, possibly split in two or more non-contiguous bands, gives also new opportunities in terms of enhanced-accuracy time delay estimation that ultimately means enhanced accuracy positioning. We will start from Section 2 with a review of the Cramér-Rao bound (CRB), and we will introduce its frequency-domain interpretation, in Section 3. Then we will discuss the opportunities for minimization of the CRB with multicarriers signals in Section 4, and later we will discuss how to optimize an MC signal format through minimization of the CRB in Section 5, to come to the fundamental part of the paper (Section 6) on Cognitive Positioning, followed by the customary Conclusions.

2 The (Modified) Cramér-Rao Bound (CRB)

The CRB is a fundamental lower bound on the variance of any estimator [1, 2] and, as such, it serves as a benchmark for the performance of actual estimators [3]. The CRB is well known and widely adopted for its simple computation, but its close-form evaluation becomes mathematically intractable when the vector of observables contains, in addition to the parameter to be estimated, also some *nuisance parameters*, i.e., other unknown random quantities whose values we are not interested in (information data, random chips of the code of a ranging signal etc.), but that concurs to shape the actual values of the observables. The MCRB for a received signal $x(t)$ embedded in complex-valued AWGN with two-sided psd $2N_0$ is found to be [6]

$$MCRB(\lambda) = \frac{N_0}{\mathbb{E}_\mathbf{u}\left\{\int_{T_{obs}} \left|\frac{\partial x(t)}{\partial \lambda}\right|^2 dt\right\}} \quad (1)$$

where λ is the (scalar) parameter to be estimated, $\mathbf{u}$ is the vector of the nuisance parameters, T_{obs} is the observation time-interval, and $\mathbb{E}_\mathbf{u}$ indicates statistical expectation wrt $\mathbf{u}$.

As stated in the Introduction, providing enhanced-accuracy location information for a wireless terminal basically amounts to providing enhanced accuracy estimates of the time-of-arrival of the data or ranging signal from/to the terminal to/from the wireless network access point. So we will focus in the following onto the issue of *Time-Delay Estimation* (TDE) for a data or ranging signal. Assuming ideal coherent demodulation (thus with the realistic assumption that during signal tracking the carrier frequency and the carrier phase are known to a sufficient accuracy), the baseband-equivalent of the received signal reduces to

$$r(t) = x(t - \tau) + n(t), \quad (2)$$

where τ is the group delay experienced by the radio signal when propagating from the transmitter to the receiver (as seen in the reference time of the receiver). The MCRB for this specific case is easily found to be

$$MCRB(\tau) = \frac{N_0}{E_\mathbf{u}\left\{\int_{T_{obs}} \left|\frac{dx(t-\tau)}{d\tau}\right|^2 dt\right\}}. \qquad (3)$$

Our criterion for optimal signal design is very simple: finding that specific waveform that, on a pre-set bandwidth and for a certain SNR, gives the minimum CRB value. We will not consider here issues related to a possible *bias* of the estimator arising in a severe multipath propagation environment, to concentrate on the main issue just stated.

3 MCRB(τ) in the Frequency Domain

The issue of the ultimate performance of TDE dates back to more than three decades, and plenty of contributions dealing with this problem can be found [7, 8]. A recent formalization of the problem that inspired our work is contained in [5], but with specific assumptions on the ciclostationarity of the signal and on the piecewise constant nature of the parameters to estimate. We intend here to review and simplify what has already been done to come to a "clean" formulation of the TDE MCRB in the frequency domain that gives much insight into the opportunities to solve the problem of signal optimization.

Assume we are receiving a generic ranging signal containing a pseudo-random ranging code **c** whose chips are considered as binary iid nuisance parameters. Our received signal $x(t)$ turns out to be a *parametric random process*, for which each sample function is a signal $\bar{x}(t)$ with finite power P_x and chip rate $R_c = 1/T_c$. We recall that the PSD of a parametric random process $x(t)$ like ours is defined to be

$$S_x(f) \triangleq \lim_{T_{obs} \to \infty} \frac{E_\mathbf{c}\{|X_{T_{obs}}(f, \mathbf{c})|^2\}}{T_{obs}} \qquad (4)$$

where $X_{T_{obs}}(f, \mathbf{c})$ is the Fourier transform of the generic sample function $x_{T_{obs}}(t)$ truncated in the time interval $[-T_{obs}/2; T_{obs}/2]$ and thus having finite energy, and where $E_\mathbf{c}\{\cdot\}$ denotes statistical expectation over the code chips. The power P_x of the (band-pass) signal $x(t)$ is

$$P_x = \frac{1}{2}\int_{-\infty}^{\infty} S_x(f) df. \qquad (5)$$

With the usual signal modeling, the MCRB is found to be

$$MCRB(\tau) = \frac{N_0}{T_{obs}\int_{-\infty}^{\infty} \frac{1}{T_{obs}} E_\mathbf{c}\left\{\left|\frac{\partial x_{T_{obs}}(t-\tau,\mathbf{c})}{\partial t}\right|^2\right\} dt} \qquad (6)$$

where we introduced the *windowed* signal, $x_{T_{obs}}(t) = x(t) \cdot \text{rect}(t/T_{obs})$, that has finite energy. Using Parseval's relation we get

$$\text{MCRB}(\tau) = \frac{N_0}{T_{obs} \int_{-\infty}^{\infty} 4\pi^2 f^2 \frac{E_c\{|X_{T_{obs}}(f)|^2\}}{T_{obs}} df} \tag{7}$$

We now adopt a crucial assumption that leads to an accurate approximation of the bound. Specifically, we assume T_{obs} very large, so that $E_c\{|X_{T_{obs}}(f)|^2\}/T_{obs} \cong S_x(f)$. Under this hypothesis,

$$\text{MCRB}(\tau) = \frac{N_0}{T_{obs} 4\pi^2 \int_{-\infty}^{\infty} f^2 S_X(f) df} = \frac{B_{eq} T_c}{4\pi^2 \cdot \frac{E_c}{N_0} \beta_x^2} \tag{8}$$

where we let $T_{obs} = N \cdot T_c$ (N very large), β_x^2 is the normalized second-order moment of the PSD of the complex signal:

$$\beta_x^2 \triangleq \frac{1}{2P_x} \int_{-\infty}^{\infty} f^2 S_X(f) df \tag{9}$$

$E_c = P_x \cdot T_c$ is the average signal energy per chip symbol of the ranging code, and finally $B_{eq} = 1/2NT_c$ is the (one-sided) noise bandwidth of a closed-loop estimator equivalent to an open-loop estimator operating on an observation time equal to NT_c. From (8), (9), we conclude that the MCRB depends on the second-order moment of the PSD of the complex signal, independent of the type of signal format (modulation, spreading, etc.) that is adopted.

Since the CRB is proportional to the inverse of the second order moment of the PSD, some considerations need to be done also on the effects of the *center of gravity* or *center frequency* of the signal spectrum. The center frequency of the signal spectrum is given by [5]

$$f_G \triangleq \frac{1}{2P_x} \int_{-\infty}^{\infty} f \cdot S_X(f) df \tag{10}$$

The (squared) Gabor bandwidth of signal x is also given by

$$(B_G)^2 = \frac{1}{2P_x} \int_{-\infty}^{\infty} (f - f_G)^2 \cdot S_X(f) df. \tag{11}$$

When the PSD is not even-symmetric, the center frequency of the PSD is not null and the normalized second moment β_x^2 is generally larger than the squared Gabor bandwidth: :

$$\beta_x^2 = f_G^2 + (B_G)^2. \qquad (12)$$

As an example, consider the bandpass received signal

$$r_{BP}(t) = \Re x(t-\tau) e^{j(2\pi f_0(t-\tau)+\varphi)} + n_{BP}(t), \qquad (13)$$

$n_{BP}(t)$ being the band-pass AWGN with power spectral density $N_0/2$. The PSD of the (bandpass) received signal is

$$S_{r_{BP}}(f) = \frac{1}{4} S_x(f - f_0) + \frac{1}{4} S_x(-f - f_0) + \frac{N_0}{2}. \qquad (14)$$

Assume also that, as is often the case in practice, $S_x(f)$ is even-symmetric and strictly bandlimited within $[-B, B]$. Instead of demodulating the signal wrt the nominal carrier frequency f_0, we may use the frequency

$$f_B = f_0 - B \qquad (15)$$

so that the resulting demodulated I/Q complex signal is

$$r(t) = x(t-\tau) e^{j(2\pi B(t-\tau)+\varphi)} + n(t), \qquad (16)$$

whose spectra is clearly asymmetric even if $S_x(f)$ is symmetric. In particular, $z(t) = x(t-\tau) e^{j(2\pi B(t-\tau)+\varphi)}$ is an *analytic signal* with no spectral components on $f < 0$. This means that $z(t) = z_I(t) + \check{z}_I(t)$ where $\check{z}_I(t)$ is the Hilbert transform of $z_I(t)$, and also means that both $z_I(t)$ and $z_Q(t) = \check{z}_I(t)$ are bandlimited to $2B$. For such situation

$$\beta_x^2 = B^2 + (B_G)^2 > (B_G)^{(2)}. \qquad (17)$$

The inevitable conclusion is that $z(t)$ is *better* than $x(t)$ for delay estimation. The price to be paid is the increased *receive* bandwidth: the receiver needs twice as much processing bandwidth wrt the case of conventional demodulation ($f_G = 0$). The most striking aspect of this computation is that the *transmit* bandwidth is *unchanged*.

GPS HW designers know very well about the classical modes of operation for a GPS receiver: the term $(B_G)^2$ applies to conventional *code tracking* on the symmetric baseband spectrum, whilst f_G^2 can be though of as relevant to *carrier navigation*, where delay estimation is performed by tracking the cycles of the carrier on the bandpass signal. On the other hand, it is known that tracking the carrier ensures high precision estimation, but with the drawback of high estimation ambiguity.

The preceding discussion shows that the *placement in frequency* of the signal spectrum can heavily affect TDE accuracy and so, positioning accuracy. This is our starting point in the discussion of the opportunities for cognitive positioning.

4 Filter-Bank Multicarrier Ranging Signals

We will now examine in detail the format of a multicarrier signal with a binary ranging code to perform (cognitive) positioning. We will assume that the chip rate of the ranging code is $R_c = 1/T_c$, and that the code repetition length L (the code period LT_c) is very large. A basic ranging MC signal can be constructed following the general arrangement of multicarrier (MC) signals: the input chip stream c_i of the ranging code is parallelized into N substreams with a MC symbol rate $R_s = R_c/N = 1/(NT_c) = 1/T_s$, where T_s is the time duration of the "slow" ranging chips in the parallel substreams. We can use a "polyphase" notation for the k-th ranging subcode ($k = 0, 1, N-1$) in the k-th substream as $c_n^{(k)} \triangleq c_{nN+k}$ where k, $0 \leqslant k \leqslant N-1$, is the subcode index for the subcarrier index, whilst n is a time index that addresses the n-th MC symbol (block) of time length $T_s = NT_c$. The substreams are then modulated onto a raster of evenly-spaced subcarriers with frequency spacing f_{sc} and the resulting modulated signals are added together to give the (baseband equivalent of the) overall ranging signal. In Filter Bank Multicarrier Modulation (FBMCM) the spectra on each subcarrier are strictly bandlimited and nonoverlapping, akin to conventional single channel per carrier (SCPC). Contrary to OFDM, orthogonality is attained in the frequency domain and holds irrespective of the observation time.

The bandwidth occupancy of an FBMCM signal can be easily calculated under the hypothesis that the chips of the ranging code are iid as is the case for a very long PN code. The power spectral density (PSD) $S_x(f)$ of $x(t)$ is

$$S_x(f) = \frac{2 \cdot P_T}{N} \sum_{k=0}^{N-1} S\left(f - k\frac{(1+\alpha)}{T_s}\right), \quad (18)$$

where

$$S(f) = \frac{E\{|c_k^{(m)}|^2\}}{T_s}|G(f)|^2 = E\{|c_k^{(m)}|^2\}G_N(f) = G_N(f) \quad (19)$$

is the PSD of each sub-stream with chip rate $1/T_s$, and where $G_N(f)$ is a Nyquist frequency-raised-cosine function with roll-off factor α and $G_N(0) = T_s$. Differently from OFDM, FBMCM do not need virtual carriers, since the spectrum is strictly bandlimited and the sampling frequency in the modulators/demodulators obey Nyquist's rule.

When T_{obs} is sufficiently large, $T_{obs} = N_m T_s$, $N_m \gg 1$, the MCRB for such a signal can be easily computed:

$$MCRB(\tau) = \frac{T_c^2}{8\pi^2 \frac{E_c}{N_0} \frac{N_m}{N} \left[\xi_g + \frac{(1+\alpha)^2}{N}\sum_k k^2\right]} \quad (20)$$

where $E_c = P_T \cdot T_c$ and ξ_g is the so-called Pulse Shape Factor (PSF), a normalized version of the Gabor bandwidth of pulse $g(t)$ [4]:

$$\xi_g \triangleq \frac{T_c^2 \cdot \int_{-\infty}^{\infty} f^2 \cdot |G(f)|^2 df}{\int_{-\infty}^{\infty} |G(f)|^2 df} \qquad (21)$$

and where $G(f)$ is the Fourier transform of the pulse $g(t)$ as in (19). We can easily show that this bound is exactly coincident with the one obtained by the general frequency-domain formulation using (8), (9) (with numerical integration) with the expression (18) for the signal PSD.

5 Towards Cognitive Positioning

CRB with uneven power allocation. A multicarrier signal naturally allows for variable allocation of signal power onto a wide bandwidth. To accommodate this feature, we only need to introduce an amplitude coefficient on each subcarrier:

$$x(t)n = \sqrt{\frac{2 \cdot P_T}{N}} \sum_{n} \sum_{k=0}^{N-1} p_k c_n^{(k)} g(t - nT_s) e^{j2\pi k(1+\alpha)t/T_s} \qquad (22)$$

where p_k^2 is the relative power weight of carrier k ($p_k \geq 0$), that satisfies

$$\sum_k p_k^2 = N \qquad (23)$$

to give the nominal total transmitted power P_T. Some p_k's can also be 0 indicating that the relevant subcarriers or even a whole subband is not being used. The relevant MCRB is found to be

$$MCRB(\tau) = \frac{T_c^2}{8\pi^2 \frac{E_c}{N_0} \frac{N_m}{N} \left[\xi_g + \frac{(1+\alpha)^2}{N} \sum_k k^2 p_k^2 \right]} \qquad (24)$$

A nice problem is now finding the *power distribution* that provides the best timing estimation accuracy, that is, minimizes (24) through maximization of

$$\sum_k k^2 p_k^2 \qquad (25)$$

with the constraint (23). When N is fixed and the spectrum is symmetric ($f_G = 0$), (25) is maximized for the optimal power scheme

$$\begin{cases} p_k = \sqrt{\frac{N}{2}} & k = \pm\frac{N-1}{2} \\ p_k = 0 & k \neq \pm\frac{N-1}{2} \end{cases} \quad (26)$$

indicating a configuration in which the power of the signal is concentrated at the edge of the bandwidth. This is a well-known results of Gabor bandwidth theory. When comparing the optimal distribution with what we get with $p_k \equiv 1$, for N very large we see that

$$MCRB_{\text{optimal}}(\tau) \cong \frac{MCRB|_{\text{flat}}(\tau)}{3} \quad (27)$$

6 And Finally: Cognitive Positioning

An MC signal with uneven, adaptive power distribution can be adopted to easily implement a *Cognitive Positioning System* (CPS) [9]. In our envisioned FBMCM scheme for positioning, the proper power allocation allows to reach the desired positioning accuracy, not only in an AWGN channel, but also in an ACGN channel (colored noise). Colored noise arises from variable levels of interference on different frequency bands. The key assumption is that such interference can be modeled as a Gaussian process. This is certainly justified in wireless networks with unregulated multiple access techniques such as CDMA and/or UWB.

Cognitive Positioning (CP) in the AWGN channel. A simple case study for a CPS starts from the assumption to use only M out of the N subcarriers in which the total assigned bandwidth is partitioned, and to use them all at the same power. The signal format is (22) with $p_k = \sqrt{N/M}$ for M subcarriers, and $p_k = 0$ for the remaining $N - M$ elements. The corresponding MCRB is minimized by a configuration in which the M subcarriers are split in two subgroups of $M/2$ subcarriers each, and placed at the edges of the bandwidth. The (optimal) MCRB can be easily computed and turns out to be

$$MCRB(\tau) = \frac{T_c^2}{8\pi^2 \frac{E_c}{N_0} \frac{N_m}{N}} \cdot \frac{1}{\left[\xi_g + \frac{(1+\alpha)^2}{M} \frac{(M+2)(M^2+4M-3MN-6N+3N^2+3)}{12}\right]} \quad (28)$$

Let us compare this result with the MCRB that applies to M contiguous carriers (symmetric around $f = 0$), that is, for the same total spectral occupancy. We only need to replace N with M in (20) and do the summation:

$$\frac{T_c^2}{8\pi^2 \frac{E_c}{N_0} \frac{N_m}{M} \left[\xi_g + \frac{(1+\alpha)^2}{12}(M^2-1)\right]} \cong \frac{T_c^2}{8\pi^2 \frac{E_c}{N_0} \frac{(1+\alpha)^2}{12} N_m M} \quad \text{(large } M\text{)}. \tag{29}$$

If we assume $M \ll N$, (28) reads

$$\cong \frac{T_c^2}{8\pi^2 \frac{E_c}{N_0} \frac{(1+\alpha)^2}{12} N_m \cdot 3N} \quad \text{(large } M\text{)} \tag{30}$$

that is considerably smaller than (29). The effect of having the two bands for positioning far apart in the frequency domain is apparent.

Cognitive Positioning in the ACGN channel. We investigate now the issue of finding that power allocation scheme that gives the minimum CRB for TDE in a Gaussian channel whose (additive) noise has a PSD $S_n(f)$. Skipping some details, our starting point is the computation of $CRB(\tau)|_{ACGN}$, namely, the CRB of TDE, as a function of the "partial" CRBs $CRB_k(\tau)$ we would get observing each subcarrier separately from the others on its own k-th subband only. By virtue of signal orthogonality we get

$$CRB(\tau)|_{ACGN} = \frac{1}{\sum_k CRB_k^{-1}(\tau)} \tag{31}$$

Doing the computations, we have

$$CRB(\tau)|_{ACGN} = \frac{1}{4\pi^2 T_{obs}(\Delta f)^3} \cdot \left(\sum_k k^2 \frac{S_x(k\Delta f)}{S_n(k\Delta f)}\right)^{-1}, \tag{32}$$

where the PSD of the transmitted signal and of the noise were considered constant across each subband. Our final result is obtained if we let $N \to \infty$ (thus $\Delta f \to df$ and $\sum \to \int$):

$$CRB(\tau)|_{ACGN} = \frac{1}{4\pi^2 T_{obs}} \cdot \left(\int_B f^2 \frac{S_x(f)}{S_n(f)} df\right)^{-1}. \tag{33}$$

Minimization of the CRB in the ACGN channel. Coming back to the problem of enhancing TDE accuracy, and sticking for simplicity to the finite-subcarriers version of the problem, we have to minimize the CRB (32) with the constraint (23) on total power. Considering that $S_x(k\Delta f)$ is proportional to p_k^2, we are to find the power distribution p_k^2 that maximizes

$$\sum_k \frac{k^2 p_k^2}{S_n(k\Delta f)} \tag{34}$$

subject to the constraints $\sum p_k^2 = N$ and, of course, $p_k \geq 0$. The optimal distribution is easily found to be

$$\begin{cases} p_k = 0 & k \neq k_M \\ p_{k_M} = \sqrt{N} \end{cases}, \quad k_M = \arg\max_k \frac{k^2}{S_n(k\Delta f)} \qquad (35)$$

that corresponds to placing all the power onto the sub-band for which the *squared-frequency to noise ratio* (SFNR) $k^2/S_n(k\Delta f)$ is maximum.

A more realistic case study for a CPS in ACGN takes also into account possible power limitations on each subcarrier that prevents from concentrating all of the signal power onto the edge subcarriers (for AWGN) or on the subcarrier with the best SFNR as above. We have thus the further constraint

$$0 \leq p_k^2 \leq P_{max} < N. \qquad (36)$$

The solution to this problem can be easily found via linear programming:

1. Order the square-frequency-to-noise-ratios $SFNR_k$ from the highest to the lowest; set the allocated power to zero; mark all carriers available
2. Find the available power as the difference between the total power N and the currently allocated power. If it's null, then STOP, else, if it's larger then P_{max}, then put the maximum power P_{max} on the available carrier with the highest SFNR; else put on the same carrier the (residual) available power
3. Update the total allocated power by adding the one just allocated, and remove the just allocated carrier from the list of available carriers. If the list is empty, then STOP, else goto 2

This gives a set of bounded-power subcarriers that gives the optimum power allocation with ACGN.

7 Conclusions and Further Works

Let us outline the main conclusions from the development above:

- The CRB for TDE is determined by the spectral shape of the observed signal in AWGN irrespective of the specific time-domain signal format
- The center frequency of the signal PSD has a strong influence on the CRB for TDE. Sideband demodulation of a bandpass signal can add a lot in terms of estimation accuracy at the expense of widening the needed processing bandwidth of the receiver, and increasing estimation ambiguity
- A multicarrier signal can automatically take advantage of the effect described in the previous item by playing with the subcarrier amplitudes and switching on/off come subcarriers (or subbands altogether) to yield better TDE accuracy
- A good candidate for an MC ranging signal is Filtered Multi-Tone (Filter-Bank Multicarrier Modulation) with variable-amplitude subcarriers

- In the presence of strong interference on the FBMCM subbands, that we can model as an ACGN channel, it is easy to find the optimum power allocation of the FBMCM radio signal on the different sub-bands to experience the minimum CRB

The fundamental issues investigated here in terms of estimation error bounds are the basis for future design of estimators that may hopefully reach such bounds. Previous work have shown that non-data aided timing estimators [10] for data communications perform not so closely to the MCRB. Design and performance assessment of code-tracking algorithms for MC ranging signals is ongoing.

References

1. H. L. Van Trees. *detection, estimation and modulation theory*. Wiley, New York, 1968.
2. H. Cramér. *Mathematical methods of statistics*. Princeton Univ. Press, New York, 1946.
3. S. M. Kay. *Fundamentals of statistical signal processing: Estimation theory*. Prentice-Hall, Englewood Cliffs, NJ, 1993.
4. U. Mengali and A. N. D'Andrea. *Synchronization techniques for digital receivers*. Plenum Press, New York, 1997.
5. T. Alberty. Frequency domain interpretation of the Cramér-Rao bound for carrier and clock synchronization. 43(2–3–4), Feb.–Apr. 1995.
6. A. N. D'Andrea, U. Mengali, and R. Reggiannini. The modified Cramér-Rao bound and its application to synchronization problems. 42(2–4):1391–1399, Feb.–Apr. 1994.
7. A.J. Mallinckrodt and T.E. Sollenberger. Optimum pulse-time determination. *IRE Trans.*, (PGIT-3):151–159, Mar. 1954.
8. M. Skolnik. *Introduction to radar systems*. McGraw-Hill, 1980.
9. H. Celebi and H. Arslan. Cognitive positioning systems. 6(12), Dec. 2007.
10. C. Saccomando F. Spalla V. Lottici, M. Luise. Non-data aided timing recovery for filter-bank multicarrier wireless communications. 54(11), Nov. 2006.

Linearisation of Transmitter and Receiver Nonlinearities in Optical OFDM Transmission

Henning Paul and Karl-Dirk Kammeyer

Abstract In this paper, linearisation of transmitter and receiver nonlinearities of an optical Intensity Modulation/Direct Detection system using digital signal processing is presented and analysed. The system performance is estimated on basis of the signal-to-interference-and-noise ratio. Furthermore, the bit error performance of a linearised system is compared to a non-linearised one.

1 Introduction

Orthogonal Frequency Division Multiplexing (OFDM) was recently proposed for optical long-haul high-speed data transmission over single-mode fibers due to the simplicity of equalization of the dispersion dominated optical channel, e.g., [1]. However, the modulator and detector components of the optical transmission system exhibit severe nonlinearities [2] which pose a challenge to OFDM. The impact of these nonlinear effects was presented in a previous work [3], where also the system performance's dependence on the setup parameters was analysed.

In this paper, linearisation of transmitter and receiver nonlinearities using predistortion functions is presented and analysed with regard to the signal-to-interference-and-noise ratio (SINR) at the receiver in dependence on setup parameters. Additionally, the bit error performance of a linearised system in dependence on the setup parameters is simulated and compared with the non-linearised system in [3].

2 System Model

The Zero-IF (Intermediate Frequency) Intensity Modulation/Direct Detection (IM/DD) system considered in this paper is able to transmit real valued signals only, thus, the OFDM signal has to be crafted in a special manner to ensure that

H. Paul (✉) and K.-D. Kammeyer
Department of Communications Engineering, University of Bremen, Germany
e-mail: {paul, kammeyer}@ant.uni-bremen.de

this requirement is fulfilled, i.e. by complex conjugate extension of subcarriers. Note that the requirement of a real-valued time domain signal does not restrict the choice of the modulation format on the subcarriers themselves, a complex valued modulation format such as Quaternary Phase Shift Keying (QPSK) is applicable and will be used for simulations later on.

For the theoretical considerations made here, the OFDM signal is modelled by a real valued, zero mean process $x(t)$ with gaussian probability density function and variance σ_x^2.

In the following, an IM/DD system in back-to-back operation with no optical filters and amplifiers, i.e. with no optical channel involved is regarded only, resulting in a system with a highly nonlinear characteristic, caused by the modulator and detector components, which will be discussed in the following.

An extension of our considerations to a scenario including the optical fiber and filters could be performed by introduction of complex valued processes and by means of, e.g. Hammerstein or Wiener models [4, 5], which are commonly used for analysis of memory nonlinearities. This extension is beyond the scope of this work, but is simplified with the approach proposed in this paper compared to the non-linearised case considered in [3].

The overall nonlinear hardware characteristic of the analysed system can be described by the expression

$$r(t) = \beta^2 \cos^2(g^0_{-\pi/2}(m \cdot s(t) + u_{\text{bias}})), \tag{1}$$

that maps a modulator input signal $s(t)$ onto a detector output signal $r(t)$. In this equation, $g^0_{-\pi/2}(\cdot)$ represents the characteristic of a hard clipping device with fixed thresholds $\vartheta_{\text{low}} = -\pi/2$ and $\vartheta_{\text{high}} = 0$, m and u_{bias} are variable hardware setup parameters, as is the power scaling factor β. The cosine characteristic is caused by a Mach-Zehnder modulator (MZM) which converts from the electrical into the optical domain, while the square operation is performed in the photo diode (PD), which performs conversion from the optical into the electrical domain by detection of the instantaneous optical power. Figure 1 shows this system, the noise term $n(t)$ depicted there represents additive noise, but is neglected for the moment.

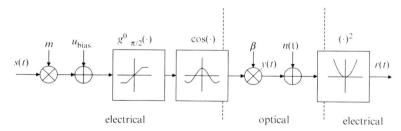

Fig. 1 Block diagram of the physical components of the transmission system

In a non-linearised system, the OFDM signal is directly used as input to the above system, i.e. $s(t) = x(t)$, but in the following considerations, the nonlinearity is supposed to be linearised employing digital signal processing at both transmitter as well as receiver side. This is accomplished by means of nonlinear functions $s(t) = \varsigma(x(t))$ at the transmitter side and $z(t) = \nu(r(t))$ at the receiving side, resulting in an overall system with output

$$z(t) = \nu(\beta^2 \cos^2(g^0_{-\pi/2}(m \cdot \varsigma(x(t)) + u_{\text{bias}}))), \qquad (2)$$

that is linear in a certain operating range.

When excited with a gaussian process $x(t)$, the system output signal $z(t)$ again is a stochastic process which generally neither is zero-mean, nor follows a gaussian distribution. This signal will be decomposed into a signal and an interference term in the next section.

2.1 Linearisation Approach

If predistortion is performed by use of an arc cosine expression

$$\varsigma(\xi) = \frac{1}{m}\left(-\arccos(m_{\text{pre}} \cdot \xi + b_{\text{pre}}) - u_{\text{bias}}\right), \qquad (3)$$

similar to the arc sine function proposed in [6] for coherent systems, and at the receiver side the square root of the detector output signal $r(t)$ is taken, i.e.

$$z(t) = \nu(r(t)) = \sqrt{r(t)}, \qquad (4)$$

the resulting overall system can be described by

$$z(t) = \beta g^1_0(m_{\text{pre}} \cdot x(t) + b_{\text{pre}}), \qquad (5)$$

with $g^1_0(\cdot)$ being a hard clipping function with lower threshold 0 and upper threshold 1.

m_{pre} and b_{pre} are user-selectable setup parameters that are introduced to control the system performance by means of operation point (b_{pre}) and drive level (m_{pre}), their exact influence on the system performance will by analysed in the next section. Note that m and u_{bias} in (1) are setup parameters as well, which are controlling drive level and operation point in the non-linearised case, but in contrast to the former ones, these are representing physical quantities, i.e. voltage gain and bias voltage. In the linearised case, these are kept fixed at an arbitrary value, but must be known to the digital preprocessing, as (3) implies.

Since both pairs of system parameters are user-controllable and determine the system performance in the linearised or non-linearised case, respectively, there exists a correspondence between them. This has to be regarded especially when these cases are supposed to be compared, as it will be the case later. Examining the MZM input signal u_{MZM} within the clipping thresholds:

$$u_{\mathrm{MZM}} = m \cdot s(t) + u_{\mathrm{bias}}, \qquad (6)$$

it evaluates for the non-predistorted case, i.e. $s(t) = x(t)$, to $u_{\mathrm{MZM}} = m \cdot x(t) + u_{\mathrm{bias}}$, while in the linearised case, it equals

$$u_{\mathrm{MZM}} = -\arccos(m_{\mathrm{pre}} \cdot x(t) + b_{\mathrm{pre}}), \qquad (7)$$

which implies that a predistortion bias b_{pre} at the MZM input will cause a physical bias

$$u_{\mathrm{bias}} = -\arccos(b_{\mathrm{pre}}), \qquad (8)$$

which can be found by Taylor series expansion of (7) around $x(t) = 0$. The linear term of this expansion is equivalent to the driving level m and evaluates to

$$m = \frac{m_{\mathrm{pre}}}{\sqrt{1 - b_{\mathrm{pre}}^2}}. \qquad (9)$$

This correspondence will be used when the performance of the linearised and the non-linearised case will be compared later on.

3 Stochastic Analysis of the System Output Process

As motivated above, the system performance of the linearised system depends on the system setup parameters m_{pre} and b_{pre}, analogue to the non-linearised system, which depends on the hardware system setup parameters. To analyse this dependence, the same approach as in [3] is used: Modelling the output

$$z(t) = \beta g_0^1(m_{\mathrm{pre}} \cdot x(t) + b_{\mathrm{pre}}), \qquad (10)$$

of the linearised system as a scaled version of $x(t)$ with an additional, uncorrelated distortion term $d(t)$ [7]

$$z(t) = \alpha_{\mathrm{lin}} \cdot x(t) + d(t), \qquad (11)$$

the scaling factor α_{lin} can be determined by means of the crosscorrelation

$$r_{XZ}(\tau) = \mathrm{E}\{x^*(t) z(t+\tau)\} = \alpha_{\mathrm{lin}} r_{XX}(\tau), \qquad (12)$$

since $x(t)$ is zero mean and $d(t)$ is uncorrelated to it. $r_{XZ}(0)$ can be calculated analytically using the definition of the expectation in (12):

$$r_{XZ}(0) = \mathrm{E}\{x^*z\}$$

$$= \int_{-\infty}^{\infty}\int_{-\infty}^{\infty} x^* z \, p_{X,Z}(x,z) dx dz \qquad (13)$$

$$= \int_{-\infty}^{\infty} x\beta g_0^1(m_{\mathrm{pre}} \cdot x + b_{\mathrm{pre}}) p_X(x) dx$$

$$= \beta \cdot \frac{m_{\mathrm{pre}}\sigma_x^2}{2}\left[\mathrm{erf}\left(\frac{1-b_{\mathrm{pre}}}{\sqrt{2}m_{\mathrm{pre}}\sigma_x}\right) - \mathrm{erf}\left(\frac{-b_{\mathrm{pre}}}{\sqrt{2}m_{\mathrm{pre}}\sigma_x}\right)\right].$$

Rearranging the last term in (12) for $\tau = 0$ using $r_{XX}(0) = \sigma_x^2$, we get

$$\alpha_{\mathrm{lin}} = \beta \cdot \frac{m_{\mathrm{pre}}}{2}\left[\mathrm{erf}\left(\frac{1-b_{\mathrm{pre}}}{\sqrt{2}m_{\mathrm{pre}}\sigma_x}\right) - \mathrm{erf}\left(\frac{-b_{\mathrm{pre}}}{\sqrt{2}m_{\mathrm{pre}}\sigma_x}\right)\right], \qquad (14)$$

where $\mathrm{erf}(\cdot)$ is the well-known error function. This factor denotes the end-to-end amplitude gain of the OFDM signal.

The system output power $\mathrm{E}\{|z(t)|^2\}$ can be calculated analytically as

$$\mathrm{E}\{|z|^2\} = \int_{-\infty}^{\infty} |z|^2 \, p_Z(z) dz$$

$$= \beta^2 \int_{-\infty}^{\infty} \left(g_0^1(m_{\mathrm{pre}} \cdot x + b_{\mathrm{pre}})\right)^2 p_X(x) dx$$

$$= \beta^2 \left[\frac{m_{\mathrm{pre}}\sigma_x}{\sqrt{2\pi}}\left(b_{\mathrm{pre}} e^{-\frac{b_{\mathrm{pre}}^2}{\sqrt{2}m_{\mathrm{pre}}\sigma_x}} - (1+b_{\mathrm{pre}}) e^{-\frac{(1-b_{\mathrm{pre}})^2}{\sqrt{2}m_{\mathrm{pre}}\sigma_x}}\right)\right.$$

$$+ \frac{m_{\mathrm{pre}}^2\sigma_x^2 + b_{\mathrm{pre}}^2 + 1}{2}\mathrm{erf}\left(\frac{1-b_{\mathrm{pre}}}{\sqrt{2}m_{\mathrm{pre}}\sigma_x}\right)$$

$$\left. - \frac{m_{\mathrm{pre}}^2\sigma_x^2 + b_{\mathrm{pre}}^2}{2}\mathrm{erf}\left(\frac{-b_{\mathrm{pre}}}{\sqrt{2}m_{\mathrm{pre}}\sigma_x}\right) + \frac{1}{2}\right], \qquad (15)$$

the power of the interference $\mathrm{E}\{|d(t)|^2\}$ thus is known using (11):

$$\mathrm{E}\{|d(t)|^2\} = \mathrm{E}\{|z(t)|^2\} - \alpha_{\mathrm{lin}}^2\sigma_x^2. \qquad (16)$$

Since for practical reasons the DC subcarrier of an OFDM system usually is not used for data transmission, only the variance $\sigma_d^2 = \mathrm{E}\{|d(t)|^2\} - \mu_d^2$ of the distortion term is of interest regarding the deterioration of the system performance. For this reason, an expression for μ_d has to be found. Since $x(t)$ is zero mean, μ_d is identical to the mean μ_z of $z(t)$ and can be calculated by

$$\mu_d = \mu_z = \mathrm{E}\{z\} = \int_{-\infty}^{\infty} z p_Z(z) dz$$

$$= \beta \int_{-\infty}^{\infty} g_0^1(m_{\mathrm{pre}} \cdot x + b_{\mathrm{pre}}) p_X(x) dx$$

$$= \beta \left[\frac{m_{\mathrm{pre}} \sigma_x}{\sqrt{2\pi}} \left(e^{-\frac{b_{\mathrm{pre}}^2}{\sqrt{2}m_{\mathrm{pre}} \sigma_x}} - e^{-\frac{(1-b_{\mathrm{pre}})^2}{\sqrt{2}m_{\mathrm{pre}} \sigma_x}} \right) + \frac{1}{2} \right. \quad (17)$$

$$\left. + \frac{b_{\mathrm{pre}} - 1}{2} \mathrm{erf}\left(\frac{1 - b_{\mathrm{pre}}}{\sqrt{2} m_{\mathrm{pre}} \sigma_x} \right) - \frac{b_{\mathrm{pre}}}{2} \mathrm{erf}\left(\frac{-b_{\mathrm{pre}}}{\sqrt{2} m_{\mathrm{pre}} \sigma_x} \right) \right].$$

Using these results, the signal-to-interference ratio (SIR) at the receiver is given by

$$\mathrm{SIR} = \frac{\alpha_{\mathrm{lin}}^2 \sigma_x^2}{\sigma_d^2} = \frac{\alpha_{\mathrm{lin}}^2 \sigma_x^2}{\mathrm{E}\{|z|^2\} - \mu_z^2 - \alpha_{\mathrm{lin}}^2 \sigma_x^2}. \quad (18)$$

Note that this expression is independent of β since the occurrences of this variable cancel in this fraction, it is a function in b_{pre} and $m_{\mathrm{pre}} \cdot \sigma_x$ only.
As soon as additive noise is introduced, i.e. $n(t) \neq 0$ in Fig. 1, the signal-to-interference-and-noise ratio (SINR) can be formulated:

$$\mathrm{SINR} = \frac{\alpha_{\mathrm{lin}}^2 \sigma_x^2}{\mathrm{E}\{|z|^2\} - \mu_z^2 - \alpha_{\mathrm{lin}}^2 \sigma_x^2 + P_n}. \quad (19)$$

This measure not only depends on the noise power P_n and the parameters b_{pre} and $m_{\mathrm{pre}} \cdot \sigma_x$, but also on the parameter β. In practical systems, a constraint has to be applied, which will be performed in the following.

4 Power Constraint

In optical transmission, the transmitting power is constrained to a value of approximately $0\,\mathrm{dBm} = 1\,\mathrm{mW}$ to avoid nonlinear effects in the optical fiber. The instantaneous electric field on the optical fiber is denoted by $y(t)$ in Fig. 1. Its second order moment $\mathrm{E}\{|y(t)|^2\}$, i.e. the optical power, is identical to $\mathrm{E}\{|z(t)|^2\}$, since $z(t) = \sqrt{r(t)} = \sqrt{|y(t)|^2} = |y(t)|$. As shown in (15), it contains a factor β^2 whose purpose is to allow adjustment of the optical power. In practical systems, this adjustment is accomplished by variation of the laser power.

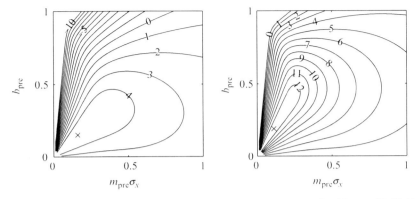

Fig. 2 Signal-to-interference-and-noise ratio in dB for noise powers $P_{opt}/P_n = 10\,\text{dB}$ (*left*), 20 dB (*right*) and varying parameters b_{pre}, $m_{pre}\sigma_x$

If the optical power is supposed to be fixed to a level P_{opt}, β has to be chosen such that

$$E\{z^2\} = P_{opt} = \beta^2 \left[\frac{m_{pre}\sigma_x}{\sqrt{2\pi}} \left(b_{pre} e^{-\frac{b_{pre}^2}{\sqrt{2}m_{pre}\sigma_x}} - (1+b_{pre})e^{-\frac{(1-b_{pre})^2}{\sqrt{2}m_{pre}\sigma_x}} \right) \right.$$
$$+ \frac{m_{pre}^2\sigma_x^2 + b_{pre}^2 + 1}{2} \text{erf}\left(\frac{1-b_{pre}}{\sqrt{2}m_{pre}\sigma_x} \right)$$
$$\left. - \frac{m_{pre}^2\sigma_x^2 + b_{pre}^2}{2} \text{erf}\left(\frac{-b_{pre}}{\sqrt{2}m_{pre}\sigma_x} \right) + \frac{1}{2} \right] \quad (20)$$

is fulfilled. Using this constraint, the SINR (19) becomes a function only depending on b_{pre}, $m_{pre}\sigma_x$ and P_{opt}/P_n. As an example, Fig. 2 shows the signal-to-interference-and-noise ratio for noise powers $P_{opt}/P_n = 10\,\text{dB}$ (left) and 20 dB (right) in dB. The maxima have been found numerically and are denoted by "x" and correspond to parameter values $b_{pre} = 0.15$, $m_{pre}\sigma_x = 0.16$ in the first and $b_{pre} = 0.19$, $m_{pre}\sigma_x = 0.11$ in the second case. The resulting system performance for these parameter pairs will be evaluated in the following by means of bit error simulations.

5 Simulation Results

The bit error performance of a linearised IM/DD optical OFDM system with 1,024 subcarriers employing QPSK modulation has been evaluated by means of Monte Carlo simulations. Due to the requirement of a real valued time domain OFDM signal, DC and Nyquist frequency subcarrier are able to convey real valued symbols only, they have been set to zero in this case. For the same reason only 511 out

Fig. 3 Bit error performance of a linearised IM/DD OFDM system for the optimal parameters $b_{pre}, m_{pre}\sigma_x$ in two operating points in dependence on the power ratio P_{opt}/P_n and bit error performance of the non-linearised system as a reference

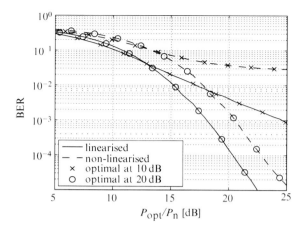

of the remaining 1,022 subcarriers are carrying independent information, while the remaining have been extended in conjugate complex fashion. Since here only the back-to-back case is considered, the length of the cyclic prefix has been set to zero. The solid lines in Fig. 3 show the bit error performance of this system for the optimal parameters $b_{pre}, m_{pre}\sigma_x$ for $P_{opt}/P_n = 10\,\text{dB}$ and $20\,\text{dB}$ as determined in the previous section. It can be seen that each one shows a lower bit error rate (BER) than the other at the corresponding power ratio P_{opt}/P_n it was optimized for. For comparison, the bit error rates of the non-linearised system in [3] with separately optimized parameters have been plotted in dashed line style. A gain of approximately 2.5 dB can be observed for $P_{opt}/P_n = 20\,\text{dB}$, for the case of 10 dB, the gain is even higher, since in the non-linearised case, an error floor is reached.

Finally, the influence of the size of the resulting linear range after linearisation is analysed by means of bit error simulations. For this purpose, b_{pre} was fixed to a value of 0.5, i.e. the operation point was established in the middle of the linear range, while for $m_{pre}\sigma_x$, values of 0.1 and 0.25 were chosen. For comparison, a non-linearised system with hardware system setup parameters u_{bias} and m chosen according to (8) and (9) was simulated. Figure 4 shows the simulated bit error rates for the linearised system in solid line style and for the non-linearised system in dashed line style. For the abscissa labelling, the optical signal-to-noise ratio (OSNR) was chosen, a measure commonly used in the optical literature, relating the overall received optical signal power to the noise power in a defined bandwidth of 0.1 nm. For calculation of the OSNR, a transmission with 42.8 Gb/s at 1,550 nm was assumed. It can be seen that the linearised and non-linearised system perform similarly for driving level $m_{pre}\sigma_x = 0.1$, with a slight advantage for the linarised system. This can be explained by the fact that in a small region around the operation point the characteristics do not deviate significantly from each other, while the linear range of the linearised system is given by $-5\sigma_x \leq x \leq 5\sigma_x$, as can be seen in Fig. 5.

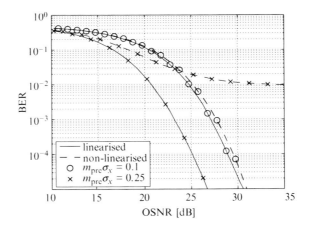

Fig. 4 Bit error performance of a linearised IM/DD OFDM system in comparison to a non-linearised system

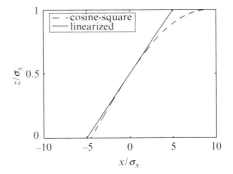

Fig. 5 Comparison of characteristics of linearised and non-linearised system for $b_{\mathrm{pre}} = 0.5$ and $m_{\mathrm{pre}}\sigma_x = 0.1$

If $m_{\mathrm{pre}}\sigma_x$ is increased to 0.25, α_{lin} is increased as well as can be verified by evaluation of (14), resulting in an improvement in the linearised case, even though the linear range is reduced to $-2\sigma_x \leq x \leq 2\sigma_x$. The non-linearised system is limited by interference introduced by its cosine-square characteristic, resulting in an error floor at a bit error rate of approximately 10^{-2} [3].

6 Conclusion

In our paper, an approach for linearisation of the cosine-square overall characteristic of an IM/DD optical OFDM system is presented. Its performance is estimated by means of the signal-to-interference-and-noise ratio, calculated by means of statistical measures of the system's output process. The required derivations were presented and also an analysis of a power constrained transmission system was introduced. Bit error simulations were performed and their results were compared to those for the non-linearised system. It was shown that linearisation by digital pre- and postprocessing is improving the system performance significantly.

References

1. Arthur J. Lowery, Liang Du, and Jean Armstrong. Orthogonal Frequency Division Multiplexing for Adaptive Dispersion Compensation in Long Haul WDM Systems. In *Proc. Optical Fiber Communication Conference (OFC)*, March 2006.
2. Govind P. Agrawal. *Fiber-Optic Communication Systems*. J. Wiley, third edition, 2002.
3. Henning Paul and Karl-Dirk Kammeyer. Modeling and Influences of Transmitter and Receiver Nonlinearities in Optical OFDM Transmission. In *Proc. 13th International OFDM Workshop 2008 (InOWo '08)*, August 2008.
4. Raviv Raich and G. Tong Zhou. On the Modeling of Memory Nonlinear Effects of Power Amplifiers for Communication Applications. In *Proc. 10th IEEE Digital Signal Processing Workshop, 2002 and the 2nd Signal Processing Education Workshop*, 2002.
5. Tong Wang and Jacek Ilow. Compensation of Nonlinear Distortions with Memory Effects in OFDM Transmitters. In *Proc. Global Telecommunications Conference 2004 (GLOBECOM '04)*, volume 4, 2004.
6. Yan Tang, Keang-Po Ho, and William Shieh. Coherent Optical OFDM Transmitter Design Employing Predistortion. *IEEE Photonics Technology Letters*, 20(11):954–956, 2008.
7. Julian J. Bussgang. *Crosscorrelation Functions of Amplitude-Distorted Gaussian Signals*. PhD thesis, Massachusetts Institute of Technology, 1952.

Multi-user OFDM Based on Braided Convolutional Codes

Michael Lentmaier, Marcos B.S. Tavares, Gerhard P. Fettweis, and Kamil Sh. Zigangirov

Abstract Braided convolutional codes (BCCs) form a class of iteratively decodable convolutional codes that are constructed from component convolutional codes. In braided code division multiple access (BCDMA), these very efficient error correcting codes are combined with a multiple access method and inherent interleaving for channel diversity exploitation into one single scheme. In this paper, we describe the BCDMA principle and present simulation results for a frequency selective Rayleigh fading channel. Results for bit interleaved coded modulation (BICM) based on turbo and LDPC codes are also given for comparison.

1 Introduction

There is a huge demand for higher data rates in next generation wireless systems. The typical multi-user scenario, where several users access the same medium and the same resources to perform communication, requires powerful and efficient broadband transmission techniques that enable the utilization of the spectrum resources with very high efficiency. *Orthogonal frequency division multiplexing* (OFDM) is a transmission technique that elegantly addresses this problem and is currently being considered as base technology for current and future wireless communication systems. Moreover, error correcting codes are indispensable elements to improve the overall capacity of communication systems.

In this paper, we present a novel multiple access technique that is based on OFDM and braided convolutional codes (BCCs) [1], denominated as *braided code division multiple access* (BCDMA) [2]. As we will discuss throughout this paper,

---

M. Lentmaier (✉), M.B.S. Tavares, and G.P. Fettweis
Vodafone Chair Mobile Communications Systems, Technische Universität Dresden,
01062 Dresden, Germany
e-mail: {michael.lentmaier, tavares, fettweis}@ifn.et.tu-dresden.de

K.Sh. Zigangirov
Department of Electrical Engineering, University of Notre Dame, Notre Dame, IN, 46556, USA
e-mail: kzigangi@nd.edu

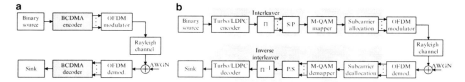

Fig. 1 Systems under study. (**a**) BCDMA system. (**b**) Bit-interleaved coded modulation (BICM) system using turbo or LDPC codes

the BCDMA scheme results in a conceptual change in the organization of modems for multiple access systems. This is due to the fact that error correction coding, interleaving for diversity exploitation, modulation and multiple access – which are generally processed by separate elements of a modem – are now concentrated in one single entity (as illustrated in Fig. 1) that is derived from the typical two-dimensional array representation of the BCCs.

2 Braided Convolutional Codes

2.1 Encoders for BCCs

BCCs are asymptotically good, turbo-like codes that can be decoded by iterative application of the BCJR algorithm [1]. Like LDPC convolutional codes they can be decoded continuously with high-speed parallel pipeline architectures and are not limited to a fixed frame length. The implementation of a rate $R = 1/3$ BCC encoder, consisting of two rate $R_{cc} = 2/3$ component convolutional encoders, is illustrated in Fig. 2(a). Characteristic for BCCs is the feedback of each parity symbol to the input of the respective other component encoder, which makes them similar to generalized LDPC codes and results in good error correction properties. Asymptotically, their free distance grows linearly with their constraint length and the bit error probability converges at least doubly exponentially to zero with decoding iterations [1, 3]. The *multiple convolutional permutors* (MCPs) [4] $\mathbf{P}^{(0)}$, $\mathbf{P}^{(1)}$ and $\mathbf{P}^{(2)}$ are fundamental elements in the construction. An MCP of multiplicity Γ is defined as an infinite diagonal-type binary matrix with a fixed number Γ of 1's in every row and column and 0's elsewhere. The width w of an MCP is defined as the maximal number of columns spanned by the non-zero elements within a row. An example of an MCP, applied to an input sequence $\mathbf{u} = (u_0, u_1, \ldots)$, is given in Fig. 2(b).

The corresponding relation between the encoded symbols is conveniently described in a two-dimensional memory array, illustrated for $\Gamma = 5$ and $w = 10$ in Fig. 3. The array is separated into three regions $\mathbb{P}^{(0)}$, $\mathbb{P}^{(1)}$ and $[\mathbb{P}^{(2)}]^T$ corresponding to the different MCPs. The white cells are used to store the individual code symbols and their positions are specified by the non-zero entries of the MCPs.

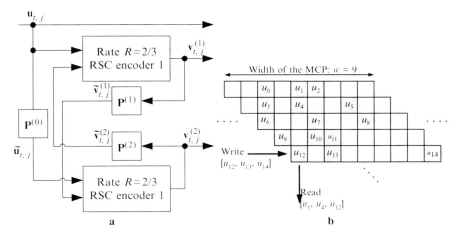

Fig. 2 (a) Encoder for rate $R = 1/3$ braided convolutional codes. (b) Example of an MCP with $\Gamma = 3$

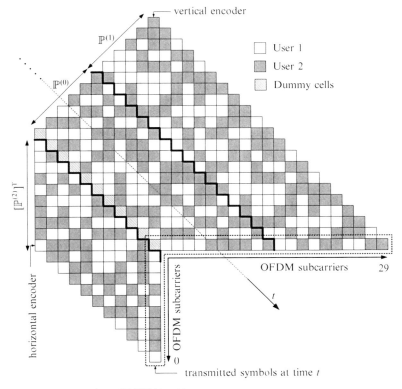

Fig. 3 Array representation of BCDMA with two users

The encoding is performed as follows. At time instant t, the information symbols $\mathbf{U}_t = (\mathbf{u}_{t,0}, \cdots, \mathbf{u}_{t,\Gamma-1})$ enter the BCC encoder and are stored in their corresponding positions in row t of $\mathbb{P}^{(0)}$. The horizontal encoder operates in this row and uses the pairs of symbols $(\mathbf{u}_{t,j}, \tilde{\mathbf{v}}_{t,j}^{(2)})$, $j = 0, \cdots, \Gamma - 1$, from regions $[\mathbb{P}^{(2)}]^{\mathrm{T}}$ and $\mathbb{P}^{(0)}$, respectively, to produce new parity symbols $\mathbf{V}_t^{(1)} = (\mathbf{v}_{t,0}^{(1)}, \cdots, \mathbf{v}_{t,\Gamma-1}^{(1)})$ to be stored in region $\mathbb{P}^{(1)}$. The symbols $\tilde{\mathbf{v}}_{t,j}^{(2)}$ are parity symbols that earlier were produced by the vertical decoder and stored in row t of the array. The vertical encoder operates analogously in column t and uses symbol pairs $(\tilde{\mathbf{u}}_{t,j}, \tilde{\mathbf{v}}_{t,j}^{(1)})$ to produce new parity symbols $\mathbf{V}_t^{(2)} = (\mathbf{v}_{t,0}^{(2)}, \cdots, \mathbf{v}_{t,\Gamma-1}^{(2)})$. The output of the BCC encoder at time t is the vector of symbols $\mathbf{V}_t = (\mathbf{V}_t^{(2)}, \mathbf{U}_t, \mathbf{V}_t^{(1)})$, whose positions are indicated by the dashed box in Fig. 3.

2.2 BCCs and M-ary Modulation Schemes

BCCs are essentially binary codes. However, we can use the array representation of the BCCs also to encode symbols of M-ary modulation schemes. In this case, each used cell of the memory array will store $q = \log_2(M)$ bits. In other words, the symbols $\mathbf{u}_{t,j}, \tilde{\mathbf{u}}_{t,j}, \mathbf{v}_{t,j}^{(1)}, \tilde{\mathbf{v}}_{t,j}^{(1)}, \mathbf{v}_{t,j}^{(2)}$ and $\tilde{\mathbf{v}}_{t,j}^{(2)}$ will be binary vectors of length q. The encoding is then performed similar to the binary case, however, each parity symbol is obtained by encoding the q bits of each input symbol sequentially. For instance, in order to obtain the parity symbols $\mathbf{v}_{t,j}^{(1)} = (v_{t,j,0}^{(1)}, \cdots, v_{t,j,q-1}^{(1)})$ from the symbols $\mathbf{u}_{t,j} = (u_{t,j,0}, \cdots, u_{t,j,q-1})$ and $\tilde{\mathbf{v}}_{t,j}^{(2)} = (\tilde{v}_{t,j,0}^{(2)}, \cdots, \tilde{v}_{t,j,q-1}^{(2)})$, the bits $u_{t,j,l}$ and $\tilde{v}_{t,j,l}^{(2)}$ enter the rate $R = 2/3$ component convolutional encoder to produce the bit $v_{t,j,l}^{(1)}$ for $l = 0$ to $q - 1$.

2.3 Decoders for BCCs

The BCCs can be decoded with a pipelined decoder architecture that enables continuous, high-speed decoding. As it can be observed in Fig. 4, the main characteristic of this kind of decoder is the fact that all I decoding iterations can be performed in parallel by $2I$ concatenated identical *processors* $\mathcal{D}_i^{(e)}$, where $i = 1, \cdots, I$ and $e = 1, 2$. Thus, after an initial decoding delay, the estimated values for the code symbols are output by the pipeline decoder at each processing cycle in a continuous manner. For the particular case of BCCs, the $2I$ parallel processors are implementing an *windowed* version of the BCJR algorithm [5, 6]. In this case, each processor operates in a finite window of length W and calculates the *a posteriori probabilities* (APP) of all code symbols (i.e., information and parity symbols).

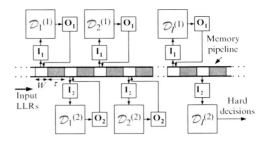

Fig. 4 Pipeline decoder for BCCs

More specifically, the decoding of BCCs using the pipeline decoder occurs as follows. At the t-th time instant, a vector of *log-likelihood ratios* (LLRs) $\mathbf{R}_t$ corresponding to the vector of code symbols $\mathbf{V}_t$ enters the memory pipeline as *a priori* information. Then, the *read-logic* $\mathbf{I}_1$ transfers the necessary data from the memory pipeline to the horizontal component code decoder $\mathcal{D}_1^{(1)}$. On its turn, the processor $\mathcal{D}_1^{(1)}$ performs the windowed BCJR decoding on the data and transfer the obtained extrinsic LLRs to the memory pipeline through the *write-logic* $\mathbf{O}_1$. Here, $\mathbf{I}_i$ and $\mathbf{O}_i$, with $i \in \{1, 2\}$, are determined by the MCPs $\mathbf{P}^{(0)}$, $\mathbf{P}^{(1)}$ and $\mathbf{P}^{(2)}$. After this, the processor $\mathcal{D}_1^{(1)}$ starts processing new data. Observe in Fig. 4 that the current memory region used by $\mathcal{D}_1^{(1)}$ is separated from the vertical decoder $\mathcal{D}_1^{(2)}$ by τ positions. This is to avoid that different decoders operate in overlapping sets of coded bits at the same time. When the extrinsic LLRs produced by $\mathcal{D}_1^{(1)}$ reaches the decoder $\mathcal{D}_1^{(2)}$, vertical decoding is performed using the extrinsic LLRs produced by $\mathcal{D}_1^{(1)}$ as *a priori* information. The new extrinsic LLRs obtained by $\mathcal{D}_1^{(2)}$ are written into the memory pipeline and $\mathcal{D}_1^{(2)}$ starts to process new data. This process repeats through all horizontal and vertical processors until the data reaches the vertical decoder $\mathcal{D}_I^{(2)}$. This last processor outputs the final LLRs and hard decision is done to determine the values of the decoded bits.

3 Braided Code Division Multiple Access

In addition to the array representation of a BCC encoder, the example shown in Fig. 3 can also be interpreted as the *subcarrier allocation pattern* of a BCDMA system for two users. In this case, we map each of the array cells at time t (i.e., cells inside the dashed box) to a subcarrier of an OFDM system. For instance, the BCDMA scheme depicted in Fig. 3 would need as basis for its realization an OFDM system with at least $N_{\text{OFDM}} = 30$ subcarriers. We assume that the white cells are associated with User 1 and gray cells are associated with User 2. The striped cells are dummy cells, they do not contain any information. If we number the OFDM subcarriers from the left to right, the symbols $\mathbb{S}_1(t) = \{\mathbf{v}_{t,0}^{(2)}, \mathbf{v}_{t,1}^{(2)}, \ldots, \mathbf{v}_{t,4}^{(2)}, \mathbf{u}_{t,0}, \mathbf{u}_{t,1}, \ldots, \mathbf{u}_{t,4}, \mathbf{v}_{t,0}^{(1)}, \mathbf{v}_{t,1}^{(1)}, \ldots, \mathbf{v}_{t,4}^{(1)}\}$, transmitted by User 1 at time instant t, will be mapped to the OFDM subcarriers $\mathbb{C}_1(t) = \{0, 5, 7, 8, 9, 10, 11, 12, 13, 18, 20, 21, 23, 26, 27\}$. Accordingly, the symbols $\mathbb{S}_2(t)$ of User 2 at time t will be mapped to the subcarriers $\mathbb{C}_2(t) = \{1, 2, 3, 4, 6, 14, 15, 16, 17, 19, 22, 24, 25, 28, 29\}$.

We can observe that the MCPs are the key components of our construction. They are responsible for defining the BCCs, as well as, the subcarrier allocation pattern of the users. In Fig. 3, $\mathbb{P}^{(0)}$, $\mathbb{P}^{(1)}$ and $[\mathbb{P}^{(2)}]^T$ denote the sets containing the MCPs $\mathbf{P}_k^{(i)}$, $i \in \{0, 1, 2\}$, $k \in \{0, \cdots, K-1\}$, of the $K = 2$ users of the system. The generalization for $K \geq 3$ is straightforward. However, a fundamental condition that the sets $\mathbb{P}^{(i)}$, $i \in \{0, 1, 2\}$, must fulfill in order to define a BCDMA system is that their elements must be orthogonal to each other. Considering the definition of an MCP, this orthogonality condition can be formulated as

$$p_{t,t'}^{(i),k_1} \cdot p_{t,t'}^{(i),k_2} = 0, \ \forall t, t' \in \mathbb{Z}^+ \text{ and } k_1 \neq k_2. \tag{1}$$

where $p_{t,t'}^{(i),k}$ is an element of the MCP $\mathbf{P}_k^{(i)}$, $i \in \{0, 1, 2\}$, associated with the k-th user of the system. This orthogonality condition implies that the different users of the BCDMA system do not share OFDM subcarriers to transmit their symbols at the time instant t. As a consequence, they do not cause interference to each other.

3.1 Spectrum Management for BCDMA Systems

In spite of having the subcarrier allocation pattern determined by the error control coding scheme, the BCDMA system is still very flexible concerning the spectrum management for the several users. Considering that the available spectrum is given by the number of OFDM subcarriers N_{OFDM}, some straightforward strategies for spectrum management are listed below:

(a) If $\lfloor N_{\text{OFDM}}/3w \rfloor < 2$ the throughput of the users can increased or reduced by increasing or reducing the multiplicities Γ, respectively. Alternatively, different rows of the MCPs can be overlapped into a single OFDM symbol to increase the throughput, or the elements of the rows of the MCPs can be spread across different OFDM symbols to decrease the throughput.

(b) If $\lfloor N_{\text{OFDM}}/3w \rfloor \geq 2$ the above mentioned techniques can also be applied. In addition, different rows of the MCPs can be placed into orthogonal sub-bands of a single OFDM symbol to increase the throughput.

Observe that by applying the spectrum managements techniques from above, the orthogonality condition in (1) must be redefined

3.2 Further Remarks

There are some points that are worth discussing when considering the constructions presented in this section. As we can observe in Fig. 3, the sub-carriers belonging to a particular user are spread in frequency domain. This is a very important feature that enables the BCDMA system to exploit the *frequency diversity* of the channel. Another very important property that belongs the BCDMA system is the *frequency hopping* experienced by the users in the succession of the time instants, i.e., the subcarriers used by a particular user are different for the succession of time instants. This last property implies that good and also bad parts of the spectrum are equally (or almost equally) distributed among the users. In other words, it means that the BCDMA system is *fair*. Additionally, if BCDMA is supposed to be deployed in a cellular system with *frequency reuse*, the frequency hopping is very important for averaging the inter-cell interference.

4 Simulation Results

In this section, we evaluate the performance of the BCDMA system compared to *bit-interleaved coded modulation* (BICM) systems [7, 8] using turbo and LDPC codes.

For the BCDMA system, the component convolutional codes have rate $R = 2/3$ and 4 states. Their generator matrices are given by

$$\mathbf{G}(D) = \begin{pmatrix} 1 & 0 & 1/(1 + D + D^2) \\ 0 & 1 & (1 + D^2)/(1 + D + D^2) \end{pmatrix}. \tag{2}$$

On the other hand, the component codes of the turbo code of the BICM system have rate $R = 1/2$ and also 4 states. Their generator matrices are given by

$$\mathbf{G}(D) = \begin{pmatrix} 1 & (1 + D^2)/(1 + D + D^2) \end{pmatrix}. \tag{3}$$

Moreover, the rate $R = 1/3$ LDPC code in our simulations is the one used in the DVB-S2 standard [9] and all MCPs, interleavers and the subcarrier allocation pattern of the BICM systems have been randomly generated.

In all systems, the underlying OFDM system consists of $N_{\text{OFDM}} = 1{,}024$ subcarriers and a *cyclic prefix* of length $N_{\text{CP}} = 16$, which is appended to the beginning of each transmitted OFDM symbol. Additionally, we use a 10-path i.i.d. frequency selective Rayleigh fading channel in our simulations that is constant over the duration of one OFDM symbol but varies independently from symbol to symbol. Moreover, in the OFDM demodulators of the systems a linear MMSE equalizer with perfect channel knowledge is used to mitigate the channel distortions.

In our simulations, the BCDMA system was observed in two different configurations, which are listed below:

1. $w = 50$, $\Gamma = 5$, QPSK modulation, block length $L = 63800$ bits.
2. $w = 50$, $\Gamma = 5$, 16-QAM modulation, block length $L = 62800$ bits.

For the BCDMA configurations from above, the BCC encoders have been terminated with $2w\Gamma\log_2(M)$ zero bits, resulting in coding rates of $R \approx 0.323$ for QPSK and $R \approx 0.312$ for 16-QAM. Moreover, $N_U = 90$ OFDM subcarriers are used in each OFDM symbol to transport user data.

The turbo coded BICM system has also been examined in different configurations. Here, an interleaver length of $N = 21600$ (resulting in terminated blocks of lengths $L = 64808$ and coding rate $R \approx 1/3$) has been combined with QPSK and 16-QAM modulations. The $R = 1/3$ LDPC code has length $L = 64800$ and has also been combined with QPSK and 16-QAM modulations. As in the case of the BCDMA system, the subcarrier allocation module of the BICM systems assigns $N_U = 90$ subcarriers to user data in each OFDM symbol.

Figure 5 shows the BER performance of the BCDMA system compared against the BICM systems. As we can observe below error rates of 10^{-3}, the BCDMA system considerably outperforms the turbo coded BICM system, which shows its typical error floor behavior in the high SNR regime. In [2], we showed that the error floor phenomenon of a BICM turbo coded system was even more evident when linear least squares equalizers are used, while the BCDMA system is quite robust against the noise enhancement of such equalizes. Here, it is worth to mention that both systems have almost the same decoding complexity. However, the BCDMA system has much more steep BER curves as an indication of its good minimum distances properties. In Fig. 5, we also have the BER curves for the LDPC coded BICM system. We can observe that the LDPC coded system outperforms the

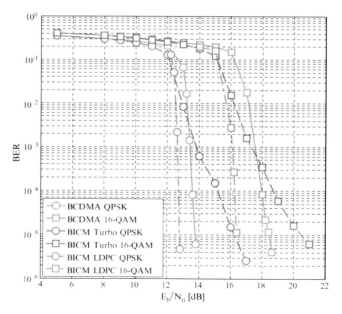

Fig. 5 BER comparison of BCDMA and BICM systems

BCDMA system. At this point, however, we should be careful since LDPC codes of rates $R < 0.5$ are usually more complex than trellis decodable codes of similar rates [10]. Moreover, if we consider the interleaving complexity of the systems in Fig. 5, we can see that the BCDMA systems with $w = 50$, in comparison to $N = 21600$ and $L = 64800$ of the BICM systems, have much more local connections in their interleavers and are, therefore, much more suitable for low complexity hardware implementations.

5 Conclusions

In this paper, we presented the general principles of a novel multiple access scheme called *braided code division multiple access* (BCDMA). Aside from providing a flexible and efficient medium access scheme for a large number of users, the application of the BCDMA concept results in a conceptual change in the design of multiple access modems. In this case, processing tasks that are traditionally considered in separate (i.e., error correction coding, interleaving for diversity exploitation, modulation and multiple access) are now combined in one single entity. Moreover, the pipeline decoding of the underlying *braided convolutional codes* (BCCs) enables very high speed VLSI implementations. Simulation results have also shown that the BCDMA system is an attractive alternative to conventional systems based on bit-interleaved coded modulation.

We consider BCDMA to be a promising research topic. Several points regarding this new technique remain open. For instance, the design of the MCPs to simultaneously maximize the error correction capabilities of the underlying BCC, frequency diversity exploitation, fairness and interference averaging property must be studied. In addition, some other issues like decoding delay reduction, VLSI implementation, rate compatibility, synchronization and channel estimation are also very interesting. Finally, we conjecture that the gap to the LDPC coded system can be reduced by proper choice of the component codes, as well as, optimization of the bit mapping scheme used in the M-ary modulations. This is our current research focus.

Acknowledgements The authors are grateful for the use of the high performance computing facilities of the ZIH at the TU Dresden.

References

1. W. Zhang, M. Lentmaier, K.Sh. Zigangirov, and D.J. Costello, Jr. Braided convolutional codes: a new class of turbo-like codes. *IEEE Trans. Inform. Theory*. submitted for publication.
2. M.B.S. Tavares, M. Lentmaier, K. Sh. Zigangirov, and G.P. Fettweis. New multi-user OFDM scheme: braided code division multiple access. In *Proc. Asilomar Conference on Signals, Systems and Computers*, Pacific Grove, CA, Oct. 2008.

3. M.B.S. Tavares, M. Lentmaier, G.P. Fettweis, and K.Sh. Zigangirov. Asymptotic distance and convergence analysis of braided protograph convolutional codes. In *Proceedings of the 46th Annual Allerton Conference on Communication, Control, and Computing*, Monticello, IL, Sep. 2008.
4. A. Jiménez Feltström, M. Lentmaier, D.V. Truhachev, and K.Sh. Zigangirov. Braided block codes. *IEEE Trans. Inform. Theory*, accepted for publication.
5. S. Benedetto, G. Montorsi, D. Divsalar, and F. Pollara. A soft-input soft-output maximum a posteriori (MAP) module to decode parallel and serial concatenated codes. *JPL TDA Progress Report*, 42(127):1–20, Nov. 1996.
6. A.J. Viterbi. An intuitive justification of the MAP decoder for convolutional codes. *IEEE J. Select. Areas Commun.*, 16(2):260–264, Feb. 1998.
7. E. Zehavi. 8-PSK trellis codes for a Rayleigh channel. *IEEE Trans. Commun.*, 40(5):873–884, May 1992.
8. G. Caire, G. Taricco, and E. Biglieri. Bit-interleaved coded modulation. *IEEE Trans. Inform. Theory*, 44(3):927–946, May 1998.
9. *European Telecommunications Standards Institute (ETSI). Digital Video Broadcasting (DVB); Second generation framing structure, channel coding and modulation systems for Broadcasting, Interactive Services, News Gathering and other broadband satellite applications; EN 302 307 V1.1.1 (2004–06).*
10. K.S. Andrews, D. Divsalar, S. Dolinar, J. Hamkins, C.R. Jones, and F. Pollara. The development of turbo and LDPC codes for deep-space applications. *Proc. IEEE*, 95(11):2142–2156, Nov. 2007.

Interference Aware Subcarrier and Power Allocation in OFDMA-Based Cognitive Radio Networks

Sami M. Almalfouh and Gordon L. Stüber

Abstract Two novel algorithms are presented for subcarrier and power allocation in OFDMA-based cognitive radio (CR) networks. In one algorithm, a transmitted power mask is imposed on different subcarriers in order to limit the interference introduced in the primary network to a specified threshold. Following this step, the subcarrier allocation problem is formulated as a multiple knapsack problem (MKP), and a search algorithm is employed to maximize the sum capacity of the CR network. The second algorithm employs a heuristic approach by assigning different non-overlapping sets of OFDMA subcarriers to different CR users according to their channel conditions. This is followed by a power allocation algorithm that maximizes the sum capacity while maintaining the interference power introduced in the primary network band below a predefined threshold. The proposed algorithms show near-optimal performance while significantly reducing the complexity of the original allocation problem which is known to be NP-hard.

1 Introduction

According to a recent report by the Federal Communications Commission(FCC) [1], radio spectrum is often under-utilized as a result of the policies and regulations governing spectrum allocation procedures. Cognitive radio (CR) has been recently suggested as a method for promoting more efficient spectrum utilization through the principles of spectrum sensing and dynamic spectrum access (DSA) [2]. Cognitive users (or secondary users) opportunistically utilize vacant channels of a nearby primary network when the primary users are inactive. There are two main constraints on the secondary users transmissions. First, the secondary users must limit the interference introduced in a neighboring primary user band below a certain level. This threshold is sometimes called the *interference temperature* level [3]. Second, the

---

S.M. Almalfouh and G.L. Stüber (✉)
School of Electrical and Computer Engineering, Georgia Institute of Technology, Atlanta, GA 30332, USA
e-mail: {sami,stuber}@ece.gatech.edu

secondary users must vacate the current channel or band they are occupying as soon as the primary users who are allocated that channel become active again.

Orthogonal frequency division multiple access (OFDMA) has been suggested as a potential access scheme for CR systems. The greatest apparent benefit from employing OFDMA in CR systems is the fact that interference to primary users can be avoided simply by "turning-off" the OFDM subcarriers that overlap with the primary users band [4]. In OFDMA, the problem of optimally allocating subcarriers and power to different users has been studied extensively and is still being investigated. In OFDMA based CR networks, the original problem of allocating OFDMA radio resources (i.e. subcarriers, power and bit loading) has an added dimension of complexity arising from the interference constraints used to protect the primary user from undue interference. The classical waterfilling algorithm for allocating power to subcarriers in OFDMA systems is not suitable in the CR context, because it does not take into consideration the protection of primary users operating in nearby bands. The basic idea in the waterfilling algorithm is to assign more power to subcarriers that experience higher channel gain, while assigning little or no power to subcarriers suffering from low channel gain. However, in the context of CR, a set of subcarriers experiencing favorable channel quality may be adjacent to a primary user band. Thus, if the waterfilling algorithm is employed, those subcarriers will typically be assigned higher power levels, which in turn will cause severe interference to the adjacent primary users. Such scenarios suggest a need for *interference aware* radio resource allocation in OFDMA-based CR networks. In [5], an iterative resource allocation approach was devised to maximize the weighted sum rate of the CR network under interference temperature constraints. A sub-optimal interference-limited subcarrier and power allocation algorithm based on Lagrangian duality theorem and frequency sharing relaxation, where multiple CR users are allowed to time-share the same subcarrier, was presented in [6]. In [7], the authors presented a suboptimal OFDM radio resource allocation algorithm that minimizes the total interference experienced by the primary user as well as the total transmitted power by the CR users. In [8], interference-limited power allocation schemes for OFDMA-based CR systems were studied for the simple case of one CR user. In this paper, we introduce two interference aware subcarrier and power allocation algorithms. In the first algorithm, a multi-level power mask is imposed on the different subcarriers so as to maintain the interference level in the primary users bands below a preset threshold; following this step a subcarrier allocation problem is formulated and an optimal or near-optimal solution is sought. The second suggested algorithm employs a two step heuristic approach, a subcarrier allocation step followed by a power allocation step, to find a near-optimal solution to the allocation problem that maximizes the sum capacity of the CR network, while the interference level in the primary users band is kept at or below a certain threshold. Our contributions are unique in the sense that we formulate the OFDMA resource allocation problem in the first algorithm as a Multiple Knapsack Problems (MKP), for which there exist many optimal and sub-optimal solutions in the literature. Moreover, the second allocation algorithm outlines a fast and near-optimal approach with linear complexity.

The remainder of this paper is organized as follows. Section 2 presents the CR system model under consideration and formulates the resource allocation problem.

Section 3 introduces two different algorithms to solve the allocation problem, the constant power subcarrier allocation algorithm as described in Section 3.1, and the joint subcarrier and power allocation algorithm as described in Section 3.2. Section 4 presents numerical examples to illustrate the performance of the two suggested algorithms. Finally, the paper is concluded in Section 5.

2 System Model and Problem Formulation

We consider a single cell spectrum sharing system consisting of two co-located CR and primary network Base Stations (BS). The CR BS serves K CR users indexed by $k \in \{1, 2, \ldots K\}$, while the primary network BS serves L Primary Users (PUs) indexed by $l \in \{1, 2, \ldots L\}$. The CR users opportunistically utilize vacant portions of the primary network bands. We assume that the vacant frequency band(s) are divided into N OFDM subcarriers indexed by $n \in \{1, 2, \ldots N\}$. Figure 1 depicts the system model described here. In this paper, we consider the CR downlink joint subcarrier and power allocation. The objective of the allocation algorithm is to maximize the sum capacity of the CR network, while limiting the level of interference introduced in the primary users band to a predefined threshold in order to protect the primary network from service degradation. Using the notation in Table 1, the allocation problem can be formulated as a constrained optimization problem as follows:

$$C = \max_{x_{kn} \in \{0,1\}, p_{kn} \geq 0} \sum_{k=1}^{K} \sum_{n=1}^{N} x_{kn} c_{kn} \qquad (1)$$

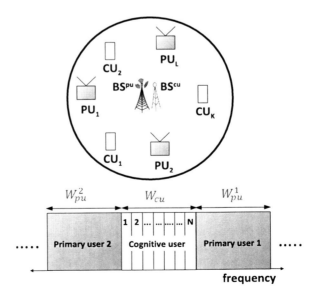

Fig. 1 Spectrum sharing model

Table 1 General notation

Notation	Description
x_{kn}	Assignment indicator, $x_{\bar{k}n} = \begin{cases} 1 & \text{subcarrier n is assigned to CR user } \bar{k} \\ 0 & \forall k \neq \bar{k} \end{cases}$
c_{kn}	Capacity of subcarrier n when assigned to user k, (5)
p_{kn}	Power allocated to user k on subcarrier n
g_{kn}	Gain of channel (subcarrier) n between the CR BS and CR user k
σ_{kn}^2	Noise power in channel (subcarrier) n assigned to CR user k
P_k	Total power budget allocated to user k
I_{kn}^l	Interference power in primary user l band contributed by subcarrier n Assigned to CR user k, (6) [9]
I_{th}^l	Maximum interference power that can be tolerated by primary user l
h_{ln}	Gain of channel (subcarrier) n between CR BS and primary user l
$\phi_n(f)$	Spectral pulse shape of subcarrier n
W_{pu}^l	Bandwidth of primary user l band
d_n^l	Spectral distance between subcarrier n and the center of primary user l band

subject to:

$$\sum_{n=1}^{N} p_{kn} \leq P_k, \ \forall\, k \in \{1,2,\ldots K\} \qquad (2)$$

$$\sum_{k=1}^{K}\sum_{n=1}^{N} I_{kn}^l \leq I_{th}^l, \ \forall\, l \in \{1,2,\ldots L\} \qquad (3)$$

$$\sum_{k=1}^{K} x_{kn} \leq 1, \ \forall\, n \in \{1,2,\ldots N\} \qquad (4)$$

where

$$c_{kn} = \log_2\left(1 + \frac{p_{kn}|g_{kn}|^2}{\sigma_{kn}^2}\right) \qquad (5)$$

$$I_{kn}^l = p_{kn}|h_{ln}|^2 \int_{d_n^l - W_{pu}^l/2}^{d_n^l + W_{pu}^l/2} \phi_n(f)\,df \qquad (6)$$

The constraint in (2) limits the power allocated to CR user k to a predefined power budget P_k, while the constraint in (3) ensures that the interference power affecting primary user l, expressed in (6), is maintained below a preset threshold I_{th}^l. The third constraint in (4) ensures that each subcarrier is exclusively assigned to one CR user.

3 Interference-Aware OFDMA Subcarrier and Power Allocation

The joint subcarrier and power allocation problem in (1)–(4) belongs to the class of Mixed Integer Nonlinear Programming (MINLP) [10], as it is comprised of binary variables x_{kn} and continuous variables p_{kn}. The optimal solution to such classes of optimization problems is known to be NP-Hard and may not be achievable in real scenarios. In the following subsections, we introduce two sub-optimal algorithms to allocate subcarriers and power to different CR users such that the sum capacity of the CR network is maximized, while the interference power at the primary user is kept below a given threshold. In Section 3.1, a constant power subcarrier allocation algorithm is considered, where a constant power level per subcarrier is set in a way such that the interference power leaked into the primary user bands is kept below the defined threshold, I_{th}^l. In Section 3.2, we introduce a sub-optimal heuristic approach to solve the original allocation problem in (1)–(4), while significantly reducing its computational complexity. Without loss of generality, we assume one primary user is active while K CR users are transmitting in a neighboring frequency band.

3.1 Constant Power Subcarrier Allocation

In the constant power subcarrier allocation scheme, a constant power level is assigned to each subcarrier such that the interference power in the primary user band is kept at or below the threshold I_{th}, i.e.,

$$\sum_{k=1}^{K}\sum_{n=1}^{N} x_{kn} \bar{p}_n |h_n|^2 \underbrace{\int_{d_n-W_{pu}/2}^{d_n+W_{pu}/2} \phi_n(f) df}_{I_n} \leq I_{th} \tag{7}$$

where $\bar{p}_n$ is the constant power level allocated to subcarrier n. Note here that the power allocated to the subcarriers is independent of the CR user k, so that p_{kn} is replaced by $\bar{p}_n$. One method for allocating $\bar{p}_n$ is to start with a low power level assigned to the subcarriers close to the primary users band, as these adjacent subcarriers contribute the majority of the interference power introduced within the primary user band. Afterwards, the power can be increased in constant increments as we move further away from the edge of the primary user band. Another method for allocating $\bar{p}_n$ to the different subcarriers is to assign less power to the subcarriers having a higher interference factor I_n, as defined in (7), thus reducing the amount of interference contributed to the primary user by those subcarriers, i.e., $\bar{p}_n \propto 1/I_n$. These two methods are discussed in some detail in [8].

Following the allocation of the different power levels $\bar{p}_n$, the original optimization problem can be reformulated as follows:

$$C = \max_{x_{kn} \in \{0,1\}} \sum_{k=1}^{K} \sum_{n=1}^{N} x_{kn} c_{kn} \tag{8}$$

subject to:

$$\sum_{n=1}^{N} x_{kn} \bar{p}_n \leq P_k \ \forall \ k \in \{1, 2, \ldots K\} \tag{9}$$

$$\sum_{k=1}^{K} x_{kn} \leq 1 \ \forall \ n \in \{1, 2, \ldots N\} \tag{10}$$

where c_{kn} is the same as in (5) with p_{kn} replaced by $\bar{p}_n$.

The optimization problem in (8)–(10) is a special case of the Multiple Knapsack Problem (MKP) [11]. The MKP describes the problem of packing K knapsacks with N objects, where each object n has a value V_n and weight W_n. The objective is to maximize the total value in all knapsacks while limiting the weight of each knapsack to a certain value, W'_k. Following the MKP analogy, the subcarrier allocation problem can be viewed as assigning N subcarriers (objects) to K CR users (knapsacks), where each subcarrier has value c_{kn} and weight $\bar{p}_n$, and each CR user has a power budget (weight limit) P_k. The difference between the MKP and the subcarrier allocation problem is that the individual subcarrier capacity (value), c_{kn}, depends on the potential CR user (knapsack) that might be assigned that particular subcarrier (object), while in the MKP the value of an object n, V_n, is independent of the potential knapsack k that might be assigned that particular object. In other words, c_{kn} depends on k while V_n is independent of k. The algorithm for finding the optimal assignment matrix $\mathbf{X} = [x_{kn}]_{K \times N}$ starts by assigning subcarrier n to the CR user k yielding the maximum c_{kn} among all CR users. If the current solution satisfies the constraint in (9), then this solution is feasible and optimal since it attains the maximum sum capacity in (8). If the current solution does not satisfy the power budget constraint in (9), the assignment matrix is modified for the CR users that violate the power limit constraint by reassigning some of the subcarriers currently assigned to CR user k to another CR user k' in such a way that the loss in sum capacity is minimized. This loss in sum capacity comes as a price for satisfying the power budget in (9). This procedure generates multiple solutions that *branch* from the original solution. The sum capacity of each new solution is calculated and these solutions are added to the pool of solutions in the initialization step, where they are sorted in descending order in terms of their corresponding sum capacities. The algorithm iterates through the available solutions until an optimal solution is achieved. The algorithm is outlined in Table 2.

Table 2 Algorithm 1: Constant Power Subcarrier Allocation

Initialization:
$\forall n \in \mathcal{N} = [1, 2, \ldots N]$
find $k_n = \arg\max_k c_{kn}$
set $\mathbf{X}^0 = [x_{kn}^0]_{K \times N}$
where $x_{kn}^0 = \begin{cases} 1 & k = k_n \\ 0 & k \neq k_n \end{cases}$
compute $C^0 = \sum_{n=1}^{N} c_{k_n n}$
Iteration
START
set $\mathbb{X} = \{\mathbf{X}^0\}, \mathbb{C} = \{C^0\}$
set $\mathbf{X}^i = \mathbb{X}(1), C^i = \mathbb{C}(1)$
remove $\mathbb{X}(1)$ from $\mathbb{X}$, and $\mathbb{C}(1)$ from $\mathbb{C}$
compute $\hat{P}_k = \sum_{n=1}^{N} \bar{p}_n x_{kn}^i$
if $\hat{P}_k < P_k \; \forall \; k \in \mathcal{K} = [1, 2, \ldots K]$
$\quad$ then $\mathbf{X}^i = [x_{kn}^i]_{K \times N}$ is optimal solution and the algorithm terminates.
else if $\hat{P}_k > P_k$ for some $k \in \mathcal{K}$
$\quad$ set $\mathcal{K}' = \{k : \hat{P}_k > P_k\}$ and $\mathcal{N}_k' = \{n : x_{kn}^i = 1\} \; \forall \; k \in \mathcal{K}'$
end if
$\forall \; k \in \mathcal{K}', \forall \; n \in \mathcal{N}_k'$
set $\mathcal{J}_k = \{n : c_{jn} < c_{kn} \forall \; j \in \mathcal{K}, j \neq k\}$
compute $\hat{P}_{\mathcal{J}_k} = \sum_{n \in \mathcal{J}_k} \bar{p}_n$
if $\hat{P}_{\mathcal{J}_k} < \hat{P}_k - P_k$ for at least one $k \in \mathcal{K}'$
$\quad$ go to *START*
else if $\hat{P}_{\mathcal{J}_k} > \hat{P}_k - P_k \forall \; k \in \mathcal{K}'$
$\quad$ compute $\Delta c_j = c_{kn} - c_{jn} \forall \; n \in \mathcal{J}_k, \forall \; k \in \mathcal{K}', \forall \; j \in \mathcal{K}$
$\quad$ set $\mathcal{M}_k = \{j' : j' = \arg\max_j \Delta c_j\}$
end if
Branching
form all branching solutions as follows:
$\mathbf{X}^B = [x_{kn}^B]_{K \times N}$, where $x_{kn}^B = \begin{cases} x_{kn}^i & \forall \; n \in \mathcal{N} \setminus \mathcal{J}_k, \forall \; k \in \mathcal{K} \\ 1 & \forall \; n \in \mathcal{J}_k, \forall \; k \in \mathcal{M}_k \end{cases}$
update $\mathbb{X}$ with the new solutions (branches)
compute the corresponding sum capacity for each solution, and update $\mathbb{C}$

3.2 Heuristic Joint Subcarrier and Power Allocation

The joint subcarrier and power allocation algorithm is comprised of two steps; the first step assigns the N subcarriers to the different CR users, where subcarrier n is assigned to CR user k having the maximum Signal to Interference Ratio (SIR), γ_{kn}, defined as

$$\gamma_{kn} = \frac{N|g_{kn}|^2}{P_k I_n} \quad (11)$$

That is, for each subcarrier $n \in [1, 2, \ldots N]$, we look for the CR user k that has the maximum γ_{kn} and assign subcarrier n to it, thus forming a set of assigned subcarriers S_k for each CR user k. In the second step, the power allocation algorithm is performed on a per user basis as follows. For each CR user k, we first arrange the subcarriers in S_k in descending order of their respective SIR values, γ_{kn}. Then, starting with the subcarrier n having the highest γ_{kn}, the power allocated to this subcarrier is increased in small fixed increments until further increase in the allocated power results in more interference power being introduced in the primary user band compared to the increase in capacity. If this is the case, the algorithm continues to the next subcarrier in S_k in the same way until the total power budget for CR user k, P_k, is consumed. The algorithm is outlined in detail in Table 3.

Table 3 Algorithm 2: Joint Subcarrier and Power Allocation

Step1: Subcarrier Assignment
$\forall n \in \mathcal{N}$
find $k_n = \arg\max_k \gamma_{kn}$
set $x_{kn} = \begin{cases} 1 & k = k_n \\ 0 & k \neq k_n \end{cases}$
set $S_k = \{n : x_{kn} = 1\} \ \forall k \in \mathcal{K}$
Step2: Power Allocation
$\forall k \in \mathcal{K}$
set $I_{th}^k = \hat{I}_{th}^k = \frac{I_{th}}{K}$
set $\hat{P}_k = P_k$
while $\hat{P}_k > 0$
    $\forall n \in S_k$
    set $p_{kn} = 0$
    while $\hat{I}_{th}^k > 0$
        set $\hat{p}_{kn} = p_{kn} + \Delta p$
        compute $u_{kn} = c_{kn} - \alpha \, I_n^k$
        if $u_{kn} \geq 0$
            set $\hat{P}_k = P_k - \Delta P$
            set $\hat{I}_{th}^k = I_{th}^k - I_{kn}$
        else if $u_{kn} < 0$
            set $\hat{p}_{kn} = p_{kn}$
            go to next subcarrier in S_k
        end if
    end while
end while

Remarks:

- c_{kn} is calculated as in (5) with p_{kn} replaced by $\hat{p}_{kn}$
- $I_n^k = \hat{p}_{kn} I_n$
- The constant α in u_{kn} has units *bits/sec/Hz/W* and can be viewed as the cost per unit power for any increase in the capacity c_{kn}.

4 Simulation Results

In this section, we introduce some numerical results for the subcarrier and power allocation algorithms introduced in this paper. We consider a wireless CR network comprised of 6 CR users and 128 subcarriers coexisting with one primary user. The channel gains g_{kn} (and h_n) are assumed to be independent and identically distributed (i.i.d) and follow a Rayleigh distribution with mean square value $E[|g_{kn}|^2] = E[|h_n|^2] = \Omega_p$ i.e., [12]

$$f_{g_{kn}}(x) = \frac{2x}{\Omega_P} \exp\left\{\frac{-x^2}{\Omega_p}\right\}, x \geq 0 \tag{12}$$

where we assume $\Omega_p = 1$. The per user power budgets are $P_1 = P_2 = 10$ dBm, $P_3 = P_4 = P_5 = 20$ dBm and $P_6 = 30$ dBm. A full response rectangular subcarrier shaping pulse is considered, i.e., $\phi_n(f) = \text{sinc}(\pi(f - f_n)T_s)^2$, where T_s is the symbol duration and f_n is the frequency of subcarrier n expressed as $n\Delta f$ and Δf is the subcarrier spacing. The primary network bandwidth is assumed to be 1 MHz. The constant power profile values ($\bar{p}_n$) used in Algorithm 1 were calculated in such a way that the allocated power to a given subcarrier is inversely proportional to the interference factor (I_n) of that subcarrier. Using the result from [8], $\bar{p}_n$ can be calculated as

$$\bar{p}_n = \frac{I_{th}}{NI_n} \tag{13}$$

The simulation results were averaged over 10, 000 independent realizations of the channel power gains. Figure 2 shows the average number of subcarriers assigned to each CR user, $\bar{N}_k$. It is clear that as the power budget of a CR user increases, it

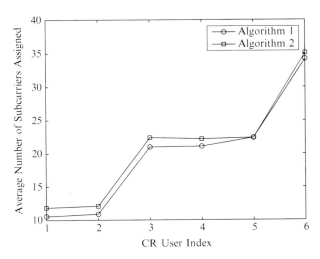

Fig. 2 Average number of subcarriers allocated to CR user k

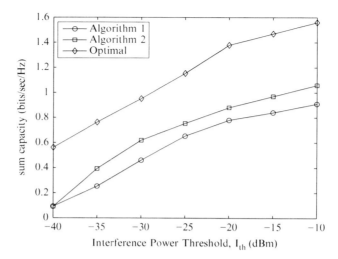

Fig. 3 CR network sum capacity against different interference power thresholds at the primary user

is allocated more subcarriers on average, and those CR users with the same power budget are assigned, on average, the same number of subcarriers. Figure 2 shows that both Algorithm 1 and Algorithm 2 yield very close subcarrier assignment solutions. In Fig. 3, the sum capacity of the CR network is plotted against different primary user interference thresholds. As expected, the sum capacity of the CR network increases as the primary user interference threshold is increased. Finally, we notice that Algorithm 2 performs slightly better than Algorithm 1. This is due to the constant power assignment in Algorithm 1, while Algorithm 2 implements both subcarrier and power allocation based on different CR users channel state information and their interference contribution to the primary user.

5 Conclusions

In this paper, we studied the problem of OFDMA subcarrier and power allocation in the context of CR networks. Two allocation algorithms were introduced. The main objective of the two allocation algorithms is to maximize the sum capacity of the CR network while maintaining the interference level introduced in the primary users bands below a predefined threshold. While the algorithms developed are sub-optimal, they still exhibit an acceptable performance compared to the optimal solution, while both algorithms significantly reduce the computational complexity of the original allocation problem, thereby making these algorithms suitable for real time systems. In future work, we will investigate fairness issues in such sub-optimal algorithms in order for each CR user to achieve the required Quality of Service (QoS) to maintain a reliable communication link.

References

1. The Federal Communications Commission (FCC). First report and order and further notice for proposed rulemaking. ET Docket No. 06-156, October 2006.
2. J. Mitola and G. Maguire. Cognitive radio: Making software radio more personal. *IEEE J. Personal Communications*, 6(4):13–18, August 1999.
3. S. Haykin. Cognitive radio: brain-empowered wireless communications. *IEEE J. Selec. Areas Commun.*, 23(2):201–220, February 2005.
4. T. Weiss and F. K. Jondral. Spectrum pooling: An innovative strategy for the enhancement of spectrum efficiency. *IEEE Commun. Mag.*, 42(3):S8–S14, March 2004.
5. A. Attar, O. Holland, M. R. Nakhai, and A. H. Aghvami. Interference-limited resource allocation for cognitive radio in orthogonal frequency-division multiplexing networks. *IET Communication*, 2(6):806–814, July 2008.
6. P. Cheng, Z. Zhang, H. H. Chen, and P. Qiu. Optimal distributed joint frequency, rate and power allocation in cognitive OFDMA systems. *IET Communications*, 2(6):815–826, July 2008.
7. T. Qin and C. Leung. A cost minimization algorithm for a multiuser OFDM cognitive radio system. In *IEEE Pacific Rim Conf on Comm, Comp and Signal Proc.*, pages 518–521, August 2007.
8. G. Bansal, J. Hossain, and V. K. Bhargava. Optimal and suboptimal power allocation schemes for OFDM-based cognitive radio systems. *IEEE Trans on Wireless Commun.*, 7(11):4710–4718, November 2008.
9. T. Weiss, J. Hillenbrand, A. Krohn, and F. K. Jondral. Mutual interference in OFDM-based spectrum pooling systems. In *IEEE Veh. Technol Conf., Spring.*, pages 1873–1877, May 2004.
10. Dimitri P. Bertsekas. *Nonlinear Programming*. Athena Scientific, Boston, MA, 1999.
11. Silvano Martello and Paolo Toth. *Knapsack Problems, Algorithms and Computer Implementations*. Wiley, 1990.
12. Gordon L. Stüber. *Principles of Mobile Communication*. Kluwer, Norwell, MA, 1996.

An Efficient ICA Based Approach to Multiuser Detection in MIMO OFDM Systems

Mahdi Khosravy, Mohammad Reza Alsharif, and Katsumi Yamashita

Abstract This paper proposes an ICA-based MIMO-OFDM system which efficiently overcomes problems inherent to ICA by using a precise and robust signal reconstruction method. It exploits the predetermined characteristics introduced to transmitted signals by a convolutional encoder at the transmitter to solve permutation indeterminacy, amplitude scaling ambiguity and phase distortion. Since, the introduced characteristics are only dependent on the convolutional code, despite the previous method, the proposed method is channel independent and robust. Moreover, the method is precise, because the accuracy of the introduced characteristics are fulfilled by an optimized convolutional code. We have compared the performance of the proposed MIMO-OFDM system with joint detection (JD) method which estimates the channels by using two training OFDM blocks. Although the JD method is a training based method, the performance of the proposed blind method is favorably comparable over slowly varying channels, and it dominates JD method over fast varying channels.

1 Introduction

Orthogonal frequency division multiplexing (OFDM) has become a popular technique for transmission of signals over wireless channels. The combination of OFDM with multiple-input multiple-output (MIMO) transceiver structure has been

---

M. Khosravy (✉) and M.R. Alsharif*
Department of Information Engineering, University of the Ryukyus, 1 Senbaru, Nishihara, Okinawa 903-0213, Japan
*Also, Adjunct Professor in University of Tehran, Iran
e-mail: k078661@eve.u-ryukyu.ac.jp, asharif@ie.u-ryukyu.ac.jp

K. Yamashita
Graduate School of Engineering, Osaka Prefecture University, 1-1 Gakuen-cho, Sakai, Osaka, Japan
e-mail: yamashita@eis.osakafu-u.ac.jp

promised as a strong candidate for future fourth generation (4G) communications [1]. Besides the advantages of a MIMO-OFDM system, deploying blind channel estimation increases the spectral efficiency of the system, since no training data is required.

The separation of multiuser signals in a MIMO-OFDM system at each frequency bin (FB) level has been represented by [2] as an instantaneous blind source separation (BSS) problem in complex domain. But the reconstruction of the multiuser data separated by BSS suffers from permutation indeterminacy, amplitude scaling ambiguity and phase distortion ambiguity which theses problems are inherent to complex BSS [3]. Furthermore, the general complexity of BSS algorithms is another important issue that should be practically modified.

This paper proposes an efficient blind multiuser detection and channel estimation technique for MIMO-OFDM systems based on independent component analysis (ICA) [4]. The proposed ICA based MIMO-OFDM system efficiently overcomes the problems of deploying ICA. To solve indeterminacies inherent to ICA, it deploys the precoding solution in [5]. To reduce the complexity of ICA algorithm it uses a concatenate structure wherein a fast BSS approach [6] is used to approximate a separating matrix as starting point of ICA.

The organization of the paper is as follows. Section 2 presents the proposed ICA based multiuser detection in a MIMO-OFDM system. Section 3 explains briefly instantaneous BSS and the used method for reducing BSS complexity. Section 4 provides the simulation results, and finally, Section 5 concludes the paper.

2 The Proposed Method

Consider a multiuser MIMO-OFDM system with M_T transmit antennas and M_R receive antennas. The symbols of each user are convolutional encoded by an FIR pre-filter at the transmitter as follows

$$S_i(kN + m) = \sum_{l=0}^{L_f-1} c(l) D_i(kN + m - l) \qquad (1)$$

where $D_i(kN + m)$ are information symbols of the ith user, and $S_i(kN + m)$ are their encoded symbols. L_f is the length of the filter and $c(.)$ are its coefficients. R_0 the auto-correlation of No.m FB track of ith encoded user symbol-block and R_1 the correlation between its mth and $m + 1$st FB tracks are respectively as follows

$$\begin{aligned} R_0 &= E\left[|S_i(kN + m)|^2\right] \\ &= c^2(0) + \cdots + c^2(L_f - 1) \end{aligned} \qquad (2)$$

$$\begin{aligned} R_1 &= E\left[S_i(kN + m)S_i^*(kN + m)\right] \\ &= c(0)c(1) + \cdots + c(L_f - 2)c(L_f - 1) \end{aligned} \qquad (3)$$

An Efficient ICA Based Approach to Multiuser Detection in MIMO OFDM Systems   49

where $E[.]$ is the expectation with respect to k. Equations (2) and (3) have been obtained in the Appendix. The pre-filter coefficients are optimized for the least possible error of average phase of R_1, that is used for removing phase distortion induced by ICA. We have proved in [5] that this optimization aim is obtained by the following criteria

$$[c(0), c(1), \cdots, c(L_c - 1)] = argmin_{c(.)} \left(\frac{\left[\sum_{l=0}^{L_c-1} c(l) \right]^2}{\sum_{l=0}^{L_c-1} c(l) c(l+1)} \right). \quad (4)$$

By solving the above optimization problem subjected to $\sum_{l=0}^{L_f-1} c(l) = 1$, the optimized coefficients are obtained as $[\frac{1}{3} \ \frac{1}{2} \ \frac{1}{6}]$. The N-length kth OFDM symbol block of each ith user

$$S_i^{(k)} = [S_i(kN), S_i(kN+1), \cdots, S_i(kN+N-1)]^T \quad (5)$$

is modulated by an N-point IDFT to

$$s_i^{(k)} = [s_i(kN), s_i(kN+1), \cdots, s_i(kN+N-1)]^T \quad (6)$$

where $i = 1, 2, \cdots, M_T$.

After adding the cyclic prefix (CP) with the length L_{cp} to avoid inter-symbol interference (ISI), the modulated signals are transmitted. The transmitted signals pass through different propagation channels and they are received by No. j antenna at the receiver. After removal of the CP and demodulation by N-point DFT, the received N-length data symbol block by the jth antenna at time k is

$$X_j^{(k)} = \sum_{i=1}^{M_T} H_{ji} S_i^{(k)} + Z_j^{(k)} \quad (7)$$

where $Z_j^{(k)}$ represents zero-mean white Gaussian noise. Because of the orthogonality among the subcarriers, H_{ji} becomes an $N \times N$ diagonal matrix of channel gains between ith transmit antenna and jth receive antenna. The received signal of mth subcarrier at the jth antenna in (7) can be rewritten as

$$X_j(kN+m) = \sum_{i=1}^{M_T} H_{ji}(m,m) S_i(kN+m) + Z_j(kN+m)$$
$$0 \leq m \leq N - 1. \quad (8)$$

For the all receive antennas ($1 \leq j \leq M_R$), we obtain

$$X(m) = \mathbb{H}(m) S(m) + Z(m), \quad (9)$$

with

$$X(m) = [X_1(kN+m), X_2(kN+m), \cdots, X_{M_R}(kN+m)]^T$$
$$S(m) = [S_1(kN+m), S_2(kN+m), \cdots, S_{M_T}(kN+m)]^T$$
$$Z(m) = [Z_1(kN+m), Z_2(kN+m), \cdots, Z_{M_R}(kN+m)]^T$$

$$\mathbb{H}(m) = \begin{pmatrix} H_{11}(m,m) & H_{12}(m,m) & \cdots & H_{1M_T}(m,m) \\ H_{21}(m,m) & H_{22}(m,m) & \cdots & H_{2M_T}(m,m) \\ \vdots & \vdots & \ddots & \vdots \\ H_{M_R 1}(m,m) & H_{M_R 2}(m,m) & \cdots & H_{M_R M_T}(m,m) \end{pmatrix}$$

It is clear From (9) that once the output signals of the DFT modulation are arranged in accordance with the index of subcarriers (m), the MIMO channel in the described system is presented as an instantaneous mixture. Therefore, the blind multiuser detection in this MIMO-OFDM system can be split into N BSS problems to obtain N un-mixing matrices related to N subcarriers ($0 \leq m \leq N-1$). Note that because of complex nature of symbols, the complex BSS should be applied over each FB track mixtures.

After solving the N complex ICA problems related to N subcarriers in (9) the separated FB tacks are obtained as

$$Y(m) = W(m)X(m) \qquad (10)$$

where $W(m)$ is the un-mixing matrix related to mth subcarrier. But even after successful separation by ICA, permutation indeterminacy, amplitude scaling ambiguity and phase distortion are new problems which necessitate user reconstruction as a post-BSS task.

It can be shown that the auto-correlation of adjacent pth and $p+1$st FB tracks recovered by complex BSS is as follows

$$E[Y_i(kN+p)Y_l^*(kN+p+1)]$$
$$= \begin{cases} 0 & \text{if } i \neq l, \\ R_1 A_{ip} A_{l,p+1} e^{(\theta_{ip} - \theta_{l,p+1})} \neq 0 & \text{if } i = l. \end{cases} \qquad (11)$$

where R_1 is a nonzero known value from (3), i and l are unknown user ownership indices, A_{ip} and $A_{l,p+1}$ are unknown scaling amplitudes, θ_{ip} and $\theta_{l,p+1}$ are unknown phase distortions. As it is seen in (11) the cross-correlation between adjacent FB tracks of the same user is a nonzero value, while it is zero for adjacent FB tracks of different users. Therefore, by doing a correlation-based grouping in the sequence from No. 1 FB to No. ($N-1$) FB, the permutation corrected multiuser symbols $\dot{Y}_i(kN+m)$ will be obtained.

After permutation alignment, the auto-correlation of the No.p FB track of ith user symbol-block will be

$$\phi_{ip}^{ip} = E[Y_i(kN+p)Y_i^*(kN+p)]$$
$$= A_{ip}^2 R_0 \qquad (12)$$

where A_{ip} is unknown amplitude scaling. So, by having R_0 from (2) the amplitude scaling of $\dot{Y}_{ip}$ can be corrected as follows

$$\ddot{Y}_i(kN+m) = \sqrt{\frac{R_0}{\phi_{im}^{im}}} \dot{Y}_i(kN+m). \qquad (13)$$

After permutation alignment and amplitude scaling correction, the phase deviation of each FB track with respect to its following one can be obtained as follows

$$\phi_{i,p+1}^{ip} = E[\ddot{Y}_i(kN+p)\ddot{Y}_{i,p+1}^*(kN+p+1)]$$
$$= e^{(\theta_{ip}-\theta_{i,p+1})} R_1 \qquad (14)$$

where θ_{ip} and $\theta_{i,p+1}$ are their unknown phase distortions. Since, the phase of R_1 is zero with the least estimation error,

$$e^{j(\theta_{ip}-\theta_{i,p+1})} = e^{j\angle \phi_{i,p+1}^{ip}} = \frac{\phi_{i,p+1}^{ip}}{|\phi_{i,p+1}^{ip}|}. \qquad (15)$$

Therefore by multiplying $\ddot{Y}_i(kN+p+1)$ by $\frac{\phi_{i,p+1}^{ip}}{|\phi_{i,p+1}^{ip}|}$, its phase deviation will be the same as $\ddot{Y}_i(kN+p)$. By doing this operation from No. 1 FB to No. $(N-1)$ FB, each FB track is adjusted to its preceding FB track which was adjusted to its prior one. So, the phase distortion of all FB tracks become θ_{i1}, that is a case similar to uncertain carrier phase in single carrier system. The same unknown resultant phase deviation for all symbols of the user can be eliminated by noncoherent detection [7]. At this point all OFDM user signals have been reconstructed, and they are ready to be transferred to user identification unit. Figure 1 shows the proposed ICA based MIMO–OFDM system.

3 Blind Source Separation

Here we face the instantaneous form of BSS. The mixing process is described as

$$x(t) = As(t) + n(t) \qquad (16)$$

where an n-dimensional vector $s(t) = [s_1(t), \cdots, s_n(t)]^T$ of independent sources is mixed by an $m \times n$ matrix A, and $x(t) = [x_1(t), \cdots, x_m(t)]^T$ is observed. BSS

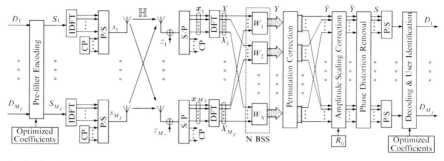

Fig. 1 The Proposed ICA based MIMO-OFDM system

problem is looking for the best $n \times m$ separation matrix W to extract signals $y(t)$ as much as possible close to unknown source signals. Without any prior knowledge of $s(t)$ and A except the independence assumption for source signals, $y(t)$ is recovered by BSS as follows

$$y(t) = Wx(t) \qquad (17)$$

where $y(t) = [y_1(t), \cdots, y_n(t)]^T$.

There are a lot of BSS methods. Here, we have deployed independent component analysis ICA with Kullback–Leibler contrast function. Because, it is known as the best performance BSS method. Since, the nature of symbols is complex, the extension of Kullback–Leibler (KL) ICA to complex domain is used [3]. For fast convergence of ICA algorithm, natural gradient learning algorithm (NGLA) [8] is used. NGLA has a hardware-friendly iterative structure that qualifies it as a better choice.

For more reducing the BSS complexity, an initial approximate to separating matrix is obtained by a low performance fast BSS method. Then the approximated matrix is given to Kullback–Leibler complex ICA as initial value. In this way NGLA starts with an initial separating matrix, and after a few iterations the high performance separating matrix is obtained. For the first step, we have used improved Stone's BSS [6] that is extended to complex domain. Stone BSS [9] is a fast method based on predictability maximization with scaling characteristics of $O(M_R^3)$. By applying two-stepped BSS, number of iterations is reduced in average from 130 to 30 iterations per each separation procedure.

4 Simulation Results

This section provides the simulation results of evaluation of performance of the proposed multiuser detection method. All results are compared with the joint detection (JD) method [10] over typical urban (TU) and hilly terrain (HT) channels. In the JD method, the channels are estimated using two training OFDM blocks. Then all user signals are recovered based on the obtained channel estimates, where zero-forcing algorithm is used for joint detection. In our evaluation, different $M_T \times M_R$ configurations of receiving two or three transmitted user signals by two, three, four or six antennas are employed. All OFDM parameters are the same for both methods. Signal constellation is DQPSK. The number of subcarriers is 64, and cycling prefix is 8. Carrier frequency and system bandwidth are respectively 0.5 GHz and 0.5 MHz.

Figures 2 and 3 demonstrate the comparison of the proposed method with JD method respectively in slowly fading ($f_d = 1.0 \times 10^{-6}$) and fast fading conditions ($f_d = 1.5 \times 10^{-4}$). f_d is the maximum Doppler frequency normalized with symbol rate. As it is seen, the performance of the proposed blind method is favorably comparable over slowly varying channels, and it is better than JD method over fast varying channels. Because the JD method can not accurately approximate the channels during the entire frame in fast variation of channels, while the proposed method does not require channel state information.

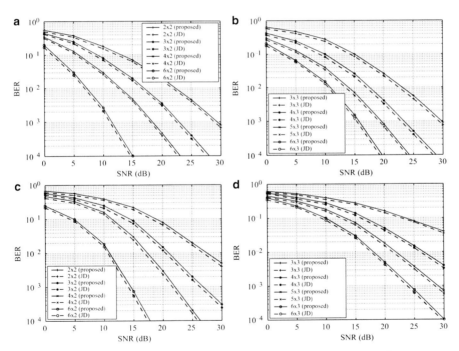

Fig. 2 BER comparison for different values of $M_R \times M_T$ while $f_d = 1.0 \times 10^{-6}$ over (**a** & **b**) tipycal urban channel, (**c** & **d**) hilly terrain channel

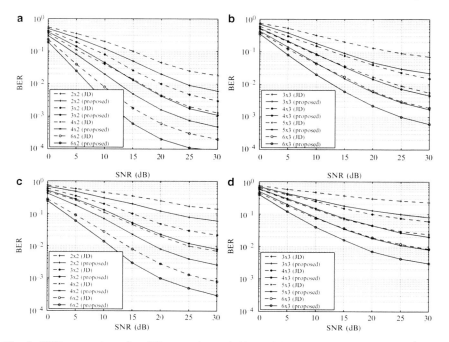

Fig. 3 BER comparison for different values of $M_R \times M_T$ while $f_d = 1.5 \times 10^{-4}$, over (**a** & **b**) typical urban channel, (**c** & **d**) hilly terrain channel

5 Conclusion

In this paper, we have proposed an efficient ICA based blind multiuser detection method for MIMO-OFDM systems. The problems inherent to complex ICA have been successfully solved in the proposed method. Furthermore, ICA complexity has been substantially reduced. Since, the coefficients of pre-filter are optimized, the introduced characteristics to transmitted symbols are accurate and the method is precise. On the other hand, it is robust and channel independent because the characteristics of the transmitted signals only depends on the optimized pre-filter coefficients. Computer simulations also demonstrates the efficiency of the proposed method.

Appendix

The auto-correlation of an FB track of encoded OFDM symbols by FIR pre-filter of (1) can be written as

$$R_0 = E\left[\left|\sum_{l=0}^{L_f-1} c(l)D_i(kN+m-l)\right|^2\right]. \quad (18)$$

where L_f and $c(.)$ are respectively the length and coefficients of the FIR pre-filter. $E[.]$ is the expectation with respect to k. Since $D_i(kN+m)$ is an information symbol modulated by an M-PSK constellation, R_0 can be written as

$$R_0 = E\left[\left|\sum_{l=0}^{L_f-1} c(l)cos\alpha_{m-l,k} + j\sum_{l=0}^{L_f-1} c(l)sin\alpha_{m-l,k}\right|^2\right]$$

$$= E\left[\left|\sum_{l=0}^{L_f-1} c(l)cos\alpha_{m-l,k}\right|^2\right] + E\left[\left|\sum_{l=0}^{L_f-1} c(l)sin\alpha_{m-l,k}\right|^2\right] \quad (19)$$

where $\alpha_{k,m-l}$ is the phase of $D_i(k,m-l)$. From the above equation, it can be obtained

$$R_0 = \sum_{l=0}^{L_f-1} c^2(l) + E\left[\sum_{\substack{l_i=0 \\ l_i \neq l_j}}^{L_f-1}\sum_{l_j=0}^{L_f-1} 2c(l_i)c(l_j)cos(\alpha_{m-l_i,k} - \alpha_{m-l_j,k})\right] \quad (20)$$

Let $\alpha_{l_i l_j} = \alpha_{k,m-l_i} - \alpha_{k,m-l_j}$. Since the used PSK modulation is without initial phase, it can be shown that $\alpha_{l_i l_j}$ is the phase of a point in the same constellation, and there is equal probability for $\alpha_{l_i l_j} = \alpha_1$ and $\alpha_{l_i l_j} = \alpha_2$, where $cos\alpha_1 = -cos\alpha_2$. Therefore, the expectation of second term of (20) is zero, and (2) is obtained.

Similarly, the correlation between m^{th} and $m+1^{st}$ FB tracks is

$$R_1 = \sum_{l=0}^{L_f-2} c(l)c(l+1) + E\left[\sum_{\substack{l_i=0 \\ l_j \neq l_i+1}}^{L_f-1}\sum_{l_j=0}^{L_f-1} c(l_i)c(l_j)e^{j(\alpha_{l_i}-\alpha_{l_j})}\right] \quad (21)$$

where $E[.]$, $c(.)$ and L_f are the same as used in (18). Since, constellation points are symmetrically located around the unit circle, and $e^{j(\alpha_{l_i}-\alpha_{l_j})}$ equals each of them with equal probability, the second term in (21) equals zero, and (3) is concluded.

References

1. Hemanth Sampath, Shilpa Talwar, Jose Tellado, Vinko Erceg and Arogyaswami Paulraj, "A fourth-generation MIMO-OFDM broadband wireless system: design, performance, and field trial results", *IEEE Commun. Mag.*, vol. 40, no. 9, pp. 143–149, Sep. 2002.
2. Chiu Shun Wong, Dragan Obradovic, Nilesh Madhu, "Independent component analysis (ICA) for blind equalization of frequency selective channels" *in Proc. 13th IEEE Workshop Neural Networks Signal Processing*, pp. 419–428, 2003.

3. Paris Smaragdis, "Blind separation of convoluted mixtures in frequency domain", *J. Neurocomputing*, vol. 22, Issues 1–3, pp. 21–34, 1998.
4. Aapo Hyvärinen, Juha Karhunen, Erkki Oja, *Independent Component Analysis*, Wiley, New York, 2001.
5. Mahdi Khosravy, Mohammad Reza Alsharif , Bin Guo, Hai Lin and Katsumi Yamashita, "A Robust and Precise Solution to Permutation Indeterminacy and Complex Scaling Ambiguity in BSS-based Blind MIMO-OFDM Receiver", *in ICA 2009*, T. Adali, Eds., *Springer-Verlags Lecture Notes in Computer Science, LNCS 5441*, pp. 670–677, 2009.
6. Mahdi Khosravy, Mohammad Reza Alsharif, Katsumi Yamashita, "A Probabilistic Short-length Linear Predictability Approach to Blind Source Separation" *in 23rd International Technical Conference on Circuits/Systems, Computers and Communications (ITC-CSCC2008)*, pp. 381–384, Yamaguchi, Japan, July, 2008.
7. Richard Van Nee, Ramjee Prasad, *OFDM for Wireless Multimedia Communications*, Artech House, 2000.
8. S. Amari, A. Cichocki and H. H. Yang,"A New Learning Algorithm for Blind Signal Separation", *Advances in Neural Information Processing Systems 1996*, Vol. 8, MIT Press, Cambridge MA, pp. 757–763, 1996.
9. James V. Stone, "Blind source separation using temporal predictability", *Neural computation*, 13(7), pp. 1559–1574, 2001.
10. Alexandros Sklavos, Tobias .E. Costa, Harald Haas, and Egon Schulz, "Joint detection in multi-antenna and multi-user OFDM systems", *in Multi-carrier Spread-Spectrum and Related Topics*, K. Fazel, S. Kaiser, Eds. Kluwer Boston, MA, 2002.

Windowing in the Receiver for OFDM Systems in High-Mobility Scenarios

Eva Peiker, Werner G. Teich, and Jürgen Lindner

Abstract Wireless connections for high-mobility scenarios lead to high demands on the applied transmission scheme. A typical scenario is the wireless communication with high speed trains, where the channel becomes fast time-variant. In this case orthogonal frequency division multiplexing (OFDM) suffers from inter-carrier interference (ICI) and therefore performance degradation occurs. We present an OFDM based system applying windowing in the receiver to fight ICI, without undertaking major modifications of the underlying conventional OFDM system. Two possibilities, how to apply windows, will be discussed in this paper.

1 Introduction

The availability of connections with high data rates to users, moving with high velocities, becomes a more and more important property of mobile systems. But a well-known problem is, that the performance of a mobile wireless system degrades severely because of Doppler spread. In fast time-variant channels orthogonal frequency division multiplexing (OFDM) suffers from inter-carrier interference (ICI), induced by Doppler spread. Without countermeasures, this leads to an increased performance degradation with increasing Doppler spread.

In the past different approaches were investigated to deal with this problem. A new hybrid transmission scheme based on a combination of M-array frequency shift keying and differential phase shift keying is introduced in [1]. Reference [2] uses soft shaping with Gauss pulses at the transmit and receive side to concentrate the ICI on adjacent subcarriers for a less complex equalization. In [3] the unused part of an oversized guard interval is utilized to reduce the ICI with the application of Nyquist windows at the receive side. In [4] we developed an approach to reduce ICI by using windowing in the receiver. The disadvantage of this approach is, that, due

E. Peiker (✉), W.G. Teich, and J. Lindner
Institute of Information Technology, University of Ulm, Albert-Einstein-Allee 43,
89081 Ulm, Germany
e-mail: {eva.peiker, werner.teich, jürgen.lindner}@uni-ulm.de

to an increased symbolperiod, the data rate decreases compared to a transmission with a conventional OFDM system using a rectangular window. In this paper, we use the approach introduced in [4], but we improve the data rate by shortening the transmit time and by exploiting Doppler diversity [8–11].

2 System Model

In [4] and [5], we have shown, that the application of windowing in the receiver leads to good performance improvements by reducing the ICI. We assumed a conventional OFDM transmitter, the only difference being an additional pre- and postfix. The receiver was modified in a way that other window functions than the common rectangular function are applied to reduce the effect of Doppler spread.

2.1 Transmitter

In comparison to a conventional OFDM system, the only difference at the transmitter of our system is an additional pre- and postfix which is appended together with the guard interval T_g. We use windows, parameterized with a roll-off factor α, ranging between 0 and 1. To avoid intersymbol interference (ISI) the symbol duration T has to be increased by a factor of $(1 + \alpha)$. Therefore the N samples coming out of the IDFT are repeated periodically in the time domain, like it is done for the samples N_g of the guard interval. The number of additional samples in pre- and postfix depends on the roll-off factor α of the used window

$$N_{prefix} = N_{postfix} = N \cdot \frac{\alpha}{2}. \qquad (1)$$

Equation (1) corresponds to the approach we used in [4] and [5]. The disadvantage of it can be seen immediately: the data rate decreases compared to the one of a conventional OFDM system with $\alpha = 0$ by the following factor

$$\frac{N_g + N}{N_g + N_{prefix} + N + N_{postfix}} = \frac{N_g + N}{N_g + N \cdot (1 + \alpha)} \qquad (2)$$

The idea of this work is to investigate the trade-off obtained by shortening the pre- and postfix. On the one hand this leads to an increased data rate r. On the other hand, cutting the window function at the receiver in general results in ICI, because the corresponding subcarriers are no longer orthogonal, even in the Doppler free case. Figure 1 gives an overview of the advanced transmission system from [4].

We modify the length of the transmit time $T_s = T_g + T_{prefix} + T + T_{postfix}$, with T_{prefix} and $T_{postfix}$ representing the length of the additional pre- and postfix, by shortening the pre- and postfix:

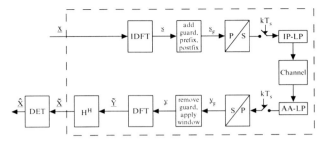

Fig. 1 Transmission system. IDFT: inverse discrete Fourier transform, DFT: discrete Fourier transform, P/S: parallel to serial conversion, S/P: serial to parallel conversion, IP-LP: interpolation lowpass, AA-LP: anti-aliasing lowpass, H^H: the Hermitian of the channel, DET: detection

$$N_{prefix} = N_{postfix} = N \cdot \kappa \cdot \frac{\alpha}{2}$$
$$\text{for } \kappa \in \left\{0, \frac{1}{8}, \frac{1}{4}, \frac{1}{2}, \frac{3}{4}, 1\right\} \tag{3}$$

A shortening factor of $\kappa = 0$ is therefore equivalent to the duration in time of a rectangular window used for a conventional OFDM transmission, which has not any additional pre- and postfix, whereas $\kappa = 1$ refers to the case we already investigated in [4] and [5].

2.2 Receiver

At the receiver (Fig. 1) the N_g samples of the guard interval are discarded as usual and the remaining ones are weighted with the used window function. Periodic repetition is one of the properties of the DFT in time as well as in frequency domain. The given sampling distance in frequency, in our case the subcarrier spacing Δf defines the repeat rate in time T_{rep}.

$$T_{rep} = T = \frac{1}{\Delta f} \tag{4}$$

This results in overlapping time samples which are added, so that the size of the DFT stays the same [3]. In the discrete domain the added samples have the distance N. The longer transmission time per symbol results in a loss in $\frac{E_b}{N_0}$. This loss is partly canceled through the signal combining gain resulting from the weighting and adding of the time samples [4].

The innovation at the receiving side, compared to [4], are the detection and the matched filtering with the hermitian of the channel H^H. Normally the detection consists of equalization and decoding. For equalization we use the block decision feedback equalizer (BDFE) [7]. Since we do not apply coding, it is not necessary to realize decoding at the receiver.

3 Windows

We use the raised cosine window (RC), the "better than" raised cosine window (BTRC), introduced in [6], the trapezoidal window and the mirrored raised cosine window (mirrored RC), introduced in [4]. The trapezoidal window is equal to the Bartlett window for $\alpha = 1$.

Figure 2 demonstrates the two effects of pulse shaping for a roll-off factor $\alpha = 1$ and no shortening of the additional pre- and postfix, i.e. $\kappa = 1$. First, the application of windowing leads to a narrower main lobe and secondly it reduces the side lobes, with the raised cosine window reducing the most. It is obvious, that windows with a stronger side lobe attenuation have wider main lobes and vice versa.

We apply the windows in two different ways compared to [4]: On the one hand, the roll-off factor α of the window is chosen in such a way, that it fits completely into the time period ($T_{prefix} + T + T_{postfix}$). An example is given in Fig. 3. This approach does not lead to additional ICI even in the Doppler free case. However, as can be seen in Fig. 3a, the window function becomes more and more similar to the rectangular one as the roll-off factor α decreases. At the same time (see Fig. 3b) the mainlobes of the corresponding spectra become wider and almost reach the width of the rectangular window. This directly implies a loss of robustness against increasing Doppler spread.

On the other hand, the roll-off factor α of the windows is chosen independently of the time period ($T_{prefix} + T + T_{postfix}$). In this case, the window functions have to be cut (see Fig. 4a) to fit into the available time period ($T_{prefix} + T + T_{postfix}$), i.e. the

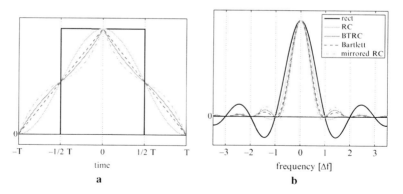

Fig. 2 (a) Rectangular window and different Nyquist windows for $\alpha = 1$ and $\kappa = 1$. (b) Corresponding spectra of the rectangular window and the Nyquist windows for $\alpha = 1$ and $\kappa = 1$

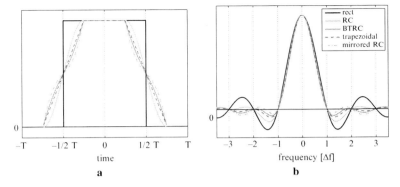

Fig. 3 (a) Rectangular window and different Nyquist windows for $\alpha = \frac{1}{2}$ and $\kappa = 1$. (b) Corresponding spectra of the rectangular window and the Nyquist windows for $\alpha = \frac{1}{2}$ and $\kappa = 1$

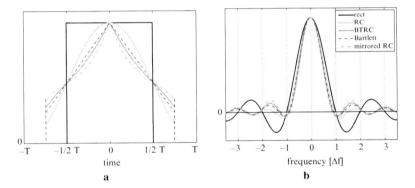

Fig. 4 (a) Rectangular window and different cut Nyquist windows for $\alpha = 1$ and $\kappa = \frac{1}{2}$. (b) Corresponding spectra of the rectangular window and the cut Nyquist windows for $\alpha = 1$ and $\kappa = \frac{1}{2}$

pre- and postfix is shortened ($\kappa < 1$). For large enough roll-off factor α, the mainlobes of the corresponding spectra (see Fig. 4b) are still narrower than the mainlobe of the rectangular window. However, compared to the uncut case (see Fig. 2b), the mainlobes are closely together. Therefore it is expected, that the different window functions show comparable robustness against increasing Doppler spread. Since the windows loose their property of being a Nyquist window, additional ICI occurs even in the Doppler free case and will degrade the performance if no proper countermeasures are taken.

Thus, in both cases, the idea is to make use of the narrower main lobe for $\alpha > 0$ in contrast to the one of the rectangular window. Consequentially, if we cut the windows, it is best to use an α-value of 1 to achieve maximum dissimilarity of the mainlobe compared to the rectangular window.

In presence of a time variant channel spectra are always spread or shifted, what, on the other hand, leads to a degradation in performance due to ICI. High side lobes result in accumulated ICI over all subcarriers whereas a wider main lobe strongly increases the ICI originating from the adjacent subcarriers. Therefore a compromise has to be found.

4 Simulation Results

For simulation we choose a two-path channel model with channel impulse response $h = [1\ 1]$. For proper pulse shaping 64 subcarriers are used and we choose a guard interval T_g of $\frac{1}{4}T$ to avoid interference between subsequent symbols. The maximum normalized frequency offset

$$\epsilon = \frac{f_{Doppler}}{\Delta f} \quad (5)$$

affects the impulse response with $[\epsilon\ 0.8\epsilon]$ and therefore Doppler spread occurs. For modulation 4QAM is chosen. Coding has not been considered in this paper. We assume perfect channel knowledge at the receiver and for equalization we use a block decision feedback equalizer (BDFE). The data rate for all the presented simulation results is the same. We adjust the length of the guard interval T_g so that the data rate becomes the same compared to the used windows.

The fast time-varying channel induces Doppler spread, which can be used as a natural source of diversity, called Doppler diversity [8–11]. Since we assume perfect channel knowledge, the maximum benefit of exploiting Doppler diversity is obtained. To distinguish between the gain achieved through application of windowing, the Doppler spread and equalization we make use of matched filter bound (MFB) simulations, where the ICI is subtracted in an ideal way. The BER is shown over $\frac{E_b}{N_0}$.

Figure 5 presents the BER resulting from a 4QAM transmission with different windows with a roll-off factor $\alpha = 1$ and full pre- and postfix, i.e. $\kappa = 1$. ϵ is chosen to be 0.5 and the BDFE is used for equalization. It is obvious that the rectangular window performs worst of all of the chosen windows, followed by the raised cosine window. The reason for this lies in the wider main lobe in the frequency domain compared to the other ones. Since the mirrored raised cosine window (Fig. 2b) has the narrowest main lobe, it performs best.

In Fig. 6 the matched filter bound of two different 4QAM transmissions with different windows is shown. On the left (Fig. 6a) the Nyquist windows are parameterized with a roll-off factor $\alpha = 1$, but the windows are shortened by $\kappa = \frac{1}{2}$, i.e. we only use half of the time for pre- and postfix and cut the windows, as it is shown in Fig. 4. For $\epsilon = 0.2$ the matched filter bound of the rectangular window performs slightly better compared to the one of the mirrored RC window. Whereas in Fig. 6b we adjust the length of the window in the time domain by choosing α in a way that the window fits completely into the available transmit time, without the need of cutting the window. Here again, the matched filter bound of the rectangular window for $\epsilon = 0.2$ performs better compared to the rectangular window. In both cases we take

Fig. 5 BER for a 4QAM transmission with different windows ($\alpha = 1$, $\kappa = 1$, $\epsilon = 0.5$) using the BDFE for equalization, no coding

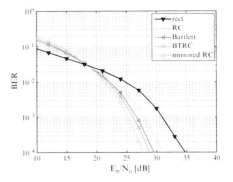

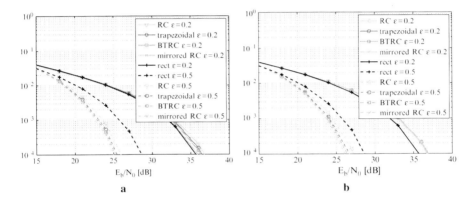

Fig. 6 BER for a 4QAM transmission to compare the matched filter bound of different windows (no coding) and different ϵ (**a**) $\alpha = 1$ and $\kappa = \frac{1}{2}$, (**b**) $\alpha = \frac{1}{2}$ and $\kappa = 1$

advantage of the narrower main lobe of the windows. But if Fig. 6a and Fig. 6b are compared, it is obvious that the matched filter bound of Fig. 6a performs better for both chosen ϵ-values compared to the one of Fig. 6b. Herefrom it can be derived, that it is better to cut the Nyquist windows and accept the additional interference, which occurs, because the spectra of the windows are no longer orthogonal even in the Doppler free case. To fight the ICI we need equalization.

The BER of a 4QAM transmission with different shortened windows ($\alpha = 1$, $\kappa = \frac{1}{2}$) for $\epsilon = 0.5$ using the BDFE for equalization is shown in Fig. 7a. The rectangular window performs worst and again it is obvious that windows with a narrower main lobe perform better than the ones with a wider main lobe (see Fig. 4). Compared to the matched filter bound of that transmission, which is shown in Fig. 6a, the performance is about $5dB$ worse.

Figure 7b presents the BER of the matched filter bound for a simulation with the BTRC window and the trapezoidal window, both with different α-values, but the same $\kappa = 0$, i.e. the windows are cut to the duration T of the rectangular window

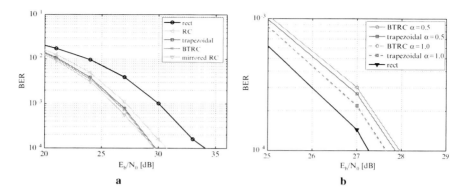

Fig. 7 (**a**) BER for a 4QAM transmission with different windows ($\alpha = 1, \kappa = \frac{1}{2}, \epsilon = 0.5$) using the BDFE for equalization, no coding. (**b**) BER for a 4QAM transmission to compare the matched filter bound of different windows ($\alpha = 1$ or $\alpha = \frac{1}{2}, \kappa = 0, \epsilon = 0.5$)

for a conventional OFDM transmission. The rectangular window performs slightly better than the BTRC window and the trapezoidal window. But, since we assume perfect channel knowledge at the moment, we expect the BTRC window and the trapezoidal window to perform better, if the channel is not perfectly known anymore.

5 Conclusions

We modified the idea of [4] by shortening the additional needed pre- and postfix by a factor of κ or by adopting the length of the window using its roll-off factor α to achieve a higher data rate. Generally, we still need additional transmit time compared to the rectangular window, which is used for a transmission over a conventional OFDM System. The simulation results show, that the performance of the cut Nyquist windows is still better compared to the rectangular window at high Doppler spreads. Up to now a two-path channel with Doppler spread and perfect channel knowledge has been used. The ICI, which is resulting from the Doppler spread, has been reduced by using windowing, but with the aspect, that the data rate decreases less compared to [4]. We used the property, that the narrower the main lobe, the less symbols of the other subcarriers are influenced. Up to now coding hasn't been considered yet, but we expect improvements. Since we assume perfect channel knowledge, we also gain high benefits because of Doppler diversity and the rectangular window still performs quite well. We expect a degradation of both, the benefit of Doppler diversity and the performance of the rectangular window, when perfect channel information is not available. But in the same step, our windows' performance should improve obviously.

Acknowledgements Some simulation results of this work were carried out by Daniel Bihlmeier within his master thesis. Thank you for the simulations.

References

1. M. Wetz, I. Perisa, W. G. Teich, J. Lindner *OFDM-MFSK with Differentially Encoded Phases for Robust Transmission over Fast Fading Channels*, Proc. 11th International OFDM Workshop, pp. 313–317, Hamburg, Germany, 30–31 August 2006.
2. S. Vogeler, P. Klenner and K. Kammeyer *Multicarrier Transmission for Scenarios with High Doppler Influence*, 10th International OFDM-Workshop, Hamburg, Germany, August 2005.
3. S. Müller-Weinfurtner and J. Huber, *Robust OFDM Reception with Near-Optimum Nyquist Window*, In Proc. of 50th IEEE Vehicular Technology Conference (VTC 1999–Fall), Amsterdam, The Netherlands, September 1999, pp. 289–293.
4. E. Peiker, J. Dominicus, W. G. Teich, J.Lindner, *Reduction of Inter-Carrier Interference in OFDM Systems through Application of Windowing in the Receiver*, 13th International OFDM-Workshop, Hamburg, Germany, August 2008.
5. E. Peiker, J. Dominicus, W. G. Teich, J. Lindner, *Improved Performance of OFDM Systems for Fast Time-Varying Channels*, ICSPCS'2008 (2nd International Conference on Signal Processing and Communication Systems), Gold Coast, Australia, December 2008.
6. P. Tan and N. Beaulieu, *Reduced ICI in OFDM Systems Using the Better Than Raised-Cosine Pulse*, IEEE Communication Letters, Vol. 8, No. 3, March 2004.
7. D. Y. Yacoub, *Spreading and Precoding for Wireless MIMO-OFDM Systems*, PhD thesis, Ulm, Germany, December 2007.
8. R. Boudreau, JLY. Chouinard and A. Yongacaglu, *Exploiting Doppler-diversity in Flat, Fast Fading channels*, Canadian Conference on Electrical and Computer Engineering, Vol. 1, 2000.
9. A.M. Sayeed, B. Aazhang, *Joint Multipath Doppler Diversity in Mobile Wireless Communications*, IEEE Transactions on Communications, May 1997.
10. R. Raulefs, A. Dammann, S. Kaiser, G. Auer, *The Doppler Spread - Gaining Diversity for Future Mobile Radio Systems*, GLOBECOM 2003, San Francisco, United States of America, December 2003.
11. T. Zemen, *OFDM Multi-User Communication Over Time-Variant Channels*, PHD thesis, Technische Universität Wien, Fakultät fr Elektrotechnik und Informationstechnik, Wien, Austria, July 2004.

Part II
Multiple-Input and Multiple-Output (MIMO)

MIMO–OFDM with Doppler Compensating Antennas in Rapidly Fading Channels

Peter Klenner, Lars Reichardt, Karl-Dirk Kammeyer, and Thomas Zwick

Abstract Multiple-input-multiple-output (MIMO) systems are well-known to improve capacity compared to single antenna transceivers. The V-BLAST scheme, where independent data streams are transmitted from several antennas, is a common representative of such a spatial multiplexing system. Originally conceived for flat-fading channels it can be easily utilized in frequency-selective channels using OFDM. However, MIMO–OFDM under high-mobility conditions requires frequent training and is exposed to intercarrier interference. In this paper, it is demonstrated via simulations that these issues can be resolved by employing Doppler compensating antenna structures at the receiver.

1 Introduction

Spatial multiplexing, also known as BLAST, increases the capacity of wireless radio links [1]. Combined with OFDM the potentially very hostile frequency-selective MIMO channel can be easily utilized, such that well-known narrow-band receivers with low complexity can be employed for spatial equalization.

However, the difficult task of channel estimation is to acquire the channel response between each pair of transmit and receive antenna. Unfortunately, the number of channel parameters, to be estimated at the receiver, increases with the number of resolvable paths and transmit antennas. For the time-invariant case a Least Squares channel estimator for OFDM with multiple transmit antennas is derived in [2]. In [3] a frequency orthogonal training scheme is proposed which commits complete OFDM symbols for training by devoting equidistant subcarriers to a particular transmitter.

P. Klenner (✉) and K.-D. Kammeyer
University Bremen, Department of Communications Engineering, 28359 Bremen, Germany
e-mail: {klenner,kammeyer}@ant.uni-bremen.de

L. Reichardt and T. Zwick
Universität Karlsruhe (TH), Institut für Hochfrequenztechnik und Elektronik,
76128 Karlsruhe, Germany
e-mail: {lars.reichardt,thomas.zwick}@ihe.uka.de

In high-mobility scenarios the channel fluctuates during signalling and the number of channel parameters is further increased. Moreover, the subcarriers are loosing their orthogonality and crosstalk between them occurs, which is commonly called intercarrier interference (ICI). Based on the central limit theorem Doppler-induced ICI can be modeled as a Gaussian process [4].

Literature contains a number of methods for ICI compensation, e.g. time-domain windowing and filtering for SINR-maximization, ICI self-cancellation methods based on transmitter-side induced redundancy and successive interference cancellation schemes. Basically these methods rely on omnidirectional antennas. However, in the mid-1960s it has been demonstrated that directional antennas on a mobile receiver can reduce the channel fading rate [5]. This effect is based on the relationship of the angle of incidence of an impinging wave and its corresponding Doppler frequency. Thus by restricting the angles of incidence to a fixed range a directional antenna reduces the Doppler spread and the fading rate of the effective channel.

In this paper the Doppler compensating effect of directive receive antennas is exploited for a V-BLAST MIMO–OFDM system to achieve high data-rates in high mobility scenarios. Simulations using statistical channel models with isotropic scattering show that directive antennas provide a superior BER performance over omnidirectional antennas. However, the BER performance of directional reception deteriorates if the wave propagation is dominated by line-of-sight components. This is evidenced by a communication scenario between two cars which is modelled using Ray-Tracing techniques.

2 System Model

The OFDM signal in the equivalent complex baseband transmitted from the tth antenna at time instant k reads

$$x(k,t) = \sqrt{\frac{\mathcal{P}}{N}} \sum_{n=0}^{N-1} d(n,t) e^{j2\pi nk/N}, \quad -N_g \leq k \leq N-1. \tag{1}$$

Thereby, the transmit power, the number of subcarriers and the length of the cyclic prefix are denoted by $\mathcal{P}$, N and N_g, respectively. The data $d(n,t)$ is drawn from an M-PSK/QAM constellation. The channel impulse response between the t-th transmit and r-th receive antenna is denoted by $h(k,\ell,t,r)$, which can model omnidirectional as well as directional reception. Details are given in the following Section 2.1. The received signal at the rth antenna reads

$$y(k,r) = \sum_{\ell=0}^{L-1} \sum_{t=0}^{N_T-1} h(k,\ell,t,r) x(k-\ell,t) + \eta(k,r). \tag{2}$$

The number of resolvable paths is given by L, additive white Gaussian noise of unit variance is denoted by $\eta(k,r)$.

2.1 Directive Receive Antennas

The Doppler spread reducing effect of directional antennas has been recognized and demonstrated in the mid-1960s [5] and is based on the correspondence between incidence angle and Doppler shift of impinging waves. Thus, by restricting the range of incidence angles a directional antenna reduces the Doppler spread, which leads to less rapidly fading channels compared to an omnidirectional antenna.

Note that the attenuation of received power from certain directions is compensated by a power gain from the main lobe, such that in an isotropic and richly scattering environment a directional antennas behaves power-neutral [5]. But in environments with directive scattering the main beam of a single directional antenna needs to be steered towards the scatterer, since otherwise no useful power will be received. Since beam steering in high-mobility scenarios is a formidable task, we follow an alternative approach and employ a number of fixed beams, pointed in such directions as to achieve equispaced Doppler subspectra. Details are described in [6].

A possible implementation of low complexity can be realized with an uniform linear antenna array (ULA) [7] oriented into the direction of motion. The array pattern of an ULA with four elements and interelement distance $\lambda/4$ is shown in Fig. 1. In particular, Fig. 1a shows the endfire configuration for which the chosen phase shift between subsequent antenna elements leads to a single main lobe pointed towards the moving direction with a half-power beamwidth of approximately $120°$. Furthermore, Fig. 1 shows the broadside configuration for which the phases between subsequent antenna elements lead to two main lobes facing in directions perpendicular to the array's alignment. Its half-power beamwidth is approximately $60°$. In conclusion, a ULA facilitates the generation of simultaneous beams pointed into distinct directions, but lacks the fine adjustment of the beamwidths to produce exactly equisized Doppler subspectra. Nevertheless the approximation due to the ULA will serve the purpose of Doppler compensation.

Uncorrelated channels are an essential prerequisite for MIMO transmission. Assuming isotropic scattering directional antennas are particularly suited to facilitate

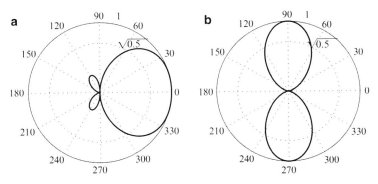

Fig. 1 Array pattern $|G(\varphi)|$ of an uniform phased linear antenna array with four elements and interelement distance $\lambda/4$, (**a**) endfire configuration, (**b**) broadside configuration

MIMO transmission under high mobility constraints since each sector receives from its mainlobe a majority of paths with different phases. Minor crosstalk between adjacent sectors due to pattern side lobes occurs, but does not lead to significant correlations.

2.2 Modelling the Impulse Response of Directional Antennas

Focusing on the receiver, the impulse response at the r-th sector antenna reads

$$h(k,\ell,r) = \frac{1}{\sqrt{N_\tau}} \sum_{\mu=0}^{N_\tau-1} G(\varphi(\mu,r)) a(\mu,r) e^{j(2\pi f(\mu,r)Tk + \phi(\mu,r))} g(\ell - \tau(\mu,r)) \quad (3)$$

and is constituted of N_τ paths, each with an incidence angle $\varphi(\mu,r)$, an amplitude $a(\mu,r)$, a delay $\tau(\mu,r)$, a phase $\phi(\mu,r)$ and Doppler frequency $f(\mu,r) = f_D \cos(\varphi(\mu,r))$, where the maximum Doppler frequency is denoted as f_D. The array pattern $G(\cdot)$ provides the directional behavior of each sector antenna. Low-pass filtering is accomplished by $g(\cdot)$. Assuming mutual uncorrelatedness between these parameter, known as WSSUS assumption, (3) is commonly used in literature to generate impulse responses according to predefined power delay profiles and Doppler spectra.

Note that the center frequencies of some sector antennas are offset from DC [6]. In fact, the channel's reduced time selectivity emerges not until this frequency offset is corrected, which is simply achieved by

$$\tilde{y}(k,r) = e^{-j2\pi f_C(r)Tk} y(k,r) \quad (4)$$

where $f_C(r)$ denotes the frequency focal point of the rth sector [6]. After removing the cyclic prefix the DFT yields the received signal in frequency domain

$$Y(n,r) = \frac{1}{\sqrt{N}} \sum_{k=0}^{N-1} \tilde{y}(k,r) e^{-j2\pi kn/N}, \quad (5)$$

which is further processed to retrieve channel estimates and eventually the transmitted data.

2.3 Ray-Tracing Simulation

Impulse responses originating from (3) lack the proper characterization of realistic scenarios and are supplemented by Ray-Tracing simulations of a typical car-to-car communication. The used Ray-Tracing simulator can be divided into three major parts, a realistic dynamic road traffic model, a comprehensive model of the environment adjacent to the road, and an accurate model for the multipath wave propagation between the transmitting and the receiving vehicles. The time-variant behavior of the channel is influenced by the movement of the transmitting and the receiving vehicles together with the dynamic behavior of other adjacent cars. For a proper characterization of the transmission channel it is therefore necessary to ensure a realistic modelling of the dynamic road traffic. The so-called Wiedemann traffic model is implemented to ensure a realistic description of traffic [8]. It considers the motion of all vehicles with different velocities and several acceleration and deceleration procedures and provides consecutive snapshots of the time-variant positions of the vehicles.

Furthermore a detailed description of the vehicle's vicinity is essential for the wave propagation modelling. The current application requires the generation of typical road traffic environments, which include the moving cars themselves, the lanes and the environment adjacent to the road. In order to generalize the approach, a stochastic model of the environment is used including the stochastic generation of both, the course of the road and the position of objects (e.g. buildings, parking cars) beside the road.

An accurate description of the multipath wave propagation in the aforementioned traffic scenarios is required to produce realistic time series of impulse responses of the channel. Ray-optical methods provide an adequate means for this purpose and are therefore implemented in the channel model. The results of this channel modelling were verified by measurements [9]. Ray optics are based on the assumption, that the wavelength is small compared to the dimensions of the objects in the simulation scenario. If this is true, different types of propagation phenomena (e.g. reflection, diffraction, scattering) can be considered. Each path is represented by a ray, which may experience consecutively several different propagation phenomena. Multiple reflections, multiple diffractions and single scattering are taken into account as propagation phenomena.

The simulation for a high-speed motorway scenario is performed. The motorway consists of six lanes, three in each direction. The traffic density is medium and the vicinity beside the motorway is vegetation, i.e., trees which are modelled as diffuse scattering objects [10]. The speed of the receiving and transmitting car is about 200 km/h resulting in a relative speed of 400 km/h. The simulation (real) time for each scenario is 10 s. In the scenario the receiving and transmitting car are driving into different directions. In the beginning of the simulation they are driving towards another, than after 1.5 s they are passing each other.

Six different antenna positions are chosen for the transmitter and the receiver. The different placements are at the front (0) and rear (1) bumper, in the left (2) and right (3) side mirror, under the car (4) and on the roof (5). The numbers are used later to identify transmitting and receiving positions.

3 Channel Estimation and Equalization

Least-squares channel estimation will be employed subsequently since it does not require any knowledge about the channel statistics (noise power, power delay profile). The details of the estimation of the average channel impulse response between all transmit antennas and the r-th receive antenna are as follows. Every Δ_f-th subcarrier contains training data, and after reception these subcarriers are stacked in the vector

$$Y_p(r) = [\ Y(0,r),\ Y(\Delta_f,r),\ Y(2\Delta_f,r),\ Y(4\Delta_f,r), \cdots\ ]^T. \qquad (6)$$

Denoting the first N_g columns in the m-th row of the DFT matrix as $\mathbf{F}(m)$ allows the definition of the matrix $\tilde{\mathbf{F}}$ as

$$\tilde{\mathbf{F}} = [\ \mathbf{F}(0)^T,\ \mathbf{F}(\Delta_f)^T,\ \mathbf{F}(2\Delta_f)^T,\ \cdots\ ]^T. \qquad (7)$$

The operator $\mathcal{D}\{\mathbf{x}\}$ produces a diagonal matrix with vector $\mathbf{x}$ on its main diagonal and allows the definition of the following matrix

$$\mathbf{M} = [\ \mathcal{D}\{\mathbf{d}(0)\},\ \mathcal{D}\{\mathbf{d}(1)\},\ \cdots,\ \mathcal{D}\{\mathbf{d}(N_T-1)\}\ ](\mathbf{I}_{N_T} \otimes \tilde{\mathbf{F}}), \qquad (8)$$

whose pseudoinverse $\mathbf{M}^+$ yields an estimate of the mean channel impulse responses

$$[\hat{\boldsymbol{h}}(0,r)^T, \hat{\boldsymbol{h}}(1,r)^T, \cdots, \hat{\boldsymbol{h}}(N_T-1,r)^T]^T = \mathbf{M}^+ \boldsymbol{Y}_p(r) \qquad (9)$$

with $\hat{\boldsymbol{h}}(t,r) = [\hat{h}(0,t,r), \hat{h}(1,t,r), \cdots, \hat{h}(L-1,t,r)]^T$. The scalars $\hat{h}(\ell,t,r)$ are Least Squares estimates for the mean of the ℓ-th channel tap, i.e., for $\bar{h}(\ell,t,r) = \frac{1}{N} \sum_{k=0}^{N-1} h(k,\ell,t,r)$. Eventually, processing the estimated channel impulse responses between every antenna pair with the DFT yields an estimate of the channel transfer functions

$$\hat{H}(n,t,r) = \sum_{\ell=0}^{N_g-1} \hat{h}(\ell,t,r) e^{-j2\pi n\ell/N}. \qquad (10)$$

Two equalization approaches will be applied, Zero-Forcing (ZF) and sphere detection (SD). The estimates of the channel transfer functions for the n-th subcarrier are collected in the matrix

$$\hat{\boldsymbol{H}}(n) = \begin{pmatrix} \hat{H}(n,0,0) & \cdots & \hat{H}(n,0,N_T-1) \\ \vdots & \ddots & \vdots \\ \hat{H}(n,N_R-1,0) & \cdots & \hat{H}(n,N_R-1,N_T-1) \end{pmatrix} \qquad (11)$$

and the corresponding symbols received on the n-th subcarrier in the vector

$$\boldsymbol{Y}_d(n) = [\ Y(n,0),\ Y(n,1),\ \cdots,\ Y(n,N_R-1)\ ]^T. \qquad (12)$$

Zero-Forcing (ZF): An ZF-estimate of the transmitted data is obtained from the channel matrix pseudoinverse $\hat{d}(n) = (\hat{H}(n))^+ Y_d(n)$ with $\hat{d}(n) = [\hat{d}(n,0), \hat{d}(n,1), \cdots, \hat{d}(n, N_T - 1)]^T$.

Sphere-Detection (SD): To facilitate sphere detection the channel matrix is transformed into an upper triangular matrix using QR decomposition, $\hat{H}(n) = QR$. Multiplying the received symbols in (12) with the matrix of eigenvectors yields

$$Q^H Y_d(n) = R\hat{d}(n) + w(n) \qquad (13)$$

The term $w(n)$ accounts for residual ICI and AWGN. Due to the triangular equation structure an efficient tree search is possible yielding a Max-APP soft value for the received bits.

4 Simulation Results

Firstly simulation results are presented for a statistical channel assuming isotropic scattering according to Jakes' model, and secondly for a car-to-car transmission scenario obtained from Ray-Tracing techniques which considers realistic propagation effects. For the statistical channel an exponentially decaying power delay profile is assumed, the time-selectivity is governed by Jakes' model with normalized maximum Doppler frequency $f_D T_s$. Bit-interleaved coded modulation is employed with a convolutional code of constraint length 7 and a fixed block-interleaver for one OFDM-symbol. Figure 2 shows the BER for a (2×2)-MIMO–OFDM system and for different Doppler frequencies. Both sectorized and omnidirectional reception suffer from larger Doppler spread. Sphere detection performs better than Zero Forcing, but due to the imperfect channel estimation the differences are rather small.

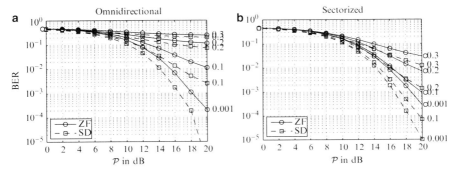

Fig. 2 Coded BER vs. transmit power $\mathcal{P}$, the curves are marked by the Doppler frequency $f_D T$, parameters: $N = 256$, $N_g = 64$, $N_R = 2$, $N_T = 2$, 16QAM, $\Delta_f = 4$

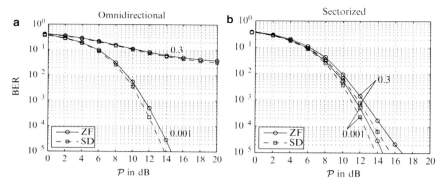

Fig. 3 Coded BER vs. transmit power $\mathcal{P}$, the curves are marked by the Doppler frequency $f_D T$, parameters: $N = 256$, $N_g = 64$, $N_R = 4$, $N_T = 2$, 16QAM

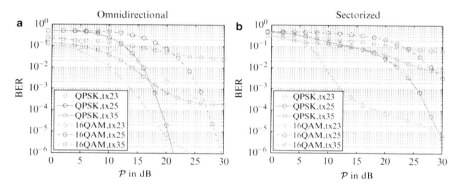

Fig. 4 Cars at 100 m distance, $N = 1024$, $N_g = 256$, $B = 64$ MHz, sphere detection, carrier frequency 5.9 GHz

In Fig. 3 the number of receive antennas is increased to $N_R = 4$. Omnidirectional reception profits from the increased receive diversity only for the low Doppler case, whereas for the high Doppler case an error floor occurs since no ICI compensation is provided. On the other hand directional reception performs superior over directional reception even at larger Doppler spread (cf. Fig. 3b).

Figure 4 shows results for a typical scenario of two approaching cars with a distance of approximately 100 m. For Fig. 4a four omnidirectional antennas at the receiving car are used at the left and right side mirror, under the car and on the roof. The best BER performance for either QPSK or 16QAM is achieved if the transmitting car employs two omnidirectional transmit antennas on the roof and the left mirror. For Fig. 4b four directional antennas are employed, positioned on the roof of the car facing towards, perpendicular and against the moving direction. Apparently none of the antenna configurations can improve over omnidirectional reception. This

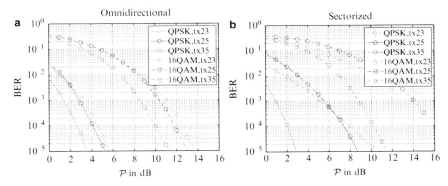

Fig. 5 Cars passing, $N = 1024$, $N_g = 256$, $B = 64$ MHz, sphere detection, carrier frequency 5.9 GHz

effect can be attributed to the lack of isotropic scattering and in particular to the sector antenna facing away from the transmitting car which does not receive useful signal power.

In Fig. 5 results are shown for the same scenario but when the cars are passing each other. Less transmit power is required for an overall BER improvement compared to the previous 100 m distance case. The most favorable transmitting antenna configuration appears to be the roof and right side mirror, which are closest to the receiving car. Using QPSK sectorized reception yields the same performance as omnidirectional reception. In every other configuration omnidirectional performs superior.

5 Conclusions

Doppler compensating antenna structures were shown to facilitate high-data rate MIMO–OFDM-transmission in isotropic scattering and rapidly fading environments. On the other hand Ray-Tracing simulations modelling car-to-car communications were used to show that directive antennas can be disadvantageous in line-of-sight scenarios. Furthermore, the possibility of creating directional antennas with a uniform linear antenna array was illustrated.

References

1. G. J. Foschini, "Layered Space-Time Architecture for Wireless Communication in a Fading Environment when using Multi-Element Antennas," *Bell Labs Tech. J.*, pp. 41–59, Autumn 1996.
2. Y. Li, N. Seshadri, S. Ariyavisitakul, "Channel Estimation for OFDM Systems with Transmitter Diversity in Mobile Wireless Channels," *IEEE J. Sel. Areas Commun.*, vol. 17, no. 3, pp. 461–471, March 1999.

3. K. F. Lee, D. B. Williams, "Pilot-Symbol-Assisted Channel Estimation for Space-Time Coded OFDM-Systems," *EURASIP Journal on Applied Signal Processing*, vol. 2002, no. 5, pp. 507–516, March 2002.
4. M. Russell, G. L. Stüber, "Interchannel Interference Analysis of OFDM in a Mobile Environment," *45th IEEE VTC*, pp. 820–824, July 1995.
5. W. C. Lee, "Preliminary Investigation of Mobile Radio Signal Fading using Directional Antennas on the Mobile Unit," *IEEE Tran. Veh. Comm.*, vol. 15, pp. 8–15, Oct. 1966.
6. P. Klenner, K.-D. Kammeyer "Doppler compensation for OFDM transmission by sectorised antenna reception," *European Transactions on Telecommunications*, vol. 19, pp. 571–579, May 2008.
7. C. A. Balanis, "Antenna Theory - Analysis and Design," 3rd ed. Wiley, New York, 2005.
8. R. Wiedemann, "Simulation des Strassenverkehrsflusses" (in German), *Tech. Rep.*, 1974.
9. J. Maurer, T. Fügen, T. Schäfer, and W. Wiesbeck, "A New Inter-Vehicle Communications (IVC) Channel Model," *VTC*, Los Angeles, 2004.
10. T. Fügen, J. Maurer, T. Kayser, and W. Wiesbeck, "Capability of 3D Ray-Tracing for Defining Parameter Sets for the Specification of Future Mobile Communications Systems," *IEEE Tr. Antennas and Propagation, Special Issue on Wireless Communications*, vol. 54, no. 11, pp. 3125–3137, Oct. 2006.

Resource Allocation Algorithms for Minimum Rates Scheduling in MIMO–OFDM Systems

Johannes Georg Klotz, Frederic Knabe, and Carolin Huppert

Abstract In this paper we present two resource allocation algorithms with minimum rate requirements for MIMO–OFDM systems, which have much a lower complexity than the optimal solution. The first method is based on a heuristic sum rate maximization algorithm using so-called eigenvalue updates. It has a very low complexity and performs quite well for high required minimum rates. The second algorithm has a higher complexity than the first one but still less than the optimal solution. It performs very well for low required minimum rates.

1 Introduction

Multiple-input-multiple-output (MIMO) systems as well as orthogonal frequency division multiplexing (OFDM) are without any doubt key elements of modern and future wireless communication systems. If the channel is known at the transmitter, broadcast techniques can be applied for transmissions from the base-station to the mobile-stations (downlink). Dirty paper coding (DPC) turned out to achieve the capacity region of the MIMO Gaussian broadcast channel [1]. An optimal iterative algorithm optimizing the sum rate was introduced in [2], but this algorithm only focuses on the total throughput and does not guarantee any quality of service. However, quite a lot of delay sensitive applications, such as voice transmission, require a minimum transmission rate to work satisfactorily. Hence, we investigate a resource allocation scenario where we try to maximize the sum rate whereas each user achieves a given minimum rate R_{min}. An optimal solution for this problem was introduced in [3] by Wunder, et al. Unfortunately, the computational complexity of this optimal algorithm is very high and thus, it is not suitable for practical applications. Therefore we introduce two new resource allocation algorithms with much less complexity. These algorithms are no longer optimal, but we show that they

---

J.G. Klotz (✉), F. Knabe, and C. Huppert
Institute of Telecommunications and Applied Information Theory,
Ulm University, Albert-Einstein-Allee 43, 89081 Ulm, Germany
e-mail: {johannes.klotz, frederic.knabe, carolin.huppert}@uni-ulm.de

still achieve quite good results. Our first algorithm, the "extended eigenvalue update algorithm" is based on a technique called "eigenvalue update" presented in [4]. The algorithm of [4] only considers a sum rate maximization and is extended in this paper to consider also minimum rate requirements. The second algorithm, which we call "rate based coding algorithm", makes use of the duality of the up- and downlink introduced in [5].

The remainder of this paper is organized as follows: In the next section we will introduce our considered system model and state the optimization problem. In Sections 3 and 4 we describe our proposed algorithms, whose simulation results are presented and discussed in Section 5. We conclude with some final remarks in Section 6.

2 System Model and Problem Formulation

We consider a MIMO–OFDM system with L subcarriers and K users. Each user has N_m antennas available at its mobile station, while at the base station N_b antennas are present. The channel matrix for each user k in subcarrier l is given by $\boldsymbol{H}^{l,k}$. We assume perfect channel knowledge at the receiver side as well as at the transmitter side. Furthermore, the average value of the additive white Gaussian noise (AWGN) at the receiver side is assumed to be $\boldsymbol{N}^{l,k} = E\left\{\boldsymbol{n}^{l,k}\boldsymbol{n}^{l,k\,H}\right\} = \boldsymbol{I}$ for all users in all subcarriers, where $\boldsymbol{n}^{l,k}$ is the noise vector of user k in subcarrier l and $E\{x\}$ denotes the expectation value of x.

The consideration of the encoding order of the DPC is crucial in the downlink scenario. In the following $\pi_l(\cdot)$ describes the user permutation such that in each subcarrier l user $\pi_l(j)$ is encoded at the jth-position, i.e., user $\pi_l(1)$ is encoded first, followed by user $\pi_l(2)$ and so on. Furthermore let $\pi_l^{-1}(\cdot)$ be the inverse of $\pi_l(\cdot)$. Hence, the achievable rate for user k in subcarrier l is given as

$$R^{l,k} = \log_2 \det\left[\boldsymbol{I} + \frac{\boldsymbol{H}^{l,k}\boldsymbol{Q}^{l,k}\boldsymbol{H}^{l,k\,H}}{\boldsymbol{I} + \sum_{n=\pi_l^{-1}(k)+1}^{K}\boldsymbol{H}^{l,k}\boldsymbol{Q}^{l,\pi_l(n)}\boldsymbol{H}^{l,k\,H}}\right], \quad (1)$$

where $\boldsymbol{Q}^{l,k}$ is the transmit covariance matrix of user k in subcarrier l. The total achievable rate for each user is hence given by

$$R^{(k)} = \sum_{l=1}^{L} R^{l,k}. \quad (2)$$

The total sum rate can be calculated by

$$R = \sum_{k=1}^{K} R^{(k)}. \quad (3)$$

We want to maximize (3) over $Q^{l,k}$, so that an overall power constraint $\sum_{k=1}^{K}\sum_{l=1}^{L} tr\{Q^{l,k}\} \leq P$ is fulfilled. Since we assume the transmission of inhomogeneous data in the described system, we also have to consider a minimum transmission rate R_{min} for the delay-sensitive data. Thus, we search for a resource allocation strategy which maximizes the system throughput provided that the minimum rate requirement is fulfilled for each user. We restrict ourselves to an equal power distribution over the subcarriers, i.e., the transmission power P_l of all subcarriers l is constant, $P_l = P/L \ \forall l$. Denoting the achievable rate of user k by $R^{(k)}$, the central problem is given by

$$\max \sum_{k=1}^{K} R^{(k)} \qquad (4)$$

subject to: $R^{(k)} > R_{min}, \ \forall k$

and $P_l = P/K \ \forall l$.

Since the transmit power is fixed we cannot guarantee that the minimum rate requirement is fulfilled for each user. If the rate $R^{(k)}$ of at least one user is smaller than R_{min}, we call this case an outage.

The proposed algorithms base on the well known Eigen beamforming technique. This technique is optimal for single user MIMO channels and is based on the singular value decomposition of the channel matrix $H^{l,k}$ of user k in carrier l:

$$H^{l,k} = U^{l,k} \Lambda^{l,k \, 1/2} V^{l,k \, H}.$$

The right singular matrices $V^{l,k}$ are used as precoding matrices. At the receiver side the signals are multiplied with the hermitian left singular matrices $U^{l,k \, H}$. Hence, the received signal can be written as:

$$y^{l,k} = U^{l,k \, H} H^{l,k} V^{l,k} \widetilde{x}^{l,k} + U^{l,k \, H} n^{l,k}$$
$$= \Lambda^{l,k \, 1/2} \widetilde{x}^{l,k} + U^{l,k \, H} n^{l,k} \qquad (5)$$

where $\widetilde{x}^{l,k}$ denotes the transmit signal for user k in carrier l. Since the matrix of eigenvalues $\Lambda^{l,k}$ is diagonal, the transmission is performed in so-called beams. The power that is assigned to such a beam is defined by the corresponding diagonal entries of the covariance matrices $P^{l,k} = E\{\widetilde{x}^{l,k} \widetilde{x}^{l,k \, H}\}$, where $Q^{l,k} = V^{l,k} P^{l,k} V^{l,k \, H}$. A beam is considered to be allocated to a carrier, if its corresponding entry in $P^{l,k}$ is non-zero. In the following, we number the beams that are allocated to one carrier as follows: The first beam of a carrier is the beam, that does not suffer from any interference, while the last beam of a carrier is disturbed by all other beams. Furthermore we denote the rate that is achieved by the bth beam of carrier l by $R_L^{l,b}$.

3 Extended Eigenvalue Update Algorithm

Our first algorithm, which is listed in Algorithm 1, uses the basic principles of the heuristic eigenvalue update algorithm for sum rate maximization presented in [4] and is therefore called "extended eigenvalue update algorithm" (EEU). In [4] the influence of a user k_{dis} using DPC on subsequent users k in carrier l is estimated by updating their eigenvalues according to

$$\Lambda^{l,k,b+1} = \Lambda^{l,k,b} - \left| \Lambda^{(l,k,b)1/2} V^{(l,k)H} V^{(l,k_{dis})}_{i_{dis}} V^{(l,k_{dis})H}_{i_{dis}} V^{(l,k)} \Lambda^{(l,k,b)1/2} \right|. \quad (6)$$

where b denotes the index of the beam that user k_{dis} uses and i_{dis} is the index of its eigenvalue. Furthermore, A_i denotes the ith row of a matrix A and $|A|$ denotes a matrix containing the absolute values of A. We use this method in the first step of the algorithm to maximize the sum rate. However, in difference to [4], where a water filling process is applied to distribute the available power to the different beams, we simply assign the same power to each beam. The number of served beams per carrier is restricted to B.

Algorithm 1 Extended eigenvalue update algorithm (EEU)

STEP I
for $l = 1$ to L **do**
  for $b = 1$ to B **do**   ▷ B denotes the maximum number of beams per carrier.
    $[k_{max}, i_{max}] = \arg\max_k \max_i \Lambda^{l,k,b}_{ii}$
    for $k = 1$ to K **do**
      update eigenvalues according to (6)
    end for
  end for
end for
Compute user rates $R^{(k)}$ and carrier rates $R^{l,b}_L$
STEP II
$b_c = B$
while $\min_k R^{(k)} < R_{min}$ and $b_c > 0$ **do**
  $\mathcal{L} = \{1, \ldots, L\}$
  while $\min_k R^{(k)} < R_{min}$ and $|\mathcal{L}| > 0$ **do**   ▷ $|\mathcal{L}|$ is cardinality of $\mathcal{L}$
    $k_{min} = \arg\min_k R^{(k)}$
    for each carrier l in $\mathcal{L}$ **do**
      $b^l_{max} = \arg\max_b \Lambda^{l,k_{min},b_c}_{bb}$
      Compute new carrier rates $\widetilde{R}^{l,b}_L$ if the b_cth beam would be used by user k_{min} with eigenvalue b^l_{max} instead
    end for
    $l_{min} = \arg\min_l \sum_{b=1}^B \left(R^{l,b}_L - \widetilde{R}^{l,b}_L \right)$ ▷ Find carrier with lowest rate loss
    $R^{l,b}_L = \widetilde{R}^{l,b}_L$ for each beam $b = 1, \ldots, B$
    $\mathcal{L} = \mathcal{L} \setminus l_{max}$
    Recompute affected user sum rates $R^{(k)}$
  end while
  $b_c = b_c - 1$
end while

The second step aims in meeting the minimum rate requirements. Therefore, previously allocated users are replaced by users that do not meet the minimum rate requirement. We replace the users that are allocated in the b_cth beam of all carriers, where we first choose $b_c = B$ and then decrease b_c as soon as there are no unchanged carriers remaining. The replacement of the b_cth beam in one carrier starts by selecting the user k_{min} with the instantaneous smallest rate. For all remaining unchanged carriers, the influence on the overall sum rate, if user k_{min} would be allocated to the carrier with its best updated eigenvalue, is computed. Finally, we choose the carrier l_{min} where the decrease of the overall sum rate is the lowest and change the allocation so that user k_{min} uses the b_cth beam of carrier l_{min}. This replacing procedure is continued until all rate requirements are fulfilled.

4 Rate Based Coding Algorithm

It has been shown in [5] that the capacity regions of the MIMO BC and the MIMO MAC are equal. We use the results of this work in our "rate based coding algorithm", which makes a resource allocation for the dual uplink channel. If we assume the reciprocal channel $\boldsymbol{H}^{l,k}{}^H$ as the uplink channel matrix, the rates that our algorithm computes are also achievable in the downlink. The advantages of doing the resource allocation in the uplink are decreased complexity and more flexibility. The achievable rate of user k in subcarrier l is given as

$$R_{MAC}^{l,k} = \log_2 \det \left[\boldsymbol{I} + \frac{\boldsymbol{H}^{l,k}{}^H \boldsymbol{Z}^{l,k} \boldsymbol{H}^{l,k}}{\boldsymbol{I} + \sum_{n=1}^{\pi_l^{-1}(k)-1} \boldsymbol{H}^{l,\pi_l(n)}{}^H \boldsymbol{Z}^{l,\pi_l(n)} \boldsymbol{H}^{l,\pi_l(n)}} \right] \quad (7)$$

in the MIMO MAC, where $\boldsymbol{Z}^{l,k}$ denotes the uplink transmit covariance matrix of user k in subcarrier l. Furthermore $\pi_l(\cdot)$ describes the user permutation in subcarrier l, where user $\pi_l(1)$ is decoded last, user $\pi_l(2)$ second to last and so on. As in the downlink case, $\pi_l^{-1}(\cdot)$ denotes the inverse of $\pi_l(\cdot)$. With the transformation given in [5] the downlink transmit covariance matrices $\boldsymbol{Q}^{l,k}$ can be computed from the uplink transmit covariance matrices $\boldsymbol{Z}^{l,k}$ so that $R_{MAC}^{l,k} = R^{l,k}$. Note that for achieving equal rates in up- and downlink the user order $\pi_l(\cdot)$ is flipped during the transformation, i.e., while user $\pi_l(1)$ suffers from interference of all other users in the downlink, it suffers from no interference in the uplink. The total rate of a user k is computed as in the downlink:

$$R_{MAC}^{(k)} = \sum_{l=1}^{L} R_{MAC}^{l,k}. \quad (8)$$

The increased flexibility of the MIMO MAC is manifested in the influence of the user order $\pi_l(\cdot)$ on the rates of the users. If the user order is changed in carrier l, the user rates also change according to the level of interference they suffer from. However, the sum of all user rates in carrier l remains the same. Thus, changing the decoding order might be used to fulfill the conditions of the optimization problem in (5) without decreasing the sum rate.

We call our algorithm "rate based coding algorithm" (RBC), because it selects the decoding order subject to the rates of the users, i.e., a user k with a low instantaneous rate $\widetilde{R}^{(k)}$ is decoded late, so that it does not suffer from too much interference and can achieve a better rate.

The algorithm, which is given in terms of pseudo code in Algorithm 2, allocates one beam to all carriers in each loop. This is repeated until the sum rate of all users can not be increased any more or until the maximum number of beams B is reached. At the beginning of each loop, we first compute the rate $R_B^{l,k,b}$, that is achieved if user k would be assigned to the bth beam of carrier l. Therefore, we assume that the beam that achieves the highest rate of all beams is used and that the power is distributed equally between the assigned beams of one carrier. It is possible that a

---

Algorithm 2 Rate based coding algorithm (RBC)

  for $b = 1$ to B **do**
    $\mathcal{L} = \{1, \ldots, L\}$
    Update rates $\widetilde{R}^{l,k}$ and $\widetilde{R}^{(k)}$ according to (7) and (8) with power $\frac{1}{b}$ per beam
    Compute achievable rates $R_B^{l,k,b}$ for all carriers l and users k
    STEP I
    while $\min_i \widetilde{R}^{(k)} < R_{min}$ and $|\mathcal{L}| > 0$ **do**
      $k_{min} = \arg\min_k \widetilde{R}^{(k)}$
      $l_{max} = \arg\max_l R_B^{l,k_{min},b}$ ▷ Find best carrier for user k_{min}
      Change user order $\pi_{l_{max}}(\cdot)$ in carrier l_{max} according to the user rates $\widetilde{R}^{(k)}$
      Update rates $\widetilde{R}^{l_{max},k}$ for carrier l_{max} and user rates $\widetilde{R}^{(k)}$
      $\mathcal{L} = \mathcal{L} \setminus l_{max}$
    end while
    STEP II
    for each carrier l in $\mathcal{L}$ **do**
      $k_{max} = \arg\max_k R_B^{l,k,b}$ ▷ Find best user for carrier l
      if user k_{max} is already assigned to carrier l **then**
        Subtract rates $\widetilde{R}^{l,k}$ achieved in carrier l from user rates $\widetilde{R}^{(k)}$
        Determine critical users with $\widetilde{R}^{(k)} < R_{min}$
        Change user order $\pi_l(\cdot)$ in carrier l according to the user rates $\widetilde{R}^{(k)}$, critical users shall be decoded last and their order shall not be changed
      else
        Do not change user order $\pi_l(\cdot)$ of carrier l, user k_{max} shall be decoded after all other users
      end if
      Update rates $\widetilde{R}^{l,k}$ for carrier l and user rates $\widetilde{R}^{(k)}$
    end for
    if $\sum_k \widetilde{R}^{(k)} > \sum_k R^{(k)}$ **then**
      $R^{(k)} = \widetilde{R}^{(k)}$ for each user $k = 1, \ldots, K$
    else
      Stop algorithm and use allocation of previous loop
    end if
  end for

beam of a user is allocated $n > 1$ times in a carrier l. In this case, this beam is assigned the power $\frac{n}{b} \cdot P_l$, where b is the current number of beams in carrier l. Note that in this case $R_B^{l,k,b}$ denotes the rate gain that is achieved by increasing the transmit power by $\frac{1}{b}$.

Each loop is divided into two steps. In the first step, the user k_{min} with the smallest instantaneous rate selects its best carrier l_{max} out of the set $\mathcal{L}$, which contains the not yet allocated carriers. Afterwards, the user order $\pi_{l_{max}}(\cdot)$ in carrier l_{max} is reordered descending by the rate of the users. The first step is repeated until all users have reached the required minimum rate or until there are no carriers remaining.

The second step selects the best possible user k_{max} for each remaining carrier l in $\mathcal{L}$. It has to be taken into account that after the first step the rate requirements are fulfilled and the second step must not violate them. This is avoided by decoding user k_{max} before all other users, so that their interference remains the same. However, this is problematic if user k_{max} is already assigned to carrier l and decoded at a later position, because the increased interference could lead to a rate lower than R_{min}. Thus, we predetermine all users that would fall below the minimum rate if the rate $\widetilde{R}^{l,k}$ that they achieve in carrier l is subtracted, decode them last, and do not change the order in between those users. The rest of the users are ordered as in the first step.

After the second step is completed for all remaining carriers, it is checked whether the allocation of the current beam would increase the overall sum rate. If not, the algorithm is stopped and the allocation of the previous loop is used. Otherwise the algorithm continues with the allocation of the next beam until the maximum number of beams B is reached.

5 Simulation Results

Our simulation scenario consists of a base station with 4 transmit antennas. It is surrounded by 20 users with signal to noise ratios equally distributed in the dB domain between 0 and 20 dB. Each user is equipped with four receive antennas. The channel is divided into $L = 64$ orthogonal subcarriers with independent Rayleigh fading. The overall transmit power is limited to $P = L$, i.e., $P_l = 1$ for all subcarriers l. All simulations results are compared to the optimal solution ("opt").

Since within the EEU the number of beams B, which will be allocated to each carrier, can be chosen arbitrarily, some properties of the algorithm can be scaled. Simulations with different choices of B have been made and in Fig. 1 the achieved average user rates R_{av} for the examined minimum rates R_{min} are shown. It can be seen, that the achieved rates grow with the number of beams B, where the increase of rates becomes smaller for a growing number of beams. Furthermore, the rate increase is larger for higher minimum rates, e.g., a selection of $B = 40$ compared to $B = 30$ achieves a rate gain of 3.2% for $R_{min} = 40$ but the rate gain for $R_{min} = 0$ is only 0.3%. Outages occurred for a selection of $B = 5$ and minimum rates $R_{min} \geq 35$ as well as for a selection of $B = 10, 15, 20$ and $R_{min} \geq 40$. Thus, at least $B = 25$ beams have to be used in order to fulfill the rate requirements for all examined

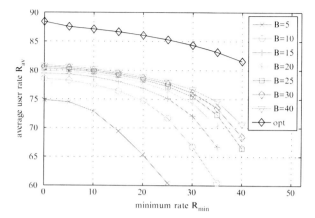

Fig. 1 R_{av} of EEU for different numbers of B

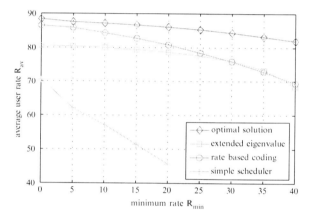

Fig. 2 R_{av} of the suggested algorithms

minimum rates R_{min}. This offers a huge flexibility and good scaling, since the allocation can be changed in small steps, i.e., if the allocation of only one beam in one carrier has changed, the effect on the sum rate of all users is very small. Therefore, the allocation can be adjusted precisely to the minimum rate requirements, which results in a low loss of sum rate for increasing minimum rates R_{min} as visible in Fig. 2. The disadvantage of such a comparatively high number of beams is the increased signaling overhead.

The RBC combines very good performance with low signaling overhead. Figure 3 shows the average number of unique beams, i.e., if a beam of one user is allocated to a carrier multiple times it is only counted once. The number of unique beams is the dominant factor for the signaling overhead. Thus, it can be stated that the signaling overhead is much less for the RBC than for the optimal solution.

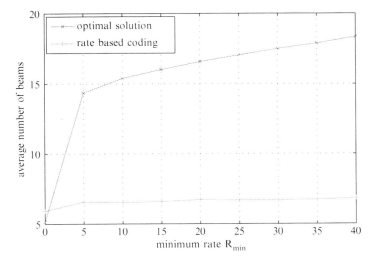

Fig. 3 Average number of unique beams per carrier

In Fig. 2 the achieved average rates per user of the two proposed algorithms are plotted for different minimum rate requirements. For comparison we added the simulation results for the optimal solution and a "simple scheduler" with only one user per carrier, that searches for the user with the instantaneous minimum rate and assigns it to its best carrier out of the not already used carriers. This is done until the minimum rate requirement is fulfilled for each user. Then the remaining unused carriers are assigned to their best users. It can be seen, that the EEU with $B = 30$ and the RBC clearly outperforms the "simple scheduler". The RBC achieves a better performance for lower minimum rates than the EEU at the cost of more complexity. Actually, it gets very close to the optimal solution for low required minimum rates. However, for high required minimum rates the EEU performs slightly better. Both presented algorithms require lower computational complexity than the iterative optimal solution.

6 Conclusion

In this paper we suggested two new heuristic resource allocation algorithms for the MIMO-OFDM downlink channel. We showed in terms of simulation results that our algorithms perform very well when compared to the optimal algorithm.

Furthermore we showed that the suggested EEU can be scaled by changing the maximum number of beams B. By choosing this value wisely a trade-off between signaling overhead and average sum rate can be made.

We also investigated the number of unique beams per carrier within the RBC, which has a significant influence on the signaling overhead. It was shown that this number is limited and almost independent of the minimum rate. Hence the overhead is substantially reduced in comparison to the optimal solution.

Acknowledgments Parts of this work were supported by the German research council "Deutsche Forschungsgemeinschaft" (DFG) under Grants Bo 867/18-1 and Bo 867/19-1.

References

1. G. Caire and S. Shamai. On the Achievable Throughput of a Multiantenna Gaussian Broadcast Channel. *IEEE Transactions on Information Theory*, 49(7):1691–1706, 2003.
2. N. Jindal, W. Rhee, S. Vishwanath, S. A. Jafar, and A. J. Goldsmith. Sum Power Iterative Water-filling for Multi-antenna Gaussian Broadcast Channels. *IEEE Transactions on Information Theory*, 51(4):1570–1580, 2005.
3. G. Wunder and T. Michel. Minimum Rates Scheduling for MIMO-OFDM Broadcast Channels. In *Proc. 9th IEEE Intern. Symp. on Spread Spectrum Techniques and Applications (ISSSTA 2006)*, Manaus, Brazil, August 2006. invited.
4. J. G. Klotz, C. Huppert, and M. Bossert. Heuristic Resource Allocation for Sum Rate Optimization in MIMO-OFDM Systems Using Eigenvalue Updates. *IEEE International Symposium on Wireless Communication Systems (ISWCS)*, October 2008.
5. S. Vishwanath, N. Jindal, and A. Goldsmith. Duality, Achievable Rates and Sum-Rate Capacity of Gaussian MIMO Broadcast Channels. *IEEE Transaction on Information Theory*, 49(10):2658–2668, Oct 2003.

Transmit Antenna and Code Selection for Mimo Systems with Linear Receivers

Ismael Gutierrez, Joan Lluis Pijoan, Faouzi Bader, and Alain Mourad

Abstract In this paper the combination of Transmit Antenna Selection with Linear Dispersion Code Selection is studied. Two optimization criteria are proposed (bit error rate minimization and throughput maximization). The performance of the proposed spatial link adaptation scheme is evaluated under low mobility environments concluding that maximum spatial diversity is achieved as well as a smooth transition between codes with low spatial multiplexing rate, high spatial diversity (suitable for low SNR), and codes with high multiplexing rate but low diversity order (suitable for high SNR) in order to maximize the overall system throughput.[1]

1 Introduction

The use of multiple antennas at the transmitter and at the receiver has demonstrated to increase the channel capacity and diversity [1]. Multiple Input Multiple Output (MIMO) systems may exploit the channel diversity and capacity by using different

---

I. Gutierrez (✉) and J.L. Pijoan
Research Group on Electromagnetism and Communications (GRECO),
Ramon Llull University,
Psg. Bonanova, 8. 08022 – Barcelona, Spain
e-mail: {igutierrez, joanp}@salle.url.edu

F. Bader
Centre Tecnològic de Telecomunicacions de Catalunya (CTTC),
Avd. Canal Olímpic, s/n. 08860 – Castelldefels, Spain
e-mail: faouzi.bader@cttc.es

A. Mourad
Advanced Technology, Standards and Regulation, Samsung Electronics Research Institute
Communications House, South Street, Staines (United Kingdom) TW18 4QE
e-mail: alain.mourad@samsung.com

[1] This work has been partially carried out in the framework of the Celtic project MOBILIA (CP5–016), and in the Spanish National project TSI-020400–2008–82.

space-time-(frequency) coding techniques. When the channel state information is available at the transmitter (CSIT), the best space time coding technique is the beamforming where the information is transmitted in the strongest eigenmode(s) of the channel and the power is allocated to each eigenmode following the water-filling principle. However, CSIT techniques require that the transmitter estimates the full channel matrix which in some scenarios might become unfeasible.

On the other hand, if CSI is not available at the transmitter, the well known Space-Time Block Coding (STBC) schemes are preferred. Hassibi and Hochwald proposed in [2] a framework where any kind of linear STBC could be analyzed, classifying such class of space-time codes as Linear Dispersion Codes (LDC). The main advantage from the LDCs is that different tradeoffs between diversity and spatial multiplexing (SM) can be achieved thanks to proper design of the code [2, 3]. As a result, different authors have proposed to use LDCs, since specific LDC codes structures are able to optimize particular channel metrics or enhance certain parameters (i.e. channel capacity, outage probability, bit error rate, etc.) [4, 5].

Nevertheless, many research efforts have been done in order to exploit the best from each approach under the concept of Partial CSIT (PCSIT). In this case, the transmitter is provided only with a small amount of information about the channel state (e.g. Frobenius channel norm, channel rank, channel condition number, etc.), thus the transmitter sets up the transmitted signal to the current channel [6, 7] (this is referred as precoding). The simplest scheme of precoding is Transmit Antenna Selection (TAS) where the best (set of) antenna(s) are selected for transmission. Actually, it has been shown that the TAS scheme gives the same diversity order than without selection at the expense of reducing the coding gain [8]. This principle has been analyzed in [9–11] for Spatial Multiplexing, Orthogonal STBCs and LDCs respectively. Yet, another well-known precoding technique which fixes the set of precoding matrices from a limited codebook (known a priori from both transmitter and receiver) has also shown to provide large capacity and link reliability improvement by using a low rate channel feedback providing uniquely the codeword index [12, 13].

As a result, the latest researches have combined both types of precoding (TAS and codebook-based) schemes for further enhance of the system performance. This paper extends the works in [14, 15] by applying both the TAS and codebook-based precoding schemes from a pure LDC perspective. Then, a spatial adaptation scheme for the downlink/uplink is developed where the proper Transmit antenna subset and LDC (from a set of predefined LDCs) are selected for each frame. Two optimization criteria (i.e. minimizing the bit error rate and maximizing the throughput) are evaluated showing that both the diversity order and the system throughput are maximized. In the following, the proposed scheme is referred as the Transmit Antenna and Code Selection (TACS).

2 System Model

The MIMO system model with M and N transmitter and receiver active antennas respectively is defined by

$$\mathbf{Y} = \sqrt{\frac{\rho}{M}} \mathbf{HS} + \mathbf{N}, \tag{1}$$

where $\mathbf{S} \in C^{M \times T}$ and $\mathbf{Y} \in C^{N \times T}$ are the transmitted and the received signals from each antenna during each channel access, and the channel matrix $\mathbf{H} \in C^{N \times M}$ is assumed constant during T periods (i.e. block fading channel model). The transmitted signal has unitary power, and the noise matrix $\mathbf{N}$ follows a circular complex Gaussian distribution with zero mean and unitary standard deviation. The Linear Dispersion Code (LDC) structure subsumes most of the previous ST codes such as the Bell-Labs Layered Architecture Space Time coding (BLAST), the Alamouti scheme, etc. [2]. Then, considering the LDC framework, the transmitted signal matrix $\mathbf{X}$ necessarily has the following structure

$$\mathbf{S} = \sum_{q=1}^{Q} (\alpha_q \mathbf{A}_q + j\beta_q \mathbf{B}_q), \tag{2}$$

where $\mathbf{A}, \mathbf{B} \in C^{M \times T}$ are the basis matrices, $\mathrm{E}\{\mathrm{tr}(\mathbf{S}^H \mathbf{S})\} = MT$, and $s_q = \alpha_q + j\beta_q$ are the complex data symbols we want to transmit with $\mathrm{E}\{s_q^* s_q\} = 1$. The number of basis matrices is Q, and the spatial multiplexing rate is Q/MT. The rate R achieved by the system is given by $R = Qn/T$ [bits/s/Hz], where n means the number of bits transmitted per complex symbol.

Then substituting (2) into (1) and applying the *vec* operator on both sides of the expression, the (real valued) system equation can be rewritten as

$$\underbrace{\begin{bmatrix} \Re(\mathbf{y}_0) \\ \Im(\mathbf{y}_0) \\ \vdots \\ \Re(\mathbf{y}_{Q-1}) \\ \Im(\mathbf{y}_{Q-1}) \end{bmatrix}}_{\triangleq \underline{y}} = \sqrt{\frac{\rho}{M}} \mathcal{H} \underbrace{\begin{bmatrix} \alpha_0 \\ \beta_0 \\ \vdots \\ \alpha_{Q-1} \\ \beta_{Q-1} \end{bmatrix}}_{\triangleq \underline{s}} + \underbrace{\begin{bmatrix} \mathbf{n}_0 \\ \mathbf{n}_0 \\ \vdots \\ \mathbf{n}_{Q-1} \\ \mathbf{n}_{Q-1} \end{bmatrix}}_{\triangleq \underline{n}} \tag{3}$$

where $\mathbf{s}$ is the real input symbols vector and $\mathbf{n}$ is the real vector noise i.i.d. components $\mathcal{N}(0, 1/2)$-distributed. The equivalent real valued channel matrix $\mathcal{H}$ is then given by

$$\mathcal{H} = \underbrace{\begin{bmatrix} \mathbf{I}_N \otimes \mathcal{A}_0 & \mathbf{I}_N \otimes \mathcal{B}_0 & \ldots & \mathbf{I}_N \otimes \mathcal{A}_{Q-1} & \mathbf{I}_N \otimes \mathcal{B}_{Q-1} \end{bmatrix}}_{2NT \times 4MNQ} \times \underbrace{\begin{bmatrix} \mathbf{I}_{2Q} \otimes \underline{h} \end{bmatrix}}_{4MNQ \times 2Q}. \tag{4}$$

with

$$\mathcal{A}_q = \begin{bmatrix} \Re\{\mathbf{A}_q\} & -\Im\{\mathbf{A}_q\} \\ \Im\{\mathbf{A}_q\} & \Re\{\mathbf{A}_q\} \end{bmatrix}_{2T \times 2M},$$

$$\mathcal{B}_q = \begin{bmatrix} -\Im\{\mathbf{B}_q\} & -\Re\{\mathbf{B}_q\} \\ \Re\{\mathbf{B}_q\} & -\Im\{\mathbf{B}_q\} \end{bmatrix}_{2T \times 2M}, \quad (5)$$

$$\underline{h} = \begin{bmatrix} h_0 \\ h_1 \\ \vdots \\ h_{N-1} \end{bmatrix}_{2MN}, \quad h_n = \begin{bmatrix} \Re\{\mathbf{h}_n\} \\ \Im\{\mathbf{h}_n\} \end{bmatrix}_{2M},$$

where $\mathbf{h}_n$ is the n-th row of the MIMO channel matrix $\mathbf{H}$.

Typically, the Maximum Likelihood (ML) detection is assumed during the LDCs design. However, it is well-known that the complexity requirements derived from such decoding techniques is extremely high ($\mathcal{O}(2^{Qn})$), making ML unaffordable for high data rates R in real implementations. Furthermore, due to the linear relationship between input and output samples observed in (3), a linear detector is enough to recover the symbols. However, the performance of such linear decoder is far from that offered by the ML. Nevertheless, one important benefit from using a linear decoder is that an equivalent channel can be estimated for each symbol, hence Adaptive Modulation and Coding (AMC) can be applied on a per symbol basis. In this paper, a linear decoder using a Minimum Mean Square Error (MMSE) equalizer is only considered for evaluation of the TACS scheme.

Then, using a linear MMSE receiver, the Effective Signal to Interference and Noise Ratio (ESINR) per each symbol q is given by

$$ESINR_q^{(MMSE)}(\mathbf{H}) = \frac{\rho}{M\left[\mathbf{H}^H\mathbf{H} + 2\rho^{-1}\mathbf{I}_{2Q}\right]^{-1}_{q,q}} - 1, \quad (6)$$

where $\mathbf{X}^{-1}{}_{q,q}$ refers to the (q,q) element from $\mathbf{X}^{-1}$, and ρ is the average Signal to Noise Ratio (SNR). Furthermore, if the mapping applied to the symbols follows a 2^n-QAM constellation, the average pairwise error probability per stream applying the Nearest Neighbor Union Bound can be obtained as follows

$$P_{e,q} \leq 1 - \left(1 - N_e(n) \cdot \mathrm{E}\left\{Q\left(\sqrt{ESINR_q(\mathbf{H})\frac{d_{min}^2(n)}{2}}\right)\right\}\right) \quad (7)$$

where $Q(x) = 0.5 \times erfc(x/2^{1/2})$, d_{min}^2 is the squared minimum distance between any two points of the constellation (assuming an unitary average transmission power), and N_e is the average number of nearest neighbors constellation points. For a 2^n-QAM modulation $d_{min}^2 = 6/(2^n - 1)$ and $N_e = 4 \times (1 - 2^{-n/2})$. In addition, in case all the symbols within the codeword apply the same modulation, the average pairwise error probability for the whole codeword is usually approximated by (assuming $P_e < 10^{-2}$)

$$P_e \leq \mathbf{Q} \cdot N_e(n) \cdot E\left\{ Q\left(\sqrt{ESINR_{\min}(\mathbf{H}) \frac{d_{\min}^2(n)}{2}} \right) \right\}, \tag{8}$$

where $ESINR_{min} = min(ESINR_0, \ldots, ESINR_{Q-1})$ [14].

3 The TACS Selection Criteria

Then, given the ESNR per stream in (6) and the average pairwise error probabilities in (7) and (8), two different optimization scenarios are studied where both the transmit antenna subset as well as the best LDC from a set of codes are selected.

In the first scenario, we consider that the same modulation is applied to all the symbols and that the rate R is fixed. In that case, and since transmission power is fixed, we are interested in selecting the transmit antenna subset and LDC code that minimizes the error rate probability (i.e. the bit error rate – BER) while the modulation that is required by each LDC is adapted in order to achieve the cited rate R. In that case, since the Q-function is monotonically decreasing as a function of the input, the optimization problem can be defined as follows

$$\max_{LDC_i, p_i} \min_q \left\{ ESINR_q(H, LDC_i, p_i) d_{\min}^2(n_i) \right\}, \tag{9}$$

where i means the LDC index and p_i the transmitting antenna subset (set of antennas that can be used according to the number of transmitter antennas M_a and the number of antennas required by the LDC). It also noted that the constellation is a function of the LDC.

In the second scenario, the optimization is performed in order to maximize the system throughput considering a certain quality of service requirement (i.e. a maximum Block Error Rate – BLER). In that case, the problem is formulated as follows

$$\max_{LDC_i, p_i, MCS_j} \min_q R\left(1 - BLER(ESINR_q)\right) \quad \text{s.t.} : BLER \leq \mu \tag{10}$$

where j means the Modulation and Coding Scheme (MCS) index that maximizes the spectral efficiency for the specific channel state subject to a maximum Block Error Rate (BLER). Actually, the selection of the optimum MCS is carried out assuming that the ESNR is the SNR that would be obtained at the receiver in case having an Additive White Gaussian Noise (AWGN) channel. Under that assumption, there is a direct mapping between each MCS and the obtained BLER for each ESNR.

4 Simulation Results

For the TACS evaluation, the downlink mode of a WiMAX TDD system has been used [16]. The number of available transmitter antennas is $M_a = \{2, 3, 4\}$ whereas the number of receiver antennas is fixed to $N = 2$. One user is simulated which is allocated one subchannel per frame. The channel follows a spatial uncorrelated Rayleigh distribution and a block fading model is assumed (flat in frequency and constant in time).

The basic set of LDC codes that we have been used for the study are: the *Single Input Multiple Output* code using a Maximum Ratio Combiner (MRC), the *Alamouti* code (referred as G2 in the plots), the *BLAST*-like codes with $M = 2$ (referred as Spatial Multiplexing, SM, in the plots) and the *Golden* code. The codeword length for all the codes is $T = 2$. Moreover, for the SM case two types of encoding have been tested named vertical encoding (SM-VE) and horizontal encoding (SM-HE). For the vertical encoding, the same MCS is used for all the symbols transmitted within the same codeword, whereas for horizontal encoding each data stream (symbol) may apply a different MCS according to the channel status. Actually, all these codes are part of the standard and can be found in [16]. Consequently, since each i-th LDC from this basic set require at most two transmitting antennas, in case $M_i < M_a$ the best set p of transmitting antennas is selected from the M_a available antennas, and since the order in which the antennas are chosen is relevant we have P_i possible transmitting antennas combinations with

$$P_i = C \binom{M_a}{M_i} = \frac{M_a!}{M_i!\,(M_a - M_i)!}. \tag{11}$$

To solve (9) or (10), an exhaustive search is performed among all the available LDC codes and P antenna sets, despite it would be very interesting testing the performance of the TACS under an incremental or decremental search as those proposed in [8].

The analysis of the TACS scheme under the first optimization criteria in (9) (uncoded BER minimization), is depicted through Figs. 1–3. Since the rate is fixed, SM is applied with vertical encoding only. Then, in (10) the reduction of the BER as a function of M_a is shown having a fixed rate $R = 8$ bits/s/Hz. An important conclusion from these results is that the TACS achieves the maximum diversity order $g_d = M_a \times N$ and also a coding gain of 3.33, 4.16 and 5 dB compared to the Alamouti scheme for $M_a = \{2, 3, 4\}$ respectively. It is observed that the SM-VE and the Golden code are the more benefited from TAS since these codes have a lower inherent diversity gain. However, the diversity order they achieve when combined with TAS is very close to the maximum. Very relevant are also the statistics of LDC selection plotted in Fig. 2, where the percentage of use of each code as a function of the average SNR is shown. It is observed that due to the high rate requirements, at low SNRs it is more beneficial to use a LDC with high multiplexing capabilities ($Q = 4$) in order to decrease the modulation order. On the other hand, as the SNR

Transmit Antenna and Code Selection for Mimo Systems with Linear Receivers

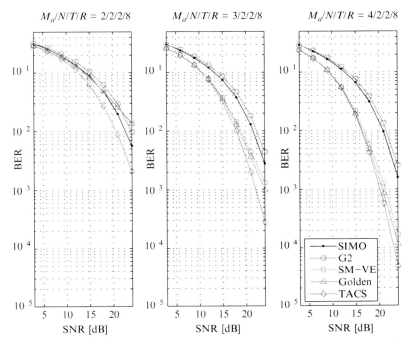

Fig. 1 BER achieved for different number of available transmitter antennas $M_a = \{2, 3, 4\}$ using the TACS scheme and the LDC basic set

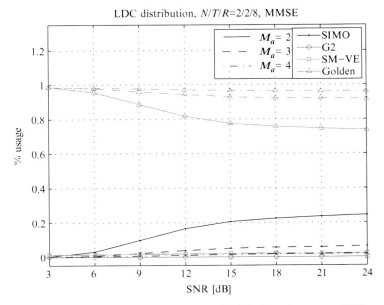

Fig. 2 Percentage of selection for each code using the TACS scheme by means of BER optimization criterion with $M_a = \{2, 3, 4\}$ and the LDC basic set

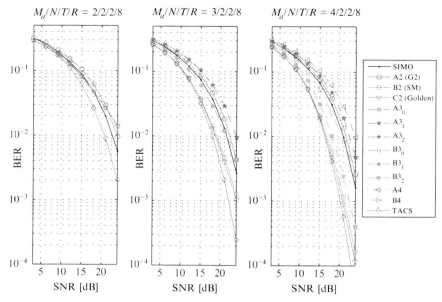

Fig. 3 BER performance of the TACS scheme with $M_a = \{2, 3, 4\}$, $N = 2$, $R = 8$ and a extended LDC set containing the STBCs defined in the IEEE 802.16 [16]

is increased the SIMO scheme is also selected (up to 25% at SNR = 24). It is then straightforward that the coding selection gives higher revenues when is applied over high SNRs values.

Afterwards, the effects of the set of different LDC available for selection is shown in Fig. 3. In this case, the LDC set is increased to all the STBCs defined in the IEEE 802.16 standard where up to four antennas can be used simultaneously for transmission. The codes are named following the notation in [16], where A/B/C-X_y means that the code requires X active transmit antennas and A codes maximize the diversity, the B codes maximize the spatial multiplexing rate, and C codes are a tradeoff between both. The index y means the different permutations of the STBC through the transmitting antennas. Comparing then the results in Fig. 3 with those in Fig. 1 it observed that for a data rate $R = 8$ bits/s/Hz the basis LDC set previously defined is enough to achieve the maximum coding gain using the TACS scheme. Actually, despite the LDC selection statistics are not here plotted, these remain unchanged compared to the case with the basic LDC set. However, it is expected that for higher data rates codes with higher spatial multiplexing rates should be more frequently selected.

Furthermore, the performance of the TACS scheme is compared for both correlated and uncorrelated subchannel across the frequency domain. For the correlated subchannel, an ITU Pedestrian A channel model is used with contiguous subcarrier permutation (Band AMC scheme in IEEE 802.16) equivalent to the block fading channel model flat in frequency previously studied. On the other hand, in the

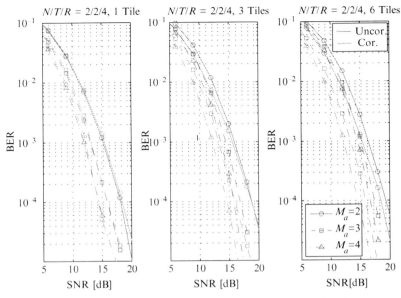

Fig. 4 Performance degradation due to frequency channel correlation as a function of the number of tiles $\{1, 3, 6\}$ per subchannel. $M_a = \{2, 3, 4\}$, $N = 2$, $R = 4$

uncorrelated case the same channel (ITU-PedA) is applied but with a subcarrier permutation following the Tile Usage of Subchannels (TUSC) scheme where one subchannel is composed by different tiles allocated at distant parts of the spectrum (one tile is given by a set of contiguous subcarriers in the frequency domain) [16]. A fixed rate $R = 4$ bits/s/Hz is chosen in this case. It is observed then in Fig.4 that under the ITU PedA channel and the Band AMC scheme the performance of the TACS is very slightly affected by channel correlation (3 dB SNR gain is obtained in all simulated cases at BER $= 10^{-3}$). In contrast, when the tiles within the subchannel are uncorrelated, the gain obtained by the TACS is lowered (only 1.5 dB when the subchannel is formed of six tiles). However, despite this reduction in the gain in case of uncorrelated tiles, the TACS scheme is able to bring another advantage that is observed in Fig. 5 where the usage of the different LDCs is depicted. It is shown that as the number of uncorrelated tiles is increased, those codes favoring the spatial diversity (i.e. the Alamouti code) are preferred or recommended. This is a very logical conclusion since for uncorrelated channels, as the number of tiles is increased it is very improbable that one antenna subset performs well across all the tiles, hence the best choice is to use the scheme with higher native diversity. We can conclude from the here above analysis that the TACS scheme is able to compensate the effects of channel correlation within the tile, and even between tiles.

Finally, the performance of TACS under the second optimization case and $M_a = 2$ is shown in Fig. 6. In this last scenario a PER bound of 1% is fixed and the MCS available are $\{4, 16\}$-QAM with turbo-coding, with coding rates varying

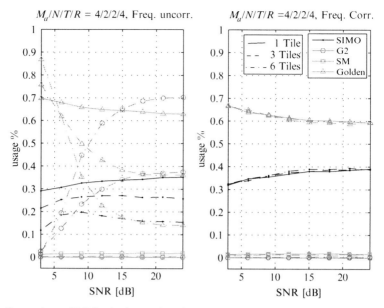

Fig. 5 Comparison of LDC selection statistic for both frequency correlated and frequency uncorrelated subchannels. $M_a = \{2, 3, 4\}$, $N = 2$, $R = 4$

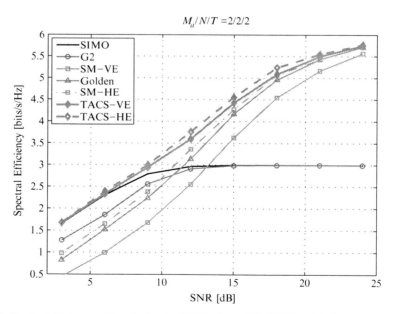

Fig. 6 Spectral efficiency achieved using the TACS scheme ($M_a/N/T = 2/2/2$) joint with AMC under the throughput maximization criterion

from 1/4 to 3/4. The MCS can be adapted on a per symbol (stream) basis. It is then shown that at low SNRs (SNR < 13 dB) the SIMO and Alamouti achieved the highest spectral efficiencies. However, as the SNR is increased, codes with higher multiplexing capacity are necessary, hence the SM and the Golden code achieve the highest spectral efficiencies. We also observed that the SM with VE implies a loss of around 2 dB compared to the Gold code, but when HE is used, the Gold code is around 0.5 dB worse than SM-HE. Above all, we observe that the TACS scheme with AMC gives the highest spectral efficiency where the highest benefit is obtained in the 8 dB < SNR < 18 dB margin where a smooth transition between both types of codes (codes with $g_s = 1$ or $g_s = 2$) is carried out.

5 Conclusions

This paper aims to fill the gap between Transmit Antenna Selection and space-time code selection. The well-defined LDC framework has been used to characterize any linear STBC. The Transmit Antenna and Code Selection (TACS) scheme is then evaluated under two optimization criteria, bit error rate minimization and spectral efficiency maximization. For the first optimization problem, it has been shown that TACS achieves the maximum diversity order as well as a remarkable SNR gain. For the second optimization problem, the TACS scheme is able increase the system throughput compared to non-adaptive techniques specially in the range of SNR between 8 to 18 dB. Furthermore, it is also concluded that the number of LDC required in the codebook is very low for data rates lower that 8 bits/s/Hz, hence the feedback required from such TACS scheme to signal the LDC and the antenna set selected is very low.

References

1. E. Telatar, "Capacity of Multi-antenna Gaussian Channels", *European Transactions on Telecommunications*, vol. 10, no. 6, pp. 851–855, Nov. 1999.
2. B. Hassibi, B.M. Hochwald, "High Rates Codes that are Linear in Space and Time", *IEEE Communications Letters*, vol. 5, no. 4, pp. 154–156, April 2001.
3. L. Zheng, D. Tse, "Diversity and Multiplexing: A Fundamental Tradeoff in Multiple Antenna Channels", *IEEE Transactions on Information Theory*, vol. 49, no. 3, pp. 1073–1096, May 2003.
4. R.W. Heath, A.J. Paulraj, "Linear Dispersion Codes for MIMO Systems Based in Frame Theory", *IEEE Transactions on Signal Processing*, vol. 50, no. 10, pp. 2429–2441, Oct. 2002.
5. R. Gohary, T. Davidson, "Design of Linear Dispersion Codes: Asymptotic Guidelines and Their Implementation", *IEEE Transactions on Wireless Communications*, vol. 4, no. 6, pp. 2892–2906, Nov. 2005.
6. J.C. Roh, B.D. Rao, "Multiple Antenna Channels with Partial Channel State Information at the Transmitter," *IEEE Transactions on Wireless Communications*, vol. 3, no. 2, pp. 677–688, March 2004.

7. D.J. Love, R.W. Heath, Jr., W. Santipach, M.L. Honig, "What is the Value of Limited Feedback for MIMO Channels?," *IEEE Communications Magazine*, vol. 42, no. 10, pp. 54–59, Oct. 2004.
8. A.B. Gershman, N.D. Sidiropoulos, "Space Time Processing for MIMO Communications", Wiley, UK, 2005.
9. R. Heath, S. Sandhu, A. Paulraj, "Antenna Selection for Spatial Multiplexing Systems with Linear Receivers", *IEEE Communications Letters*, vol. 5, no. 4, pp. 142–144, April 2001.
10. K.T. Phan, C. Tellambura, "Capacity Analysis for Transmit Antenna Selection Using Orthogonal Space-Time Block Codes," *IEEE Communications Letters*, vol. 11, no. 5, pp. 423–425, May 2007.
11. D. Deng, M. Zhao, J. Zhu, "Transmit Antenna Selection for Linear Dispersion Codes Based on Linear Receiver", *Proceedings of Vehicular Technology Conference*, VTC 2006-Spring.
12. X. Pengfei, G.B. Giannakis, "Design and Analysis of Transmit-Beamforming Based on Limited-Rate Feedback," *IEEE Transactions on Signal Processing*, vol. 54, no. 5, pp. 1853–1863, May 2006.
13. D.J. Love, R.W. Heath, Jr., "Multimode Precoding for MIMO Wireless Systems," *IEEE Transactions on Signal Processing*, vol. 53, no. 10, pp. 3674–3687, Oct. 2005.
14. R.W. Heath, D.J. Love, "Multimode Antenna Selection for Spatial Multiplexing Systems with Linear Receivers", *IEEE Transactions on Signal Processing*, vol. 53, no. 8, pp. 3042–3056, Aug. 2005.
15. A. Sezgin, E.A. Jorswieck, E. Costa, "LDC in MIMO Ricean Channels: Optimal Transmit Strategy with MMSE Detection", *IEEE Transactions on Signal Processing*, vol. 56, no. 1, pp. 313–328, Jan. 2008.
16. IEEE Standard for Local and Metropolitan Area Networks, Part 16: Air Interface for Fixed and Mobile Broadband Wireless Access Systems, *IEEE Std 802.16e^{TM}-2005*, Feb. 2006.

Space-Time-Frequency Diversity in the Next Generation of Terrestrial Digital Video Broadcasting

Armin Dammann

Abstract DVB-T2 is an emerging terrestrial digital video broadcasting standard. Compared to DVB-T, which is widely deployed at the present time, DVB-T2 promises performance gains because of improved coding, modulation and multiple antenna technologies. Alamouti Space-Frequency coding in DVB-T2 provides additional frequency diversity, especially in multipath propagation environments. Coding and interleaving is done with sufficient depth such that temporal diversity can be exploited as well. In this paper we investigate the achievable frequency diversity gains by Alamouti Space-frequency coding and the additional achievable temporal diversity gains by applying a standard conformable multi antenna Doppler diversity scheme. Simulation results for the TU6 radio propagation channel model show achievable diversity gains up to about 8 dB.

1 Introduction

Digital Video Broadcasting (DVB) [1] is a family of standards, developed by the international DVB Project. These standards provide digital broadcast services over satellite (DVB-S), cable (DVB-C) and terrestrial (DVB-T) sender networks. Published first in 1997, DVB-T has become the most widely used digital terrestrial TV standard currently deployed in more than 35 countries [2].

The DVB Project is working on the next generation of terrestrial DVB (DVB-T2) [3]. The technical specification of DVB-T2 is available as a *DVB BlueBook* [4] and has been sent to the European Telecommunications Standards Institute (ETSI) for publication as a formal standard. DVB-T2 aims to achieve a capacity improvement of 30...50% compared to DVB-T. Increased capacity and further available UHF and VHF spectrum by switch-off of analogue TV services allow to offer high data rate services such as high definition TV (HDTV) in these bands. The UK regulator

---

A. Dammann
German Aerospace Center (DLR), Institute of Communications and Navigation,
Oberpfaffenhofen, 82234 Wessling, Germany
e-mail: Armin.Dammann@DLR.de

Ofcom intends to assign one nationwide DVB-T2 multiplex for multichannel HDTV at the end of 2009, which will most likely be the first deployment of DVB-T2.

Besides providing high data rate services to stationary receivers, DVB-T2 also aims to work for portable and mobile receiver devices such as laptop computers, car or mobile phone receivers. For such multipath radio propagation environments investigations for DVB-T have already shown performance gains achievable by the use of frequency diversity enhancing techniques such as cyclic delay diversity [5]. The DVB-T2 specification optionally includes spatial diversity processing in form of Alamouti coding [6]. Additionally to spatial diversity, the signal structure of DVB-T2 is suitable to exploit both frequency and temporal diversity. This is in contrast to DVB-T whose signal structure does not allow to exploit temporal diversity offered by mobile radio propagation environments. This paper aims to provide results on the achievable time/frequency diversity performance gains for DVB-T2 by applying the standard conformable, diversity increasing multi-antenna technology Discontinuous Doppler diversity (DDoD) [7] in combination with Alamouti Space-Frequency coding as defined in DVB-T2.

2 The DVB-T2 System

DVB-T2 shows significant differences in forward error correction (FEC) coding, modulation and framing compared to DVB-T. The next subsections provide a brief introduction into the physical layer elements of DVB-T2 and point out the major improvements compared to DVB-T.

2.1 Coding, Modulation and Framing Structure

Figure 1 shows a simplified block diagram of a DVB-T transmitter for input mode 'A', where one Physical Layer Pipe (PLP), i.e. one bit stream, is transmitted in one physical channel. The information bits of the PLP are encoded using a serial concatenation of an outer BCH and an inner an LDPC code. This forward error correction (FEC) scheme has been adopted from the satellite standard DVB-S2. The

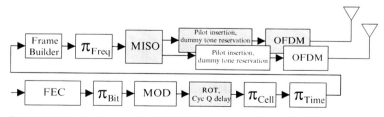

Fig. 1 Simplified block diagram of a DVB-T2 transmitter for single PLP transmission. Grey blocks are optional

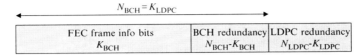

Fig. 2 FEC frame structure

Table 1 Code parameters for normal FEC frames ($N_{LDPC} = 64800$)

LDPC code rate	K_{BCH}	$N_{BCH} = K_{LDPC}$	BCH t-error correction
1/2	32 208	32 400	12
3/5	38 688	38 880	12
2/3	43 040	43 200	10
3/4	48 408	48 600	12
4/5	51 648	51 840	12
5/6	53 840	54 000	10

BCH codebits are fed into the LDPC encoder without interleaving. Both the BCH and LDPC code are systematic codes. Figure 2 shows the principle structure of a FEC frame. DVB-T2 specifies LDPC codes with different lengths and code rates. There are two different LDPC code lengths, which are $N_{LDPC} = 16200$ for short FEC frames and $N_{LDPC} = 64800$ for normal FEC frames. For $N_{LDPC} = 64800$, Table 1 shows the associated BCH code parameters N_{BCH} (codelength), K_{BCH} (code dimension) and t (number of correctable bit errors). For short FEC frames, a further code rate of 1/4 is specified. Such FEC frames are dedicated for protecting sensitive signalling data.

The FEC frame bits are interleaved (π_{Bit}) in two steps. The first stage interleaves the parity bits of the LDPC code. In a second step a column twist interleaver is used. In principle this is a block interleaver. The bit stream is written into the interleaver matrix column-wise. The start position for writing into columns is twisted (shifted) for each column. The bits are read out of the matrix row-wise.

The interleaved code bits are modulated (MOD) using a QAM modulation alphabet. The subsequent block *ROT, Cyc Q delay* is optional. In this entity, the QAM symbols – called cells – are rotated, i.e. phase shifted. For this reason, the information content of a rotated QAM symbol is contained in both its real and imaginary part. The imaginary parts are then shifted cyclically within one FEC frame. With this procedure, information is spread over two cells and, therefore, provides diversity.

The resulting cells are again interleaved (π_{Cell}) using pseudo random interleaving permutations. The interleaving depth, i.e. the interleaving permutation lengths, of the cell interleaver π_{Cell} is one FEC frame.

The time interleaver π_{Time} provides the highest depth and, therefore, most of diversity in time direction. π_{Time} may interleave up to 1,023 FEC frames. The frame building entity arranges the resulting cell stream into OFDM frequency domain symbol blocks. Each of these blocks are again interleaved by a subcarrier (frequency) interleaver π_{Freq}, which is basically a pseudo random interleaver with an interleaving depth of one OFDM symbol. After that, the OFDM symbols are optionally Alamouti encoded. In DVB-T2 Alamouti coding is in frequency direction. For each TX branch, pilots are inserted and a set of tones are reserved for PAPR reduction purposes. Finally the OFDM symbols are transformed into time domain with subsequent insertion of the guard interval.

2.2 Major Differences Between DVB-T and DVB-T2

The targeted capacity improvement of 30...50% compared DVB-T has been achieved by modifications respectively changes in coding and modulation technology. A big part of this gain is to be caused by the application of a more powerful FEC scheme, which is a concatenation of a BCH and an LDPC code in DVB-T2. For this reason, the application of 256-QAM and, therefore, higher spectral efficiencies have become possible. Further parts of the gain come from the reduction of the relative guard interval overhead by extending the set of possible FFT transformation lengths up to 32k. Also the pilot structure is more flexible in DVB-T2 and allows to reduce the pilot overhead compared to DVB-T. Table 2 summarizes the main differences in the transmission modes of DVB-T and DVB-T2.

The DVB-T standard defines frames which consist of 68 OFDM symbols and superframes containing four DVB-T frames. However, the information bit stream is continuously encoded by the memory six inner convolutional encoder. There is no termination of the convolutional code at the end of a frame. DVB-T2 follows a more block oriented and flexible framing approach. A DVB-T2 frame starts with a preamble section, which contains signalling data for instance, followed by the data OFDM symbols. The number of DVB-T2 frames within a DVB-T2 superframe is a configurable parameter. At its end, DVB-T2 superframes can contain so called future extension frames (FEFs). As their name imply, such frames are intended for flexibility in further extensions of the system.

Table 2 Comparison of transmission parameters for DVB-T and DVB-T2

	DVB-T	DVB-T2
FEC	Inner convolutional code, outer Reed-Solomon code	Inner LDPC code, outer BCH code
Modulation	4-QAM, 16-QAM, 64-QAM	4-QAM, 16-QAM, 64-QAM, 256-QAM optional constellation rotation with cyclic Q delay
Interleaving direction	Frequency	time/frequency
FFT size	2k, 8k, 4k (for DVB-H)	1k, 2k, 4k, 8k, 16k, 32k
Guard interval	1/4, 1/8, 1/16, 1/32	1/4, 19/256, 1/8, 19/128, 1/16, 1/32, 1/128
Scat. pilot overhead	8%	1%, 2%, 4%, 8%
Cont. pilot overhead	2.6%	0.35%

3 Doppler Diversity

Additionally to the Alamouti Space-Frequency coding (SFC) scheme, defined in DVB-T2, we apply a TX- antenna diversity technology called *Discontinuous Doppler diversity (DDoD)* [7]. This TX antenna diversity scheme is standard conformable, which means that it can be applied on top of already fixed standard without changing them. DDoD artificially enlarges the Doppler spread using multiple TX antennas and, therefore, increases temporal diversity offered to the system.

Let $s(k)$ be a stream of consecutive OFDM symbols in time domain, i.e. the complex valued baseband signal after OFDM modulation (the inverse FFT) and guard interval insertion. The length of one such OFDM symbol is $N_{\text{OFDM}} = N_{\text{FFT}} + N_{\text{G}}$, where N_{FFT} and N_{G} denote the FFT length and the length of the guard interval respectively. This signal is copied into N_{T} TX-antenna branches, normalized in order to keep the overall TX power independent of N_{T} and finally multiplied by a TX branch specific function $\gamma_i(k)$. Thus, the transmitted signals are

$$s_i(k) = \frac{1}{\sqrt{N_{\text{T}}}} \cdot s(k) \cdot \gamma_i(k), \quad i = 0, \ldots, N_{\text{T}} - 1. \quad (1)$$

We choose time domain signals

$$\tilde{\gamma}_i(k) = \exp(j 2\pi \Delta f_i \cdot T \cdot k) \quad (2)$$

as complex valued exponential functions. $T = \frac{1}{N_{\text{FFT}} \cdot \Delta f_{\text{SC}}}$ is the system sampling time and can be calculated from the subcarrier spacing Δf_{SC} and the FFT length N_{FFT}. This introduces a Doppler shift of Δf_i for each TX antenna branch and in total an artificial Doppler spread when receiving the superimposed signals from all TX antennas. This effect can be described by an effective single-input single-output (SISO) channel. Figure 3 shows the Doppler spectrum of that resulting channel. Directly using $\gamma_i(k) = \tilde{\gamma}_i(k)$ leads to the Doppler Diversity (DoD). DoD has the drawback of increasing intercarrier interference (ICI) because of additional introduced time variance within the duration of the OFDM symbols. Since temporal diversity is an effect which is obtained from changes of the fading processes between different OFDM symbols, the idea of the discontinuous variant of Doppler

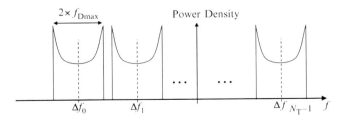

Fig. 3 The effect of Discontinuous Doppler Diversity on the Doppler spectrum

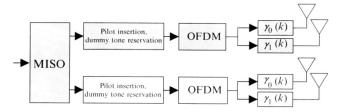

Fig. 4 Application of Discontinuous Doppler Diversity to DVB-T2

diversity is to keep the artificial Doppler increasing functions $\gamma_i(k)$ constant within the duration of an OFDM symbol and change their values at the transition from one OFDM symbol to the next one. This leads to a discontinuously sampled version of the original functions $\tilde{\gamma}_i(k)$ which can be described as

$$\gamma_i(k) = \tilde{\gamma}_i(N_{\text{OFDM}} \cdot (k \div N_{\text{OFDM}})) \\ = \exp(j2\pi \Delta f_i \cdot T \cdot N_{\text{OFDM}} \cdot [k \div N_{\text{OFDM}}]), \quad (3)$$

where '÷' means an integer division with positive remainder. When selecting Δf_i symmetrically around zero, the effective maximum Doppler spread of the effective channel is

$$f_{\text{Dmax}}^{\text{eff}} = \underbrace{\frac{\max_i\{\Delta f_i\} - \min_i\{\Delta f_i\}}{2}}_{=\Delta f_{\max}} + f_{\text{Dmax}}, \quad (4)$$

where f_{Dmax} is the maximum Doppler spread of the original channel's fading processes.

Figure 4 shows the application of DDoD to DVB-T2. For each Alamouti encoded TX antenna branch, we insert additional two DDoD branches. The artificial Doppler functions $\gamma_i(k)$, $i = 1, 2$ are identical for each Alamouti branch. Thus, the additionally inserted Doppler spread, assumed that we select the Doppler shifts symmetrically around zero is

$$\Delta f_{\max} = \frac{|\Delta f_0 - \Delta f_1|}{2} = \Delta f_0 = \Delta f_1. \quad (5)$$

4 Results

For our investigations we use the 6-path *Typical Urban* (TU6) multipath Rayleigh fading channel model [8] with a maximum path delay of 5 μs and a Jakes Doppler spectrum form. For comparison we additionally use an independent Rayleigh (IR) fading channel. For the IR channel, the fading coefficients are mutually uncorrelated in both frequency and time direction. We apply a frequency domain simulation approach, i.e. we implement the channel in frequency domain as a set of parallel flat fading channels. Thus, the received symbols in frequency domain are modelled as

$$R_{k,\ell} = H_{k,\ell} \cdot S_{k,\ell} + N_{k,\ell}, \tag{6}$$

where $S_{k,\ell}$ is the transmitted symbol at subcarrier k of OFDM symbol ℓ. $H_{k,\ell}$ and $N_{k,\ell}$ denote the frequency domain fading coefficients and additive white Gaussian noise (AWGN) respectively. With this simulation approach we do not observe intercarrier interference (ICI). Note, we obtain a signal-to-interference level of SIR = 20 dB at Doppler bandwidth of 10% of the subcarrier spacing, i.e., $f_{\text{Dmax}}/\Delta f_{\text{SC}} = 0.1$. For all the simulations, the SIR is sufficiently low such that the frequency domain simulation approach is justified.

The average subcarrier SNR is defined as

$$\text{SNR} = \frac{E\{|H_{k,\ell}|^2\} \cdot E\{|S_{k,\ell}|^2\}}{E\{|N_{k,\ell}|^2\}} = \frac{E\{|S_{k,\ell}|^2\}}{E\{|N_{k,\ell}|^2\}} \tag{7}$$

and does not depend on the subcarrier or OFDM symbol index. $E\{.\}$ denotes the expectation operator.

For simulations we use normal FEC frames (64,800 bits), a code rate of 3/4, 16-QAM, 8k FFT length and a guard interval of 1/32. Applying pilot pattern PP3, these parameters provide 6,494 data subcarriers per OFDM symbol. The time interleaving depth selected for our simulations is ten FEC frames, which corresponds to 25 OFDM symbols. At the receiver, we use symbol-by-symbol maximum a-posteriori QAM demodulation in the logarithmic domain (Log-MAP) [9], LDPC belief propagation decoding with a maximum of 50 iterations and hard decision bounded minimum distance BCH decoding.

Figure 5 shows the bit error rates (BER) versus the average subcarrier SNR at the receiver for the TU6 channel, one TX antenna (SISO) and various Doppler frequencies. For a normalized Doppler bandwidth of $f_{\text{Dmax}}/\Delta f_{\text{SC}} = 0.01$ the observable Doppler diversity gain compared to the static case ($f_{\text{Dmax}} = 0$) is about 0.8 dB. Increasing to $f_{\text{Dmax}}/\Delta f_{\text{SC}} = 0.1$ provides an SNR gain of approx 4.4 dB at BER = 10^{-4}. For comparison we have included the BER graph for uncorrelated fading in time direction. This offers the maximum temporal diversity with a maximum achievable Doppler diversity gain of 5.7 dB compared to $f_{\text{Dmax}} = 0$. Note, the temporal uncorrelated case can be approached by applying DDoD as described in Section 3. The maximum achievable overall diversity gain in both time and frequency direction can be observed from the IR graph.

Equivalently to the SISO case, Fig. 6 shows the achievable Doppler diversity gains for Alamouti SFC (MISO). Compared to SISO, Alamouti SFC yields a gain of 4 dB for the static case and 2.2 dB for uncorrelated fading in time direction. The SNR difference at BER = 10^{-4} between temporal uncorrelated fading and $f_{\text{Dmax}} = 0$ is 3.9 dB. This maximal achievable Doppler diversity gain is lower compared to SISO since Alamouti SFC already increases and exploits frequency domain (multipath) diversity.

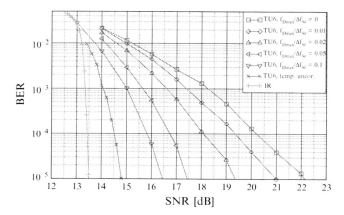

Fig. 5 BER after BCH decoding for DVB-T with different Doppler bandwidths $f_{\text{Dmax}}/\Delta f_{\text{SC}}$, 1 TX antenna (no Alamouti SFC), 8k FFT, 16-QAM, code rate R=3/4, guard interval length $N_{\text{G}}/N_{\text{FFT}} = 1/32$

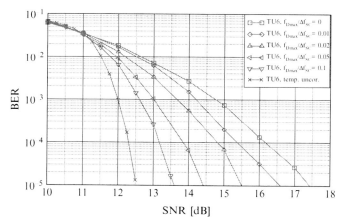

Fig. 6 BER after BCH decoding for DVB-T with different Doppler bandwidths $f_{\text{Dmax}}/\Delta f_{\text{SC}}$, 2 TX antennas (Alamouti SFC)8k FFT, 16-QAM, code rate R=3/4, guard interval length $N_{\text{G}}/N_{\text{FFT}} = 1/32$

For both configurations SISO and MISO Alamouti we additionally apply the standard conformable DDoD scheme as described in Section 3. We compare the additionally achievable Doppler diversity gains with the static case (see $f_{\text{Dmax}} = 0$ graphs in Figs. 5 and 6). The Doppler bandwidth of the original channel is $f_{\text{Dmax}} = 0$. Using DDoD, the artificially inserted Doppler bandwidth is increased from $\Delta f_{\text{max}}/\Delta f_{\text{SC}} = 0.0025$, which is 0.25% of the subcarrier spacing, up to 10%. Note, with DDoD we increase Doppler diversity but do not insert any additional ICI. Figure 7 shows the additionally achievable Doppler diversity gains. Compared to the SISO case, where we do not use Alamouti SFC, the application of 2-TX-antenna

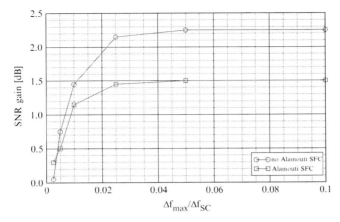

Fig. 7 SNR gain at a BER of 10^{-4} of Discontinuous Doppler Diversity versus the normalized inserted Doppler spread $f_{max}/\Delta f_{SC}$, 8k FFT, 16-QAM, code rate R=3/4, guard interval length $N_G/N_{FFT} = 1/32$

DDoD provides SNR gains of up to 2.5 dB. It can be observed that no further gains can be achieved when increasing the artificial Doppler bandwidth δf_{max} beyond 5% of the subcarrier spacing. Even in combination with Alamouti SFC there is an additional Doppler diversity gain of 1.5 dB. Note, the combination of Alamouti and DDoD uses 4 TX antennas in total as outlined in Fig. 4.

5 Summary

In this paper we have briefly introduced the emerging terrestrial video broadcasting standard DVB-T2. We outlined major differences compared to the currently deployed standard DVB-T with respect to coding, interleaving, modulation and framing. Compared to DVB-T the coding and interleaving structure of DVB-T2 is capable of exploiting Doppler diversity. Therefore, we have investigated the application of the Doppler diversity increasing TX-antenna scheme Discontinuous Doppler Diversity in combination with Alamouti Space-Frequency coding as specified in DVB-T2. Simulation results have shown significant diversity gains in the order of several dBs.

References

1. Ulrich H. Reimers. DVB — The family of international standards for digital video broadcasting. *Proceedings of the IEEE*, 94(1):173–182, January 2006. ISSN: 0018-9219, doi: 10.1109/JPROC.2005.861004.
2. DVB-T fact sheet, June 2008. Online available at http://www.dvb.org.

3. DVB-T2 fact sheet, August 2008. Online available at http://www.dvb.org.
4. Digital Video Broadcasting Project. *Framing structure, channel coding and modulation for a second generation digital terrestrial television broadcasting system (DVB-T2)*, June 2008.
5. Armin Dammann and Stefan Kaiser. Standard conformable antenna diversity techniques for OFDM and its application to the DVB-T system. In *Proceedings IEEE Global Telecommunications Conference (GLOBECOM 2001), San Antonio, TX, USA*, pages 3100–3105, November 2001.
6. Siavash M. Alamouti. A simple transmit diversity technique for wireless communications. *IEEE Journal on Selected Areas in Communications*, 16(8):1451–1458, October 1998.
7. Armin Dammann and Ronald Raulefs. Increasing time domain diversity in OFDM systems. In *Proceedings IEEE Global Telecommunications Conference (GLOBECOM 2004), Dallas, TX, USA*, volume 2, pages 809–812, November 2004.
8. COST 207 Management Committee. *COST 207: Digital Land Mobile Radio Communications, Final Report*. Commission of the European Communities, 1989.
9. Patrick Robertson, Emmanuelle Villebrun, and Peter Höher. A comparison of optimal and sub-optimal MAP decoding algorithms operating in the log domain. In *Proceedings IEEE International Conference on Communications (ICC 1995), Seattle, USA*, volume 2, pages 1009–1013, June 1995.

Part III
Channel Estimation & Characterization

Robust 2-D Channel Estimation for Staggered Pilot Grids in Multi-Carrier Systems: LTE Downlink as an Example

Muhammad Danish Nisar, Wolfgang Utschick, and Josef Forster

Abstract The paper extends the recently proposed concept of finite dimensional robust 2-D MMSE channel estimation to the staggered pilot grids in multi-carrier systems. The proposed scheme is particularly useful once the receiver has no knowledge of the channel correlation function and it intends to pursue the MMSE channel estimation for improved performance. We build a minimax optimization setup, as a function of the pilot grid structure, that leads to the Least Favourable (LF) channel correlation sequence, which is then employed by the receiver to obtain the maximally robust channel estimation coefficients guaranteeing the best worst-case performance. For staggered pilot grid, such as the one found in LTE Downlink specifications, we present three implementation variants for the maximally robust 2-D MMSE channel estimation and compare their performance in terms of coded BER.

1 Introduction

Multi-carrier systems typically employ the transmission of known pilot sub-carriers to enable pilot-aided channel estimation at the receiver. However, since the transmission of pilot sub-carriers lead to a reduction in effective data throughput, the pilot sub-carriers are desired to be distributed over the time frequency grid in a regular but sparse manner. The pilot sub-carrier spacing along time and along frequency directions are typically selected by keeping in view the channel time variation and frequency selectivity scenarios that are expected in the system [1, 2]. In order to be able to grasp the fine channel variations at the receiver, staggered pilot grids are often employed in practical systems such as LTE Downlink [3]. In staggered pilot grids, the successive set of pilot sub-carriers in time are staggered along

---

M.D. Nisar (✉) and W. Utschick
Technical University Munich, Associate Institute for Signal Processing (MSV), Munich, Germany

M.D. Nisar and J. Forster
Nokia Siemens Networks, Radio Access, Algorithms & Simulations, Munich, Germany
e-mail: mdanishnisar@ieee.org, utschick@tum.de, josef.forster@nsn.com

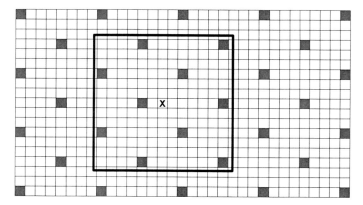

Fig. 1 Staggered Pilot Grid. Time along horizontal and frequency along vertical direction. Blue squares represent pilot positions, so that we have $\Delta_T = 4$ and $\Delta_F = 6$ in this illustration. The observation window (thick bordered rectangle) is shown with $K_T = 2$ and $K_F = 1$ with respect to the crossed data CFR position

the frequency direction. Precisely speaking, this implies a certain shift in the frequency index of the pilot sub-carriers that are successively transmitted along time. An example is shown in Fig. 1 with staggering equal to half the pilot frequency separation.

At the receiver, *Least Squares (LS)* estimates of the *Channel Frequency Response (CFR)* coefficients are obtained at the pilot positions followed by linear or more sophisticated interpolation to estimate the CFR at the data sub-carriers. Among the various interpolation techniques, the 2-D MMSE estimator leads to the minimum estimation MSE. However, it requires the knowledge of the channel correlation function which may not be accurately available in a practical system. This necessitates the need of a robust 2-D MMSE filter that employs a fixed channel correlation spectrum but guarantees a good worst-case performance as well. Such a robust estimator has been proposed in [4] for the case of an infinite number of pilot observations along time, whereby the authors suggest to use a rectangular correlation spectrum for robust estimation performance. For the practically relevant case of a finite number of observations, we recently proposed in [5] to employ the LF correlation sequence that results from a minimax based semi-definite optimization procedure to have a robust performance.

In this paper, we extend the proposed Maximally Robust 2-D MMSE channel estimator to multi-carrier systems with staggered pilot grids, and compare three variants of the concept in terms of their estimation performance. Simulation results are presented for LTE downlink systems that employ a staggered pilot grid.

Notation. The operators $E[\bullet]$, $|\bullet|^2$, $(\bullet)^*$, $(\bullet)^H$, $\text{vec}(\bullet)$ stand for expectation, absolute value square, complex conjugate, hermitian and vectorization (vertically stacking the columns of a matrix below each other) respectively, while ι denotes the imaginary unit.

2 Robust 2-D Channel Estimation

We consider a multi-carrier system with pilot sub-carriers distributed over the time-frequency grid in a periodic manner. Pilot spacings along time and frequency are labeled with Δ_T and Δ_F respectively, and successive pilots are assumed to be staggered by $\Delta_F/2$ along frequency as shown in Fig. 1. Let K_T and K_F denote the number of pilot channel estimates on either side of the data position of interest, along the time and the frequency directions respectively, to be employed for estimation. The rectangular observation window (marked in Fig. 1 with a rectangle around the crossed data position of interest) thus formed is labeled with a matrix H and let the total number of pilot CFR estimates falling inside this observation window be N_p, then the LS estimates of pilot CFRs can be arranged in the observation vector as

$$\tilde{h}_p = h_p + \eta, \tag{1}$$

where $\eta \in \mathbb{C}^{N_p}$ denotes the pilot channel estimation error vector. Now with $w^H \in \mathbb{C}^{1 \times N_p}$ denoting the vector containing 2-D filter coefficients, the estimation MSE between the CFR $H_{f,t}$ at fth frequency and tth time index, and its MMSE estimate $\hat{H}_{f,t} = w^H \tilde{h}_p$ obtained as $\mathrm{E}[|H_{f,t} - \hat{H}_{f,t}|^2]$, can be written as

$$\varepsilon(w, \{r_H(i,j)\}) = r_H(0,0) + w^H \left(R_{h_p} + R_\eta \right) w - w^H r_{h_p} - r_{h_p}^H w, \tag{2}$$

where $\{r_H(i,j)\}$ denotes the unknown 2-D channel correlation sequence[1] defining both $R_{h_p} = \mathrm{E}\left[h_p h_p^H \right]$ and $r_{h_p} = \mathrm{E}\left[h_p H_{f,t}^* \right] \in \mathbb{C}^{N_p}$ while the noise covariance matrix is assumed to be $R_\eta = \sigma_\eta^2 I_{N_p}$. Minimization of the MSE w.r.t the filter coefficients yields the well known MMSE solution,

$$w_{\mathrm{MMSE}} = \left(R_{h_p} + R_\eta \right)^{-1} r_{h_p}, \tag{3}$$

leading to the minimum MSE attained (cf. (2))

$$\varepsilon(w_{\mathrm{MMSE}}, \{r_H(i,j)\}) = r_H(0,0) - r_{h_p}^H \left(R_{h_p} + R_\eta \right)^{-1} r_{h_p} \tag{4}$$

In the sequel, we look for a maximally robust 2-D MMSE estimator based on a fixed 2-D channel correlation sequence that promises the best worst case performance.

2.1 Maximally Robust Estimator

Intuitively speaking, to find the *Maximally Robust* (MR) estimator we first maximize the MSE (cf. (2)) over the set of all valid 2-D channel correlation sequences to arrive

---

[1] Note that the wide sense stationarity of the random process $H_{f,t}$ is assumed through out this paper so that the correlation function is independent of the indices f and t.

at the worst case scenario and then minimize the resultant MSE via optimization for the filter to finally arrive at the MR estimator, i.e.

$$\min_{\boldsymbol{w} \in \mathbb{C}^{N_\text{p}}} \max_{\{r_H(i,j)\} \in \mathcal{U}_{r_H}^1} \varepsilon(\boldsymbol{w}, \{r_H(i,j)\}), \tag{5}$$

where $\mathcal{U}_{r_H}^1$ denotes the set of all valid 2-D channel correlation sequences with a bounded L$_1$ norm, i.e. $r_H(0,0) \leq \beta$ and with bandwidth restrictions of $\omega_{\text{t,max}}$ and $\omega_{\text{f,max}}$ on their time and frequency correlation spectra in accordance with the Doppler frequency and the channel delay spread respectively.

We now employ a theorem [6, 7] on the equivalence (saddle point nature) of the finite dimensional minimax and max-min problems under the convexity and compactness constraints on the uncertainty class $\mathcal{U}_{r_H}^1$ and the linearity constraint on the estimator. Consequently, given that the prerequisites are fulfilled, the original minimax problem in (5) can be reformulated into the equivalent max-min problem,

$$\max_{\{r_H(i,j)\} \in \mathcal{U}_{r_H}^1} \min_{\boldsymbol{w} \in \mathbb{C}^{N_\text{p}}} \varepsilon(\boldsymbol{w}, \{r_H(i,j)\}). \tag{6}$$

In essence, the problem of finding the maximally robust estimator is casted into the one of finding the *Least Favorable (LF)* 2-D correlation sequence. Note that the minimization problem in (6) is nothing else than the conventional MMSE optimization problem leading to the following residual problem (cf. (3) and (4)),

$$\max_{\{r_H(i,j)\} \in \mathcal{U}_{r_H}^1} r_H(0,0) - \boldsymbol{r}_{h_\text{p}}^\text{H} \left(\boldsymbol{R}_{h_\text{p}} + \boldsymbol{R}_\eta\right)^{-1} \boldsymbol{r}_{h_\text{p}}. \tag{7}$$

Thus, we arrived from a minimax optimization setup down to a pure maximization problem. Since the objective function in (7) is monotonically increasing in $r_H(0,0)$, the maximization is reached once $r_H(0,0) = \beta$, so that we actually need to minimize the subtractor

$$\min_{\{r_H(i,j)\} \in \tilde{\mathcal{U}}_{r_H}^1} \boldsymbol{r}_{h_\text{p}}^\text{H} \left(\boldsymbol{R}_{h_\text{p}} + \boldsymbol{R}_\eta\right)^{-1} \boldsymbol{r}_{h_\text{p}}. \tag{8}$$

where $\tilde{\mathcal{U}}_{r_H}^1$ is identical to $\mathcal{U}_{r_H}^1$ except that $r_H(0,0) = \beta$. Transforming the problem into epigraph notation [8, p. 75] by introduction of a slack variable t and then employing the Schur complement positive semidefiniteness theorem,[2] the optimization problem reduces to,

$$\min_{t, \{r_H(i,j)\} \in \tilde{\mathcal{U}}_{r_H}^1} t \quad \text{s.t.} \quad \begin{bmatrix} t & \boldsymbol{r}_{h_\text{p}}^\text{H} \\ \boldsymbol{r}_{h_\text{p}} & \boldsymbol{R}_{h_\text{p}} + \boldsymbol{R}_\eta \end{bmatrix} \succeq 0. \tag{9}$$

---

[2] Given a matrix $\boldsymbol{M} = \begin{bmatrix} \boldsymbol{A}_{11} & \boldsymbol{A}_{12} \\ \boldsymbol{A}_{21} & \boldsymbol{A}_{22} \end{bmatrix}$, and the Schur complement of $\boldsymbol{A}_{22}$ as $\boldsymbol{S}_{22} = \boldsymbol{A}_{11} - \boldsymbol{A}_{12}\boldsymbol{A}_{22}^{-1}\boldsymbol{A}_{21}$, we have $\boldsymbol{M} \succeq 0$ if and only if $\boldsymbol{A}_{11} \succ 0$ and $\boldsymbol{S}_{11} \succeq 0$.

Next we decompose the $\{r_H(i,j)\} \in \tilde{\mathcal{U}}_{r_H}^1$ constraint into individual analytical constraints. The positive semidefiniteness property of the finite length correlation sequence can be expressed in terms of positive semidefiniteness of the channel correlation matrix, $\boldsymbol{R}_H = \mathrm{E}\left[\mathrm{vec}(\boldsymbol{H})\,\mathrm{vec}(\boldsymbol{H})^{\mathrm{H}}\right]$. In order to incorporate constraints on the bandwidths of time and frequency correlation sequence, we use a theorem [9] on the existence and uniqueness of band-limited positive semidefinite extensions. It allows the bandwidth constraints to be expressed as positive semidefinite constraints on Toeplitz matrices constructed from the sequence obtained after filtering the original correlation sequence with the filter

$$u(m) = e^{\iota(\omega_h+\omega_l)/2}\delta(m-1) + e^{-\iota(\omega_h+\omega_l)/2}\delta(m+1) \\ - 2\cos((\omega_h-\omega_l)/2)\delta(m) \tag{10}$$

To this end, bandlimitedness of time and frequency correlation sequences can be assured by positive semidefinite constraints on the Toeplitz matrices $\boldsymbol{R}_T$ and $\boldsymbol{R}_F$ constructed from the filtered sequences [5]. Thus the uncertainty class constraint can be equivalently described by following positive semidefiniteness constraints,

$$\{r_H(i,j)\} \in \tilde{\mathcal{U}}_{r_H}^1 \iff \boldsymbol{R}_H \succeq 0, \boldsymbol{R}_T \succeq 0, \boldsymbol{R}_F \succeq 0, r_H(0,0) = \beta \tag{11}$$

The overall optimization problem can finally be posed by incorporating (11) into the optimization problem (9) as,

$$\min_{t, \{r_H(i,j)\}} t \quad \text{s.t.} \quad \begin{bmatrix} t & \boldsymbol{r}_{h_p}^{\mathrm{H}} \\ \boldsymbol{r}_{h_p} & \boldsymbol{R}_{h_p} + \boldsymbol{R}_\eta \end{bmatrix} \succeq 0, \quad r_H(0,0) = \beta, \\ \boldsymbol{R}_H \succeq 0, \quad \boldsymbol{R}_T \succeq 0, \quad \boldsymbol{R}_F \succeq 0. \tag{12}$$

Thus, we end up in a semidefinite optimization problem with a linear cost function, an equality constraint, and a few positive semidefiniteness constraints. As such, the problem that can be solved via any semidefinite problem solver like SeDuMi [10]. The solution of this problem yields the LF 2-D CFR correlation sequence $\{r_H^{\mathrm{LF}}(i,j)\}$ which can then be used for the computation of the *maximally robust (MR) 2-D MMSE estimation filter* coefficients, i.e.

$$\boldsymbol{w}_{\mathrm{MMSE}}^{\mathrm{MR}} = \left(\boldsymbol{R}_{h_p}^{\mathrm{LF}} + \boldsymbol{R}_\eta\right)^{-1}\boldsymbol{r}_{h_p}^{\mathrm{LF}}. \tag{13}$$

with $\boldsymbol{r}_{h_p}^{\mathrm{LF}}$ and $\boldsymbol{R}_{h_p}^{\mathrm{LF}}$ as defined earlier. The superscripts $(\bullet)^{\mathrm{LF}}$ emphasize that they are based on the optimized LF correlation sequence.

Note that the optimization procedure (12) needs to be carried out only once for a set of grid parameters such as pilot spacings. The resultant LF correlation sequence or the corresponding estimator coefficients can therefore be computed offline. As such, the computational complexity of the estimation process itself is analogous to that of any other 2-D MMSE estimator. It should be noted that we employ the approach presented in [11] to handle the grid edge effects.

2.2 Implementation Variants

Optimal Strategy. Owing to the relative staggering in successive pilot symbols, the number of pilot sub-carriers to be used for interpolation may vary in alternate blocks. As an example, consider the data position marked with a cross in Fig. 1, the number of pilot observations to be employed for interpolation is $2K_T + 1$ for the pilot-containing block to the left, but is $2K_T$ for the block to the right. This complicates the structure of r_{h_p} and R_{h_p} and needs special care in the implementation of (12) and (13). The scheme is optimal in the sense that it incorporates the staggering of pilot blocks in the formulation of robust estimation problem unlike the proceeding schemes.

De-Staggering based Strategy. A sub-optimal work around, to avoid the staggering-related structural complications, as hinted above, is to artificially de-stagger the grid at the receiver. This involves basically a linear interpolation along the frequency direction in staggered blocks to determine the LS estimates of data sub-carriers at the non-staggered positions. Afterwards, we may treat these LS estimates as that of virtual pilot sub-carriers and remove the original staggered pilot observations. This leaves the grid with same pilot spacings, but with the staggering eliminated, resulting into simpler implementations of (12) and (13). Discarding of the original (staggered) pilot estimates and their replacement by de-staggered pilot estimates results into sub-optimality of the scheme.

Compensated Staggering based Strategy. Another alternative to artificially eliminate staggering at the receiver, is to employ linear interpolation along frequency in all the pilot blocks to create virtual pilot sub-carriers in between the original pilot sub-carriers leading to an effective reduction of pilot spacing along frequency to $\tilde{\Delta}_F = \Delta_F/2$. We increase $\tilde{K}_F = 2K_F$ so as to make a fair comparison with other approaches. Note that by this compensation of staggering we created observations with correlated noise, against our assumption, so that this approach is also sub-optimal.

Linear Interpolation based Strategy. As a benchmark, the simplest scheme that we employ for estimation of data CFRs is the linear interpolation along frequency followed by linear interpolation along time. This serves as the performance lower bound for other more sophisticated interpolation schemes.

3 Simulation Results

We provide simulation results for the LTE downlink system [3] that employs a staggered pilot grid. Owing to the proposed frequency hopping at slot boundaries, we consider estimation of CFRs at each slot independently. This leaves us with only two pilot containing blocks in the observation window, i.e. we have $K_T = 1$ with a time spacing of $\Delta_T = 4$ and with the pilot sub-carrier frequency spacing of $\Delta_F = 6$.

The pilot blocks are staggered along frequency by $\Delta_F/2 = 3$. Various other parameters have been selected according to the LTE specifications for the high data rate 20 MHz band [3]. Among the estimation parameters, we choose $K_F = 1$ for simplicity. For the case of compensated staggering based strategy described above we have $\tilde{\Delta}_F = 3$ and $\tilde{K}_F = 2$. We employ the standard frequency domain MMSE equalizer and a turbo decoder to show the overall system performance.

A comparison of the performance of the three alternatives for robust channel estimation described above is presented along with that of the simple linear interpolation along frequency and time. Simulations are carried out for an allocation of 6 Resource Blocks in downlink. 16-QAM and 64-QAM modulation schemes are used in conjunction with a rate 4/5 and 2/3 turbo code respectively.

At the benchmark level of 10^{-2} coded BER, for 16-QAM scenario in Fig. 2a we observe that performance of simple linear interpolation suffers about 2.5 dB as compared to the optimal scheme. On the other hand, the performance of the three MMSE based robust channel estimation variants are within 0.3 dB of each other, implying an improvement in excess of 2.2 dB over the linear interpolation approach. Among the robust estimation variants, the performance gap increases at lower benchmark levels. For instance at the coded BER of 10^{-3}, the de-staggering based implementation suffers a degradation of about 2.5 dB as compared to the optimal strategy while the degradation for compensated staggering based strategy is only 0.5 dB.

Similarly for 64-QAM scenario in Fig. 2b at the coded BER of 10^{-2}, linear interpolation looses about 5 dB, while the de-staggering based strategy looses about 3 dB, as compared to the optimal robust channel estimation. The performance of compensated staggering based strategy remains within 1 dB of the optimal scheme.

Hence, we conclude from the above simulation results that besides the optimal scheme whereby we incorporate the staggered structure of pilot grid in the semi-definite optimization procedure, the compensated staggering based robust channel estimation offers a nice trade off between the (offline) optimization implementation complexity and the final system performance.

4 Conclusion

The paper applied the principle of finite observation based robust 2-D MMSE channel estimation to multi-carrier systems with staggered pilot grids, such as the one specified in LTE downlink specifications. After deriving a general framework of robust 2-D channel estimation, three implementation variants were described. Simulation results presented for an LTE downlink system indicate that besides the staggering based optimal robust channel estimation, robust estimation applied to the grid after compensation of staggering offers a suitable performance complexity trade off.

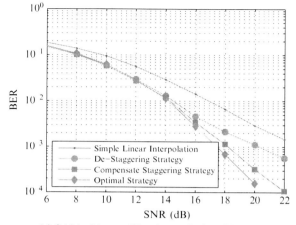

a 16-QAM with rate 4/5 code at velocity of 30 kmph

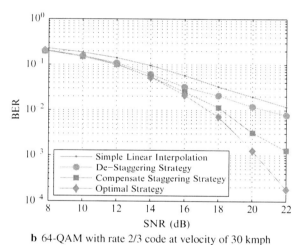

b 64-QAM with rate 2/3 code at velocity of 30 kmph

Fig. 2 Performance comparison of different strategies for robust channel estimation in case of staggered pilot grid. Coded BER at velocity 30 kmph

References

1. R. Nilsson, O. Edfors, M. Sandell, and P. O. Borjesson. An analysis of two-dimensional pilot-symbol assisted modulation for OFDM. *IEEE International Conference on Personal Wireless Communications*, pages 71–74, Dec 1997.
2. G. Auer and I. Cosovic. On pilot grid design for an OFDM air interface. *Wireless Communications and Networking Conference, 2007*, pages 2172–2177, March 2007.

3. Standardization Committee 3GPP. Evolved universal terrestrial radio access: Physical channels and modulation, 36.211. *Online, http://3gpp.org*, 2008.
4. Y. Li, L. J. Cimini, and N. R. Sollenberger. Robust channel estimation for OFDM systems with rapid dispersive fading channels. *IEEE Transactions on Communications*, 46:902–915, July 1998.
5. M. Danish Nisar, W. Utschick, and T. Hindelang. Robust 2-D Channel estimation in multi-carrier systems. *Submitted to the IEEE Transactions on Wireless Communications*.
6. V. N. Soloviov. Towards the theory of Minimax-Bayesian estimation. *Theory Probab. Appl.*, 44:739–754, 2000.
7. Frank A. Dietrich. *Robust Signal Processing for Wireless Communications*. Springer, 1st edition, 2008.
8. Stephen P. Boyd and L. Vandenberghe. *Convex Optimization*. Cambridge University Press, 2004.
9. K. S. Arun and L. C. Potter. Existence and Uniqueness of Band-Limited, Positive Semidefinite Extrapolations. *IEEE Trans. on Acoustics, Speech and Signal Processing*, 48:547–549, Mar 1990.
10. J. F. Sturm. Using SeDuMi 1.02, A MATLAB toolbox for optimization over symmetric cones. *Optimization Methods and Softwares*, 11–12:625–653, 1999.
11. M. Danish Nisar, W. Utschick, and T. Hindelang. Robust 2-D Channel estimation for multi-carrier systems with finite dimensional Pilot Grid. *IEEE International Conference on Acoustics, Speech and Signal Processing*, Apr, 2009.

Block-IFDMA – Iterative Channel Estimation Versus Estimation with Interpolation Filters

Anja Sohl and Anja Klein

Abstract In this paper, Discrete Fourier Transform (DFT) precoded Orthogonal Frequency Division Multiple Access (OFDM) with block-interleaved subcarrier allocation per user is considered which is denoted as Block Interleaved Frequency Division Multiple Access (B-IFDMA). The B-IFDMA parameter design is affected on the one hand by considerations to reduce the energy consumption by entering a micro sleep mode. These considerations recommend a transmission on a high number of subcarriers and a small number of successive symbols. On the other hand, in terms of channel estimation, an increasing number of successive symbols is beneficial to reduce the pilot symbol overhead. The paper contributes to the choice of the B-IFDMA system parameters, i.e., blocksize in frequency domain and number of successive symbols in time domain, that are investigated in consideration of channel estimation performance and pilot symbol overhead. Further on, a novel iterative Decision Directed Channel Estimation with Wiener filtering is proposed that requires only one pilot carrying B-IFDMA symbol in time domain and, thus, reduces the pilot symbol overhead especially if a small number of successive symbols is transmitted. In this paper, this novel channel estimation algorithm is introduced and compared to Wiener interpolation filtering and additionally to a low-complexity channel estimation approach.

1 Introduction

Within the ongoing research activities of future mobile radio systems, discrete fourier transform (DFT) precoded orthogonal frequency division multiple access (OFDMA) is a candidate multiple access scheme for non-adaptive transmission.

For DFT precoded OFDMA, there exist different possibilities of allocating subcarriers to a certain user. In this paper, block interleaved frequency division multiple access (B-IFDMA) is considered, where the data symbols of a specific user are transmitted on several blocks of K_f adjacent subcarriers that are equidistantly

---

A. Sohl (✉) and A. Klein
Technische Universität Darmstadt, Communications Engineering Lab, 64283 Darmstadt, Germany
e-mail: {a.sohl,a.klein}@nt.tu-darmstadt.de

S. Plass et al. (eds.), *Multi-Carrier Systems & Solutions 2009*,
Lecture Notes in Electrical Engineering 41,
© Springer Science+Business Media B.V. 2009

distributed over the available bandwidth. B-IFDMA is a generalization of the well-known interleaved frequency division multiple access (IFDMA), where a block consists of a single subcarrier [1, 2]. Due to the blockwise subcarrier allocation, B-IFDMA exhibits higher robustness against carrier frequency offsets than IFDMA and, at the same time, maintains the advantage of high frequency diversity [3]. In terms of pilot assisted channel estimation (PACE), interpolation in Frequency Domain (FD) is applicable for B-IFDMA within each block of K_f adjacent subcarriers. Thus, B-IFDMA benefits from a lower pilot symbol overhead compared to IFDMA if the distance between neighboring blocks is larger than the coherence bandwidth [4].

Furthermore, B-IFDMA supports an additional user separation via a time division multiple access (TDMA) component. I.e., during one TDMA slot, each user is assigned to a specific set of K_t successive B-IFDMA symbols. For each user, this opens up the possibility to enter a micro sleep mode and to achieve considerable energy savings if K_t is small compared to the interval between consecutive TDMA slots [3]. Thus, it is desirable to achieve a certain data rate by the transmission of a small number of successive B-IFDMA symbols and the allocation of a high number of subcarriers to each user [3].

On the other hand, considering PACE, the application of interpolation filters in time domain (TD) and FD, which is a favorable technique to estimate the channel for the non-pilot carrying symbols and subcarriers, involves the usage of at least two B-IFDMA symbols within the K_t successive symbols in TD and at least two subcarriers within each block of K_f subcarriers in FD for pilot transmission. Hence, in TD the pilot symbol overhead increases if less successive symbols are transmitted and a TDMA slot with a high number of successive B-IFDMA symbols that are assigned to a user would be beneficial. In FD, interpolation filtering is only possible if there are more than two subcarriers per block ($K_f > 2$).

In order to reduce the pilot symbol overhead for transmission with a small number K_t of successive symbols assigned to a user, we propose a novel iterative Decision Directed Channel Estimation with Wiener filtering (iterative DDCE+WF) in TD for B-IFDMA which avoids the need of a second pilot carrying symbol in TD while achieving estimation performances comparable to Wiener interpolation.

In this paper, the novel iterative DDCE+WF is compared with a conventional Wiener interpolation filter and with an alternative approach requiring only one pilot carrying B-IFDMA symbol. The optimal choice of the B-IFDMA system parameters, i.e., number K_f of subcarriers per block and number K_t of successive symbols, is investigated in consideration of channel estimation performance and pilot symbol overhead.

2 System Model

In this section, a system model for B-IFDMA will be derived. In the following, all signals are represented by their discrete time equivalents in the complex baseband. Further on, $(\cdot)^T$ denotes the transpose and $(\cdot)^H$ the Hermitian of a vector or a matrix.

Assuming a system with U users, the data symbols transmitted by a user with index u at symbol rate $1/T_s$ are grouped into a vector $\mathbf{d}^{(u)}(i) = (d_0^{(u)}(i),\ldots,d_{Q-1}^{(u)}(i))^T$ with index i, $i = 0,\ldots,K_t - 1$, denoting the index of the B-IFDMA symbol. The data symbols $d_q^{(u)}(i)$ can be taken from the alphabet of a modulation scheme like Phase Shift Keying or Quadrature Amplitude Modulation, that is applied to coded or uncoded bits. Let $\mathbf{F}_N$ and $\mathbf{F}_N^H$ denote the matrix representation of an N-point DFT and an N-point inverse DFT (IDFT) matrix, respectively, where $N = U \cdot Q$ is the number of available subcarriers in the system. The assignment of the data symbols $d_q^{(u)}(i)$ to the user specific set of Q subcarriers can be described by a Q-point DFT precoding matrix $\mathbf{F}_Q$, an $N \times Q$ mapping matrix $\mathbf{M}^{(k)}$ and an N-point IDFT matrix $\mathbf{F}_N^H$ [5]. The mapping matrix $\mathbf{M}^{(u)}$ has to describe the allocation of Q/K_f blocks each consisting of K_f adjacent subcarriers to the user with index u. $\mathbf{M}^{(u)}$ is given by its elements $M^{(u)}(n,q)$ in the n-th row and q-th column as

$$M^{(u)}(n,q = r + s \cdot K_f) = \begin{cases} 1 & n = s \cdot \frac{NK_f}{Q} + r + u \cdot K_f \\ 0 & \text{else} \end{cases}, \quad (1)$$

with $n = 0,\cdots,N-1$, $r = 0,\ldots,K_f - 1$, and $s = 0,\ldots,Q/K_f - 1$. Thus, the resulting ith, $i = 0,\cdots,K_t - 1$, B-IFDMA symbol of user u at chip rate $1/T_c = U/T_S$ is given by

$$\mathbf{x}^{(u)}(i) = \mathbf{F}_N^H \cdot \mathbf{M}^{(u)} \cdot \mathbf{F}_Q \cdot \mathbf{d}^{(u)}(i). \quad (2)$$

In order to avoid intersymbol and inter-carrier interference, a Cyclic Prefix is inserted in-between successive B-IFDMA symbols. The resulting B-IFDMA signal is transmitted over a channel with impulse response $\mathbf{h}^{(u)}(i)$ and $L < Q$ non-zero coefficients $h_l^{(u)}(i)$, $l = 0,\cdots,L-1$, at chip rate. The channel is assumed to be time-invariant during the transmission of one B-IFDMA symbol and the transmission over this multipath channel can be described by a flat fading channel for each allocated subcarrier in FD. With $H_q(i)$ denoting the complex channel coefficient, $D_q(i)$ the DFT of the transmitted data symbols and $V_q(i)$ the Additive White Gaussian Noise (AWGN) on the subcarrier with index q and the symbol with index i, the received values on each subcarrier in FD can be described by

$$R_q(i) = H_q(i) \cdot D_q(i) + V_q(i), \quad q = 0,\cdots,Q-1. \quad (3)$$

3 Channel Estimation

In the following, different appropriate methods of channel estimation in FD and TD are described for B-IFDMA. The proposed new iterative DDCE+WF is introduced as an approach to estimate the channel in TD iteratively. For simplicity and without loss of generality, the user index u is omitted in the sequel.

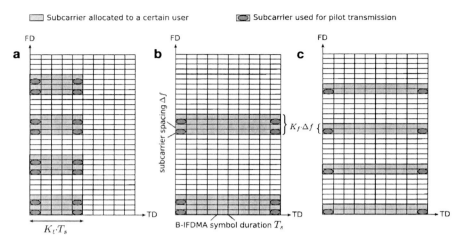

Fig. 1 Exemplary subcarrier allocations and pilot arrangements for B-IFDMA (**a**) $K_t = 5$, $Q = 16$, $K_f = 4$, (**b**) $K_t = 10$, $Q = 8$, $K_f = 4$, (**c**) $K_t = 10$, $Q = 8$, $K_f = 2$

In order to estimate the channel at the receiver, P_t symbols with index $\iota = 0, \cdots, P_t - 1$ are used to transmit pilot symbols on a subset consisting of P_f out of Q subcarriers in FD [4]. An estimate of the channel transfer factor of the pilot carrying subcarrier with index $\kappa = 0, \cdots, P_f - 1$ in the corresponding B-IFDMA symbol with index ι is determined by a Least-Squares (LS) estimation and, thus, given by

$$\hat{H}_\kappa(\iota) = \frac{R_\kappa(\iota)}{P_\kappa(\iota)}, \qquad (4)$$

with $P_\kappa(\iota)$ the pilot symbol transmitted on the κ-th pilot carrying subcarrier in the ι-th pilot carrying B-IFDMA symbol. The LS-estimates $\hat{\mathbf{H}}(\iota) = [\hat{H}_0(\iota), \cdots, \hat{H}_{P_f - 1}(\iota)]$ are exploited to estimate the channel transfer factors of the remaining, non-pilot carrying subcarriers and symbols. First, the estimation is performed in FD and then used to estimate the channel in TD. Depending on the block size K_f in FD and the number K_t of successive B-IFDMA symbols, different pilot arrangements and, thus, different channel estimation approaches are applicable. In Fig. 1, the B-IFDMA signal is given in TD and FD for exemplary sets of parameters and various pilot arrangements.

3.1 Frequency Domain

Wiener Interpolation Filter. For the application of the well-known Wiener Interpolation Filter in FD at least two subcarriers within each block of K_f subcarriers are used for pilot transmission, cf. Fig. 1a, b. In general, the distance between neighboring blocks is larger than the coherence bandwidth and, thus, interpolation is only possible within each block of subcarriers. I.e., an estimate of the channel transfer coefficients of the non pilot-carrying subcarriers within each block can be obtained by filtering with a Wiener filter according to [6]. With $b_{\kappa,q}$ the Wiener filter coefficients determined to minimize $E\{|H_q(\iota) - \tilde{H}_q(\iota)|^2\}$, the channel estimate in FD is given by

$$\tilde{H}_q(\iota) = \sum_{\kappa=1}^{P_f K_f/Q} b_{\kappa,q} \cdot \hat{H}_\kappa(\iota) . \qquad (5)$$

Repetition. For the application of Repetition in FD only one pilot symbol is required within each block of K_f subcarriers in FD, as it is depicted in Fig. 1c. The channel variations are assumed to be negligible within a certain fraction of the coherence bandwidth and the LS-estimate of the nearest pilot carrying subcarrier is used for equalization. Thus, the vector $\tilde{\mathbf{H}}(\iota)$ containing the channel estimates for each allocated subcarrier is given by

$$\tilde{\mathbf{H}}(\iota) = [\underbrace{\hat{H}_{\kappa=1}(\iota), \hat{H}_{\kappa=1}(\iota), \cdots}_{K_f\text{-times}}, \cdots, \underbrace{\hat{H}_{\kappa=P_f-1}(\iota), \hat{H}_{\kappa=P_f-1}(\iota)}_{K_f\text{-times}}] . \qquad (6)$$

Subcarrierwise Least-Squares. For Subcarrierwise LS in FD each allocated subcarrier is used for pilot transmission, i.e. $P_f = Q$, and the channel transfer factors $\hat{H}_\kappa(\iota)$, $\kappa = 0, \cdots, Q-1$, of the allocated subcarriers are determined by LS-estimation, i.e. $\hat{H}_\kappa(\iota) = \frac{R_\kappa(\iota)}{P_\kappa(\iota)}$ and $\tilde{\mathbf{H}}(\iota) = [\hat{H}_0(\iota), \cdots, \hat{H}_{Q-1}(\iota)]$. In this paper, the LS estimation on each subcarrier is included to provide a reference compared to the channel estimation approaches with reduced number of pilot symbols.

3.2 Time Domain

Wiener Interpolation Filter. For the application of Wiener Interpolation Filter in TD at least two pilot carrying B-IFDMA symbols, i.e. $P_t \geq 2$, are required in TD as it is shown in Fig. 1. In this paper, the first and the last B-IFDMA symbol are chosen to transmit pilot symbols. With $c_{\iota,i}$ the Wiener filter coefficients determined to minimize $E\{|\mathbf{H}(i) - \tilde{\mathcal{H}}(i)|^2\}$, the channel estimate in TD is given by

$$\tilde{\mathcal{H}}(i) = \sum_{\iota=1}^{P_t} b_{\iota,i} \cdot \tilde{\mathbf{H}}(\iota) . \qquad (7)$$

Iterative DDCE + Wiener Filtering. For the application of the proposed new iterative DDCE+WF only one pilot carrying B-IFDMA symbol is required in TD and, thus, the iterative DDCE+WF is feasible where the application of Wiener Interpolation Filtering fails. In Table 1, the basic method of the iterative DDCE+WF is outlined. It is applied in TD after the channel transfer factors have been determined

Table 1 Iterative DDCE + Wiener Filter

---
1. Initializing Estimates
  (a) Wiener Interpolation Filtering / Repetition of $\hat{\mathbf{H}}(0)$ in FD
      $\rightarrow \tilde{\mathbf{H}}(0)$
  (b) Equalization with $\tilde{\mathbf{H}}(0)$
      Estimation of transmitted symbols $\rightarrow \hat{\mathbf{D}}(1)$
  (c) $\tilde{\mathbf{H}}(1) = \mathbf{R}(1)/\hat{\mathbf{D}}(1)$
  (d) Wiener Filtering of $[\tilde{\mathbf{H}}(0)\,\tilde{\mathbf{H}}(1)] \rightarrow \tilde{\mathcal{H}}(1)$

For $i = 1,\ldots, K_t - 1$
2. Equalization with $\tilde{\mathcal{H}}(i)$
    Estimation of transmitted symbols $\rightarrow \hat{\mathbf{D}}(i+1)$

3. Decision Directed Channel Estimation
    $\tilde{\mathbf{H}}(i+1) = \mathbf{R}(i+1)/\hat{\mathbf{D}}(i+1)$

4. Wiener Filtering
    Wiener Filtering of $[\tilde{\mathcal{H}}(i)\,\tilde{\mathbf{H}}(i+1)] \rightarrow \tilde{\mathcal{H}}(i+1)$

---

for each subcarrier (e.g. of the first symbol) via Repetition or Wiener interpolation. For B-IFDMA, the DDCE principle described in [7] leads to a high noise amplification with increasing number K_t of B-IFDMA symbols. Therefore, we propose a new algorithm where a Wiener filter is applied to the decision directed estimates and the filtered estimates are used iteratively for DDCE again.

Repetition. Repetition is the alternative approach to iterative DDCE + WF and requires only one pilot symbol, i.e. $P_t = 1$, within the K_t B-IFDMA symbols in TD. Here, the channel variations are assumed to be negligible within a certain fraction of the coherence time and the LS-estimate $\tilde{\mathbf{H}}(\iota)$ of the pilot carrying symbol is used for equalization of the non-pilot carrying symbols, i.e. $\tilde{\mathcal{H}}(i) = \tilde{\mathbf{H}}(\iota)$ for $i = 0, \cdots, K_t - 1$.

4 Performance Analysis

In this section, the performance of different combinations of channel estimation approaches in TD and FD are investigated for velocities of $v = 20$ km/h and $v = 50$ km/h, respectively. Figures 2 and 3 show the Mean Square Error (MSE) between the estimated and the true channel transfer function, i.e. MSE $= \sum_{i=0}^{K_t-1} \|\tilde{\mathcal{H}}(i) - \mathbf{H}(i)\|^2/(Q \cdot K_t)$, as a result of 500 channel realizations. The MSE is depicted in dependency of the Signal-to-Noise Ratio (SNR) E_s/N_0, i.e. energy per symbol over noise power, and for the parameters given in Table 2. The presented results already include the differing pilot symbol overhead for the particular estimation approach

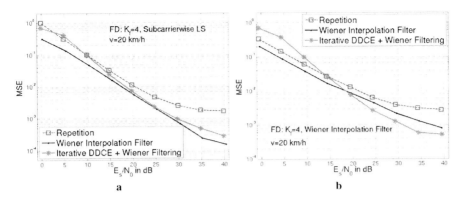

Fig. 2 Comparison of Repetition, Wiener Interpolation Filter and iterative DDCE + Wiener Filtering in TD for (**a**) Subcarrierwise LS estimation in FD and $v = 20$ km/h, (**b**) Wiener Interpolation Filter in FD and $v = 20$ km/h

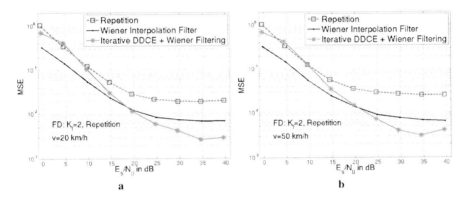

Fig. 3 Comparison of Repetition, Wiener Interpolation Filter and iterative DDCE + Wiener Filtering in TD for (**a**) interpolation with Repetition in FD for $v = 20$ km/h, (**b**) interpolation with Repetition in FD for $v = 50$ km/h

as an SNR degradation [8]. Further on, the results are valid for the assumption that the channel correlations in TD and FD are known to the Wiener Filters.

In Fig. 2a, results are presented for $K_f = 4$ and subcarrierwise LS channel estimation in FD, i.e. the Q allocated subcarriers in the pilot carrying symbol are used for pilot transmission. In TD, the channel is estimated with a Wiener Interpolation Filter, Repetition and iterative DDCE+WF at a velocity of $v = 20$ km/h. The iterative DDCE+WF clearly outperforms the channel estimation with Repetition for SNR > 15dB and reaches the performance of the Wiener Interpolation Filter for SNR-values between 15 and 30 dB.

In Fig. 2b, results are presented for the same parameters as in Fig. 2a but for channel estimation with a Wiener Interpolation Filter in FD. I.e., only two subcarriers

Table 2 Simulation parameters

Carrier Frequency	3.7 GHz
Bandwidth	40 MHz
Total No. of Subcarriers	1024
Subcarrier Spacing Δf	39.1 kHz
No. Q of Subcarriers per user	16
No. K_t of successive B-IFDMA symbols	10
Guard Interval	3.2 μs
Channel	WINNER SCM, Urban Macro
Coherence bandwidth	$B_{coh} < 20 \cdot \Delta f$

out of a block consisting of $K_f = 4$ subcarriers are used for pilot transmission, cf. Fig. 1b. It can be seen that the performance of the Wiener Interpolation Filter and Repetition in TD is degraded for SNR > 15 dB as a result of the reduced number of pilot symbols in FD. The interpolation error that is caused in FD is propagated by the Wiener Interpolation Filter and Repetition in TD. In contrast, the iterative DDCE+WF shows only slight performance degradations compared to the result in Fig. 2a. Due to symbol detection, updating channel estimation and subsequent Wiener Filtering in each iteration for iterative DDCE+WF, the propagation of the interpolation error in FD is mitigated and can be nearly avoided for a velocity of $v = 20$ km/h.

Figure 3 shows the MSE for the case that each block in FD consists of $K_f = 2$ subcarriers and only one subcarrier per block is used for pilot transmission, cf. Fig. 1c. As there is only one pilot symbol available within the coherence bandwidth, the channel is estimated by Repetition in FD.

In Fig. 3a, the results are presented for $v = 20$ km/h. It shows that the performance of all three estimation approaches in TD is strongly degraded due to the inaccurate channel estimation in FD and exhibits an error floor for high SNR-values. Nevertheless, the performance of the iterative DDCE+WF is least affected in comparison to Wiener Interpolation and Repetition and, thus, outperforms even the Wiener Interpolation Filter for SNR > 18 dB.

Fig. 3b shows the same results for $v = 50$ km/h. It can be seen, that the performance of the Wiener Interpolation Filter and iterative DDCE+WF is hardly affected by the increasing velocity and only the performance of Repetition in TD suffers from a higher error floor.

4.1 Parameter Selection for B-IFDMA

The results presented in Figs. 2 and 3 show that for the proposed channel estimation approaches, the performance of the estimation in FD strongly influences the overall estimation performance. Dependent on the choice of the channel estimation approach in FD and, thus, dependent on the coherence bandwidth and blocksize K_f in FD, the application of Wiener Interpolation Filter, Repetition or iterative DDCE+WF in TD lead to different estimation performances. In Table 3,

Table 3 Parameter selection

FD	$2^{\beta-1} \cdot 10 \cdot \Delta f \leq B_{\text{coh}} < 2^\beta \cdot 10 \cdot \Delta f,$ ($\beta=1,2,3,4,\ldots$)			
	$K_f \leq 2^\beta$		$K_f > 2^\beta$	
	Repetition		**Wiener**	
	$K_t < \frac{K_{\text{coh}}}{12}$	$K_t \geq \frac{K_{\text{coh}}}{12}$	$K_t < \frac{K_{\text{coh}}}{17}$	$K_t \geq \frac{K_{\text{coh}}}{17}$
TD	SNR $\geq$ 20 dB else	**Wiener**	SNR $\geq$ 18 dB else	**Wiener**
	DDCE **Wiener**		**DDCE** **Wiener**	

an overview is given for reasonable choices of channel estimation approaches in FD and TD in dependency of the coherence bandwidth, the coherence time and the B-IFDMA parameters K_f and K_t. Table 3 is deduced from simulation results that are not shown in this paper due to limited space. In the following, K_{coh} denotes the number of B-IFDMA symbols that can be transmitted within the coherence time $T_{\text{coh}} = \frac{8.1 \cdot 10^{-2}}{v \text{ in m/s}}$ m. Wiener denotes the Wiener Interpolation Filter and DDCE denotes the iterative DDCE+WF.

5 Conclusion

In this paper, different channel estimation approaches in FD and TD and their various combinations have been investigated for the case that the data symbols of a user are transmitted on K_t successive B-IFDMA symbols. A new iterative DDCE+WF that requires only one pilot carrying B-IFDMA symbol in TD has been introduced and compared to the Wiener Interpolation Filter. It came out that the performance of iterative DDCE+WF is even better than performance of the Wiener Interpolation Filter for certain parameters. The iterative DDCE+WF is preferable to Wiener Interpolation Filtering if the channel estimation performance in FD is poor. For a channel estimation in FD that suffers from interpolation errors and velocities up to $v = 70$ km/h it is feasible to transmit within a small number K_t of consecutive symbols compared to the coherence time and use iterative DDCE+WF. The better the channel estimation performance in FD and the higher the velocity, the better is the performance for Wiener Interpolation especially for high SNR-values.

References

1. U. Sorger, I. De Broeck, and M. Schnell. IFDMA - A New Spread-Spectrum Multiple-Access Scheme. In *Proc. ICC*, pages 1013–1017, Atlanta, Georgia, USA, June 1998.
2. T. Frank, A. Klein, E. Costa, and E. Schulz. IFDMA - A Promising Multiple Access Scheme for Future Mobile Radio Systems. In *Proc. PIMRC*, Berlin, Germany, September 2005.

3. Tommy Svensson, Tobias Frank, David Falconer, Mikael Sternad, Elena Costa, and Anja Klein. B-IFDMA - A Power Efficient Mutliple Access Scheme for Non-frequency-adaptive Transmission. In *Proc. IST Mobile & Wireless Communications Summit*, Budapest, Hungary, July 2007.
4. A. Sohl, T. Frank, and A. Klein. Channel estimation for block-IFDMA. In *Proc. International Multi-Carrier Spread Spectrum Workshop*, Herrsching, Germany, May 2007.
5. T. Frank, A. Klein, E. Costa, and A. Kuehne. Low complexity and power efficient space-time-frequency coding for OFDMA. In *Proc. IST Mobile & Wireless Communications Summit*, Mykonos, Greece, June 2006.
6. P. Hoeher, S. Kaiser, and P. Robertson. Two-dimensional pilot-symbol aided channel estimation by wiener filtering. In *Proc. of ICASSP*, Munich, Germany, April 1997.
7. Jianjun Ran, Rainer Gruenheid, Hermann Rohling, Edgar Bolinth, and Ralf Kern. Decision-directed Channel Estimation Method for OFDM systems with high velocities. In *Proc. 57th IEEE VTC*, April 2003.
8. A. Sohl, T. Frank, and A. Klein. Channel estimation for DFT precoded OFDMA with block-wise and interleaved subcarrier allocation. In *Proc. International OFDM Workshop*, Hamburg, Germany, August 2006.

A Study on Channel Estimation Using EM Algorithm for Mobile WiMAX Systems

Masahiro Fujii, Yasufumi Kawahara, Yu Watanabe, and Makoto Itami

Abstract Mobile WiMAX system has recently been popularized as one of high speed wireless communication techniques. It is, however, necessary to accurately estimate the time-variant channel in order to equalize the received signal from the moving vehicular at high speed. In this paper, we propose a novel channel estimation method using the Expectation-Maximization (EM) algorithm for the uplink PUSC (Partial Usage of SubChannels) mode of the mobile WiMAX system in time-varying multi-path channels. The proposed algorithm estimates the channel frequency responses among the data signal subcarriers based on the estimates of the channel frequency responses on the pilot signal subcarriers. We assume that the estimate of the channel response at the data signal subcarrier is given by a linear combination, whose combination coefficients are determined using the EM algorithm, of the estimates of the channel responses among the pilot signal subcarriers. By computer simulations, we evaluate performances of the proposed algorithm in comparison with the conventional linear interpolation algorithm.

1 Introduction

Mobile WiMAX system is just beginning to be put to practical use for high speed broadband wireless access systems, and the service in Japan will start in 2009. In mobile WiMAX standard as amended in IEEE 802.16e, Orthogonal Frequency Division Multiple Access (OFDMA) is adopted for the physical layer [1, 2]. In mobile WiMAX OFDMA PHY, subcarriers of an OFDM signal are divided into subsets

---

M. Fujii (✉), Y. Kawahara, and Y. Watanabe
Department of Information Systems Science, Graduate School of Engineering,
Utsunomiya University, 7-1-2 Yoto, Utsunomiya, Tochigi, 321-8585, Japan
e-mail: {fujii,kawahara,yu}@is.utsunomiya-u.ac.jp

M. Itami
Department of Applied Electronics, Tokyo University of Science, 2641 Yamazaki, Noda, Chiba, 278-8510, Japan
e-mail: itami@te.noda.tus.ac.jp

of subcarriers. The subset called subchannel is minimum transmission unit that can be allocated to a user. The subchannelization scheme is helpful for scalability of the system and multiple access by OFDM method. Moreover, there are two types of subcarrier permutations for mobile WiMAX profile. One of them is a distributed subcarrier permutation such as Fully Usage SubChannels (FUSC) and Partial Usage SubChannels (PUSC) in order to obtain frequency diversity effect and average inter-cell interference. Another one is an adjacent subcarrier permutation such as Adaptive Modulation and Coding (AMC). Using these subcarrier permutations for the uplink and the downlink, we can improve the system throughput and support the system scalability. Mandatory subcarrier permutations of the mobile WiMAX are PUSC, FUSC and AMC for the down link and PUSC and AMC for the uplink.

It is necessary to estimate channel state information at the receiver in order to apply the receiver with coherent detection in fading channel. Especially, we need to estimate it with high accuracy when the mobile terminal moves rapidly for the uplink PUSC mode of the mobile WiMAX systems. In this paper, we proposed a new channel estimation method using Expectation-Maximization (EM) algorithm [3] for the support of mobility in the fast-fading channel environment.

2 Signal Model

In this paper, we focus on a channel estimation algorithm for the uplink PUSC permutation mode of the mobile WiMAX system. In the uplink PUSC, a subchannel is made up of six groups of four adjacent subcarriers and these groups of four subcarriers are modulated with a mix of data and pilot subcarriers over three OFDM symbols. A tile structure spans over three OFDM symbols (in time) of four subcarriers (in frequency) illustrated in Fig. 1. $M_p = 4$ pilot symbols and $M_d = 8$ data symbols are, respectively, assigned on the subcarriers of four corners and on the remaining subcarriers.

The received signal vector, $\underline{y}_p = [y_{p;0}, \cdots, y_{p;M_p-1}]^T$, on the pilot subcarriers is given by

$$\underline{y}_p = x_p \underline{h}_p + \underline{\eta}_p \quad (1)$$

where $x_p = \text{diag}[x_{p;0}, \cdots, x_{p;M_p-1}]$, $\underline{h}_p = [h_{p;0}, \cdots, h_{p;M_p-1}]^T$ and $\underline{\eta}_p = [\eta_{p;0}, \cdots, \eta_{p;M_p-1}]^T$, respectively, denote the pilot symbol diagonal matrix which is known at the receiver, the channel frequency response vector and the additive white Gaussian noise vector. When we cannot use a priori information on the channel fading such as a statistic characteristic among time-frequency domain, the optimum estimate vector of the channel on the pilot subcarriers is given by

$$\hat{\underline{h}}_p = x_p^{-1} \underline{y}_p. \quad (2)$$

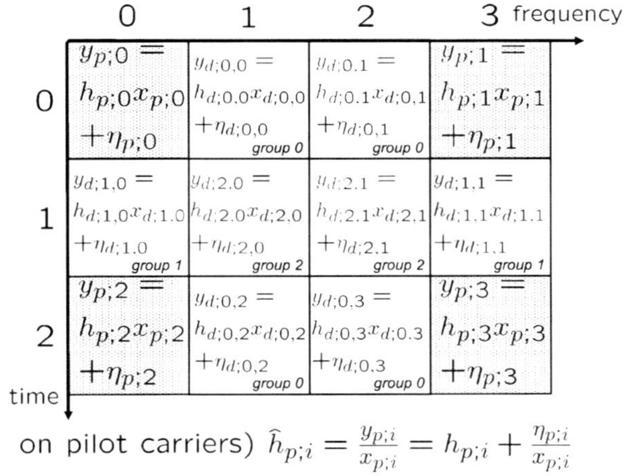

	0	1	2	3 frequency
0	$y_{p;0} =$ $h_{p;0}x_{p;0}$ $+\eta_{p;0}$	$y_{d;0,0} =$ $h_{d;0,0}x_{d;0,0}$ $+\eta_{d;0,0}$ group 0	$y_{d;0,1} =$ $h_{d;0,1}x_{d;0,1}$ $+\eta_{d;0,1}$ group 0	$y_{p;1} =$ $h_{p;1}x_{p;1}$ $+\eta_{p;1}$
1	$y_{d;1,0} =$ $h_{d;1,0}x_{d;1,0}$ $+\eta_{d;1,0}$ group 1	$y_{d;2,0} =$ $h_{d;2,0}x_{d;2,0}$ $+\eta_{d;2,0}$ group 2	$y_{d;2,1} =$ $h_{d;2,1}x_{d;2,1}$ $+\eta_{d;2,1}$ group 2	$y_{d;1,1} =$ $h_{d;1,1}x_{d;1,1}$ $+\eta_{d;1,1}$ group 1
2	$y_{p;2} =$ $h_{p;2}x_{p;2}$ $+\eta_{p;2}$	$y_{d;0,2} =$ $h_{d;0,2}x_{d;0,2}$ $+\eta_{d;0,2}$ group 0	$y_{d;0,3} =$ $h_{d;0,3}x_{d;0,3}$ $+\eta_{d;0,3}$ group 0	$y_{p;3} =$ $h_{p;3}x_{p;3}$ $+\eta_{p;3}$

time

on pilot carriers) $\hat{h}_{p;i} = \frac{y_{p;i}}{x_{p;i}} = h_{p;i} + \frac{\eta_{p;i}}{x_{p;i}}$

Fig. 1 Uplink PUSC tile

Let us suppose that M_d data subcarriers are divided into M_g groups and $M_d = \sum_{m=0}^{M_g-1} M_{d;m}$ holds where $M_{d;m}$ denotes the number of elements for the m-th data symbol group. In this paper, we assume that $M_g = 3$, $M_{d;0} = 4$, $M_{d;1} = 2$ and $M_{d;2} = 2$ illustrated in Figure 1. Then, the m-th received signal vector on the data subcarriers $\underline{y}_{d;m} = [y_{d;m,0}, \cdots, y_{d;m,M_{d;m}-1}]^T$ is expressed as

$$\underline{y}_{d;m} = \boldsymbol{x}_{d;m}\underline{h}_{d;m} + \underline{\eta}_{d;m} \qquad (3)$$

where $\boldsymbol{x}_{d;m}$, $\underline{h}_{d;m}$ and $\underline{\eta}_{m,d}$ are, respectively, the data symbol diagonal matrix, the channel frequency response vector and the additive white Gaussian noise vector. Using the known matrix $u_m \in \mathbb{C}^{M_{w;m} \times M_{d;m}}$ and the unknown vector $\underline{w}_m = [w_{m,0}, \cdots, w_{m,M_{w;m}-1}]^T$ at the estimator, an estimate vector of the channel frequency response vector $\underline{\hat{h}}_{d;m}$ for the m-th data subset is given by

$$\underline{\hat{h}}_{d;m} = \boldsymbol{u}_m^T \underline{w}_m \qquad (4)$$

where $u_m = [\underline{u}_{m,0}, \cdots, \underline{u}_{m,M_{d;m}-1}]$ and $\underline{u}_{m,n} = [u_{m,n,0}, \cdots, u_{m,n,M_{w;m}-1}]^T$. The known vector $\underline{u}_{m,n}$ is given by

$$\underline{u}_{m,n} = \boldsymbol{v}_m \underline{\hat{h}}_p \qquad (5)$$

using the prearranged interleaving matrix $\boldsymbol{v}_m$ whose elements are equal to 0 or 1 and the estimate vector $\underline{\hat{h}}_p$ in (2). A relationship between $\underline{\hat{h}}_{d;m}$, $\underline{w}_m$ and $\underline{\hat{h}}_p$ is shown in Fig. 2 corresponding to Fig. 1. Then, we can rewrite the received signal vector $\underline{y}_{d;m}$ by

$$\underline{y}_{d;m} = \boldsymbol{x}_{d;m}\underline{\hat{h}}_{d;m} + \underline{\zeta}_{d;m} \qquad (6)$$

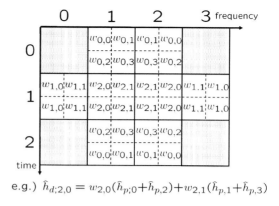

Fig. 2 The weighting coefficient assignment for the proposed algorithm

e.g.) $\hat{h}_{d;2,0} = w_{2,0}(\hat{h}_{p;0}+\hat{h}_{p,2})+w_{2,1}(\hat{h}_{p,1}+\hat{h}_{p,3})$

where $\underline{\zeta}_{d;m} = \underline{\eta}_{d;m} + \mathbf{x}_{d;m}(\underline{h}_{d;m} - \hat{\underline{h}}_{d;m})$. In this paper, we assume that $\underline{\zeta}_{d;m}$ is a zero-mean complex-valued Gaussian random variable vector with covariance matrix $E\{\underline{\zeta}_{d;m_1}\underline{\zeta}_{d;m_2}\} = \sigma^2_{\zeta_{d;m_1}}\delta_{m_1,m_2}\mathbf{I}$. Then, the conditional pdf of $\underline{y}_{d;m}$ given $\mathbf{x}_{d;m}$, $\underline{w}_m$ and $\sigma^2_{\zeta_{d;m}}$ is

$$p\left(\underline{y}_{d;m}|\mathbf{x}_{d;m}, \underline{w}_m, \sigma^2_{\zeta_{d;m}}\right) = \prod_{n=0}^{M_{d;m}-1} p\left(y_{d;m,n}|x_{d;m,n}, \underline{w}_m, \sigma^2_{\zeta_{d;m}}\right) \quad (7)$$

where

$$p\left(y_{d;m,n}|x_{d;m,n}, \underline{w}_m, \sigma^2_{\zeta_{d;m}}\right) = \frac{1}{\pi\sigma^2_{\zeta_{d;m}}} \exp\left(-\frac{|y_{d;m,n} - x_{d;m,n}\underline{u}^T_{m,n}\underline{w}_m|^2}{\sigma^2_{\zeta_{d;m}}}\right). \quad (8)$$

3 Channel Estimation Algorithm

Maximizing the conditional pdf $p(\mathbf{x}_{d;m}, \underline{y}_{d;m}|\theta_m = \{\underline{w}_m, \sigma^2_{\zeta_{d;m}}\})$ of the parameter θ_m in order to obtain estimate vector of $\underline{h}_{d;m}$ using (4), we can obtain the optimum estimate of the channel frequency response vector $\underline{h}_{d;m}$. It is, however, difficult to find the closed form solution for the estimation because we need to estimate θ_m and $\mathbf{x}_{d;m}$ jointly. Our purpose of the study is to maximize the objective function $f(\theta_m)$ in the presence of hidden parameter $\mathbf{x}_{d;m}$ given as follows.

$$\hat{\theta}_m = \arg\max_{\theta_m \in \Theta_m} f(\theta_m) \quad (9)$$

where

$$f(\theta_m) = \ln \sum_{\boldsymbol{x}_{d;m} \in \mathbb{X}^{M_{d;m}}} p(\boldsymbol{x}_{d;m}, \underline{y}_{d;m} | \theta_m)$$

$$= \ln \sum_{n=0}^{M_{d;m,n}-1} \sum_{x_{d;m,n} \in \mathbb{X}} p(x_{d;m,n}, y_{d;m,n} | \theta_m), \quad (10)$$

Θ_m is the set of possible θ_m and $\mathbb{X}$ denotes the set of possible data symbol. However, it is hard directly to obtain a closed form solution for the ML estimator using (10) because it involves the logarithm of a summation. Let us introduce a function $B(\theta_m; \theta_m^{(i)})$ of θ_m given $\theta_m^{(i)}$ that is an estimate of θ_m is given by

$$B(\theta_m; \theta_m^{(i)}) = \sum_{n=0}^{M_{d;m}-1} \sum_{x_{d;m,n} \in \mathbb{X}} p(x_{d;m,n} | y_{d;m,n}, \theta_m^{(i)}) \ln \frac{p(x_{d;m,n}, y_{d;m,n} | \theta_m)}{p(x_{d;m,n} | y_{d;m,n}, \theta_m^{(i)})}. \quad (11)$$

Since Jensen's inequality gives $B(\theta_m; \theta_m^{(i)}) \leq f(\theta_m)$ with equality if $\theta_m = \theta_m^{(i)}$, $B(\theta_m; \theta_m^{(i)})$ is the lower bound of $f(\theta_m)$. We can obtain a new estimate $\theta_m^{(i+1)}$ of θ_m in the form

$$\theta_m^{(i+1)} = \arg \max_{\theta_m \in \Theta_m} B(\theta_m; \theta_m^{(i)}) \quad (12)$$

because $f(\theta_m^{(i)}) = B(\theta_m^{(i)}; \theta_m^{(i)}) \leq B(\theta_m^{(i+1)}; \theta_m^{(i)}) \leq f(\theta_m^{(i+1)})$ holds. The inequality is evidence that $\theta_m^{(i+1)}$ is a better estimate than $\theta_m^{(i)}$ under the condition that $f(\theta_m)$ is a monotonic function of θ_m. We can find the asymptotic ML estimate by updating the estimate of θ_m from $\theta_m^{(i)}$ to $\theta_m^{(i+1)}$ by iteration. As larger iterated i be, $\theta_m^{(i)}$ will converge on the value of θ_m that achieves at least the local maximum of the objective function $f(\theta_m)$. Equation (11) and (12) are, respectively, called the Expectation step and Maximization step and the iterative procedure is called the EM algorithm. Supposing that $p(x_{d;m,n} | \theta_m) = p(x_{d;m,n})$, the lower-bound function can be rewritten as

$$B(\theta_m; \theta_m^{(i)}) = -M_{d;m} \ln \sigma_{\zeta_{d;m}}^2 - \frac{1}{\sigma_{\zeta_{d;m}}^2} \sum_{n=0}^{M_{d;m}-1} \sum_{x_{d;m,n} \in \mathbb{X}} \left| y_{d;m,n} - x_{d;m,n} \underline{u}_{m,n}^T \underline{w}_m \right|^2$$

$$\times p(x_{d;m,n} | y_{d;m,n}, \theta_m^{(i)}) + \text{const.} \quad (13)$$

using (8). In order to find the value of θ_m that maximizes $B(\theta_m; \theta_m^{(i)})$, it is necessary to satisfy $\partial B / \partial \underline{w}_m = \underline{0}$ and $\partial B / \partial \sigma_{\zeta_{d;m}}^2 = 0$. We can obtain new estimate $\theta_m^{(i+1)} = \{\underline{w}_m^{(i+1)}, (\sigma_{\zeta_{d;m}}^2)^{(i+1)}\}$ of θ_m by solving simultaneous equations. Consequently, $\underline{w}_m^{(i+1)}$

and $(\sigma_{\zeta_{d;m}}^2)^{(i+1)}$ that are elements of $\theta_m^{(i+1)}$ can exactly be expressed as

$$\begin{cases} \underline{w}_m^{(i+1)} \\ = \left(\sum_{n=0}^{M_{d;m}-1} \underline{u}_{m,n}^* \underline{u}_{m,n}^T \sum_{x_{d;m,n} \in \mathbb{X}} |x_{d;m,n}|^2 p(x_{d;m,n}|y_{d;m,n}, \theta_m^{(i)}) \right)^{-1} \\ \times \left(\sum_{n=0}^{M_{d;m}-1} y_{d;m,n} \underline{u}_{m,n}^* \sum_{x_{d;m,n} \in \mathbb{X}} x_{d;m,n}^* p(x_{d;m,n}|y_{d;m,n}, \theta_m^{(i)}) \right) \\ (\sigma_{\zeta_{d;m}}^2)^{(i+1)} \\ = \frac{1}{M_{d;m}} \sum_{n=0}^{M_{d;m}-1} \sum_{x_{d;m,n} \in \mathbb{X}} |y_{d;m,n} - x_{d;m,n} \underline{u}_{m,n}^T \underline{w}_m^{(i+1)}|^2 p(x_{d;m,n}|y_{d;m,n}, \theta_m^{(i)}) \end{cases} \quad (14)$$

Using (8), the conditional pdf of $x_{d;m,n}$ given $y_{d;m,n}$ and $\theta_m^{(i)}$ in (14) can be represented as

$$p(x_{d;m,n}|y_{d;m,n}, \theta_m^{(i)}) = C \exp\left(-\frac{\left| x_{d;m,n} - \frac{y_{d;m,n}}{\underline{u}_{m,n}^T \underline{w}_m^{(i)}} \right|^2}{\frac{(\sigma_{\zeta_{d;m}}^2)^{(i)}}{|\underline{u}_{m,n}^T \underline{w}_m^{(i)}|^2}} \right) \quad (15)$$

where C is a normalized factor to satisfy $\sum_{x_{d;m,n} \in \mathbb{X}} p(x_{d;m,n}|y_{d;m,n}, \theta_m^{(i)}) = 1$ and we note that $p(x_{d;m,n}|\theta_m^{(i)}) = p(x_{d;m,n})$. The conditional pdf is also updated by updating the estimate of θ_m from $\theta_m^{(i)}$ to $\theta_m^{(i+1)}$ because it depends on the estimate of $\theta_m^{(i)}$. The EM algorithm increases the objective function $f(\theta_m)$ by repeating the updating procedures as necessary and guarantees to converge to the local maximum of the function. We can obtain the estimate $\hat{\theta}_m$ when the algorithm converges to the local maximum of $f(\theta_m)$. We stop the EM algorithm after I iterations and obtain $\hat{\theta}_m = \theta_m^{(I)}$.

4 Numerical Examples & Conclusions

In this section, by computer simulations, we evaluate channel estimation error of the proposed estimator in comparison with the conventional one. In contrast with our proposed algorithm with dynamically-reconfigurable weighting coefficient vector $\underline{w}_m$, a conventional algorithm makes use of a prearranged $\underline{w}_m$ illustrated in Fig. 3 [4] corresponding to Fig. 2. In order to evaluate the estimation error performance, we introduce the Mean Squared Estimation Error (MSEE) defined as

$$\text{MSEE} = \frac{\text{E}\{\|\underline{h}_{d;m} - \hat{\underline{h}}_{d;m}\|^2\}}{\text{E}\{\|\underline{h}_{d;m}\|^2\}}. \quad (16)$$

In our simulations, we use the system parameters based on uplink PUSC mode of the mobile WiMAX system [1]. We assume the channel impulse model based on the vehicular test environment channel B of ITU-R M. 1225 [5]. We set the staring value

$\underline{w}_m^{(0)}$ of the EM algorithm to the prearranged coefficient vector of the conventional algorithm illustrated in Fig. 3.

Figure 4 shows MSEE curves as functions of E_b/N_0 in case that $f_d T_s = 0.02$ where f_d and T_s, respectively, denote maximum Doppler frequency and one OFDMA symbol period. We assume that the channel is deep frequency selective and fast fading in ITU-R vehicular channel B and the maximum number of EM iteration is equal to 20. Such deep fading channel results in a saturation of the MSEE curve using the conventional estimator as functions of E_b/N_0 because the simple linear interpolation do not work well. On the other hand, the proposed algorithm can decrease the estimation error although the computational complexity increase by the iterative process of the EM algorithm.

Next, Fig. 5 shows MSEE performances as function of the number of iterations of EM algorithm under the same condition in Fig. 4. These curves are MSEE performances of the proposed algorithm in case that E_b/N_0s are 10, 20 and 30[dB].

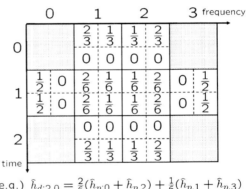

Fig. 3 The weighting coefficient assignment for the linear interpolation

e.g.) $\hat{h}_{d;2,0} = \frac{2}{6}(\hat{h}_{p;0} + \hat{h}_{p,2}) + \frac{1}{6}(\hat{h}_{p,1} + \hat{h}_{p,3})$

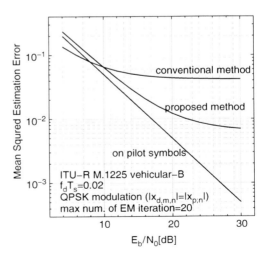

Fig. 4 E_b/N_0 v.s. Mean squared estimation error

Fig. 5 The number of EM iterations v.s. Mean squared estimation error

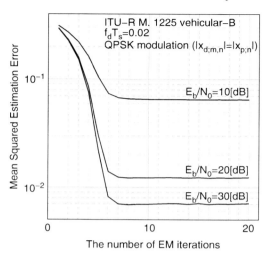

When the number of EM iterations is larger than 7, the MSEEs converge the constant value and the proposed estimator achieves the local maximum of the objective function $f(\theta_m)$. The convergence value depends on the initial value of θ_m. We can improve the MSEE performance by using diverse initial value although the procedure causes high computation complexity.

5 Conclusions

In this paper, we proposed a novel channel estimation method using the Expectation-Maximization (EM) algorithm for the uplink PUSC (Partial Usage of SubChannels) mode of the mobile WiMAX system in time-varying multi-path channels. The proposed algorithm estimated the channel frequency responses among the data signal subcarriers based on the estimates of the channel frequency responses on the pilot signal subcarriers. The estimate of the channel response at the data signal subcarrier was given by a linear combination, whose combination coefficients were determined using the EM algorithm, of the estimates of the channel responses among the pilot signal subcarriers. By computer simulations, we evaluated performances of the proposed algorithm in comparison with the conventional simple linear interpolation algorithm. The proposed estimator using the dynamical weighting coefficients outperformed the conventional estimator using the prearranged coefficients. We are studying how to reduce the computational complexity of the proposed estimator with a slight performance degradation.

References

1. WiMAX Forum, "Mobile WiMAX-Part 1: A Technical Overview and Performance Evaluation," June, 2006.
2. L. Nuaymi, *WiMAX Technology for Broadband Wireless Access*, John Wiley & Sons Ltd, The Atrium, Southern Gate, Chichester, West Sussex, PO19 8SQ, England, 2007.
3. T. M. Moon, "The Expectation-Maximization Algorithm," IEEE Signal Processing Mag., vol. 13, no. 6, pp. 47–60, Nov., 1996.
4. Altera Corporation, "Channel Estimation & Equalization for WiMAX," Application Note 434, May, 2007.
5. Rec. ITU-R M. 1225, "Guidelines for evaluation of radio transmission technologies for IMT-2000," 1997.

A Technique for Correcting Residual Frequency Offset in OFDM Systems

Young Mo Chung

Abstract Orthogonal frequency division multiplexing (OFDM) systems are very sensitive to carrier frequency offset (CFO) and Schmidl and Cox Algorithm (SCA) was proposed for the carrier synchronization. However, there remains an estimation error with the SCA, resulting in a residual frequency offset (RFO). In this paper, an attempt to reduce a CFO estimation error by exploiting the samples in the cyclic prefix (CP) has been made and an RFO estimation with reduced complexity is proposed. An analysis for the error variances of the proposed CFO estimations is given and the performance of the RFO estimation is investigated by computer simulations. The results show that the proposed CFO estimators provide better performance than the SCA in terms of the error variance and the RFO estimation obtained with 250 subcarrier symbols provides almost equivalent BER performance to the perfect synchronization.

1 Introduction

Orthogonal frequency division multiplexing (OFDM) system has received a great attention for its high spectrum efficiency and robustness to the multipath fading. However, OFDM system requires a very accurate frequency synchronization since OFDM system is very sensitive to carrier frequency offset (CFO). The CFO comes from the mismatch between the transmitter and receiver oscillator. The CFO generates an interchannel interference (ICI), which degrades bit error rate (BER) performance.

So far, several methods on the synchronization for OFDM have been proposed [1–7]. In [1], Schmidl and Cox algorithm (SCA) was proposed for carrier synchronization method employing a data aided scheme. The SCA requires low overhead bits for synchronization and possesses a low complexity. However, it is known

---

Y.M. Chung
Department of Information and Communications Engineering, Hansung University,
389 Samsun-Dong 2Ga, Sungbook-Gu, Seoul 136-792, Korea
e-mail: ymchung@hansung.ac.kr

that the SCA does not provide sufficiently accurate estimate for a frequency offset. The estimation error results in a residual frequency offset (RFO) which rotates the received constellation and causes bit errors in coherently demodulated OFDM systems.

Attempts have been made to correct the RFO. In [2], an RFO is estimated by observing a block of OFDM symbols. The estimated residual correction factor is updated on block basis, resulting in a delay in demodulated symbols. On the other hand, in [3], an RFO is estimated by observing the phase difference between two consecutive OFDM symbols at the output of the fast Fourier transform (FFT) block while the RFO is compensated at the input of the FFT block. By doing the estimation and compensation in different domains, there exists a mismatch in time between the measurement and correction, which causes correction errors. Recently, in [4], an iterative scheme using maximum likelihood (ML) and expectation-maximization (EM) is proposed for an RFO estimation. This method provides a good BER performance. However, EM algorithm is known to require lots of computations.

In this paper, first, attempts to reduce a CFO estimation error have been made. Two CFO estimations exploiting the samples in the cyclic prefix (CP) are proposed by assuming that the channel delay spread is shorter than the guard interval. The performances of the proposed estimators are investigated analytically and compared with that of the SCA. Next, a simple and practical method for correcting an RFO is proposed. The angle of the rotation at each subchannel is measured by comparing the symbol with its demodulated value. By taking an average of the angles, the angle and RFO estimates are obtained. In OFDM systems with a large number of subcarriers, the number of computations required for comparing and averaging the entire subchannel symbols could be large. In our approach, we use fewer subchannel symbols than the entire subchannel symbols for estimating the RFO. The performance of the proposed RFO estimator is investigated through computer simulations.

2 Signals and Offset Correction Method

2.1 OFDM Signals

A sampled complex baseband OFDM symbol can be expressed as

$$x_{l,n} = A \sum_{k=0}^{K-1} S_{l,k} e^{j2\pi kn/N}, \quad 0 \leq n \leq N-1, \tag{1}$$

where A is an amplitude, $S_{l,k}$ is the complex data symbol at the kth subcarrier of the lth OFDM symbol, K is the number of subcarriers, and N is the number of samples in a useful OFDM symbol duration T. The CP of length N_g is added to form an OFDM symbol sequence as $\{x_{l,N-N_g}, x_{l,N-N_g+1}, \cdots, x_{l,N-1}, x_{l,0}, x_{l,1}, \cdots, x_{l,N-1}\}$. The channel is assumed to be slow fading and have a finite length impulse response with N_c samples. We let $N_m = N_g - N_c$ and assume $N_m > 0$. In addition, a correct frame and timing synchronization is assumed. Then, the samples of the received signal are expressed as

$$y_{l,n} = A \sum_{k=0}^{K-1} H_{l,k} S_{l,k} e^{j2\pi(k+\delta)n/N} + w_{l,n}, \quad -N_m \leq n \leq N-1, \quad (2)$$

where $H_{l,k}$ is the transfer function at the kth subchannel, $w_{l,n}$ is the additive white Gaussian noise (AWGN), and δ is the CFO normalized to the subchannel spacing. Note that (2) is valid for $(N + N_m)$ samples since we assumed that $N_g > N_c$.

2.2 Schmidl and Cox Algorithm (SCA)

In [1], at the beginning of a synchronization frame, two OFDM symbols are assigned for synchronization. The first OFDM symbol is used for both the symbol timing and CFO estimation. The second one is used only for the CFO, especially for the estimation of frequency offset larger than the subchannel spacing. The fractional part of the normalized frequency offset, which is an offset smaller than the subchannel spacing, is estimated with the first training symbol. The integer part of the normalized frequency offset is estimated with both the training symbols. A CFO estimate is obtained by adding these two parts of the offset.

The first training symbol is composed of two identical half symbols, which is made by giving a pseudo noise (PN) symbol sequence at even frequencies and zeros at odd frequencies. It is noted that the two training symbols have the CP with N_g samples. The length of the half symbol is denoted as L and thus $L = N/2$. When the channel is slow fading, in the first training symbol, sample pairs with L samples apart have phase difference of $\phi = \pi\delta$ due to the CFO. Thus, the rotated angle due to the offset can be estimated by

$$\hat{\phi}_i = \tan^{-1}\left(\sum_{m=0}^{L-1} y_{1,i+m}^* y_{1,i+m+L}\right), \quad (3)$$

where $y_{1,n}$ stands for the first training OFDM symbol in a synchronization frame. Note that i indicates the starting sample index of the L pairs and is given by $-N_m, -N_m + 1, \cdots, 0$.

The normalized frequency offset estimate is given by

$$\hat{\delta}_i = \frac{\hat{\phi}_i}{\pi}. \quad (4)$$

For $-\pi < \hat{\phi}_i \leq \pi$, it is seen that $-1 < \hat{\delta}_i \leq 1$. This shows that only a CFO smaller than the subchannel spacing can be estimated with (3). A CFO larger than the subchannel spacing is estimated with both the first and second training symbols [1].

The offset estimation error is given by $\epsilon_i = \hat{\delta}_i - \delta$. In [1, 5], it is shown that $E[\epsilon_i] = 0$ and

$$E[\epsilon_i^2] = \frac{\sigma^2}{\pi^2 E[\sum_{m=0}^{L-1} |r_{i+m}|^2]}, \qquad (5)$$

where $\sigma^2 = E[|w_{l,n}|^2]$ and $r_{i+m} = A \sum_{k=0}^{K-1} H_{1,k} S_{1,k} e^{j2\pi(k+\delta)(i+m)/N}$. By assuming $H_{1,k}$ and $S_{1,k}$ are independent and average channel gain $E[|H_{1,k}|^2] = H^2$, we obtain

$$E\left[\sum_{m=0}^{L-1} |r_{i+m}|^2\right] = LKA^2H^2S^2, \qquad (6)$$

where $S^2 = E[|S_{1,k}|^2]$. Then, the variance of the offset estimation error is expressed as $E[\epsilon_i^2] = \frac{1}{\pi^2 L \cdot SNR}$ as in [1], where $SNR = \frac{KA^2H^2S^2}{\sigma^2}$.

The integer part of the normalized CFO is estimated by using the ratio of the first and second training symbol and the performance of this estimator is analysed in [1]. The integer part estimator of the frequency offset is known to provide a good estimate. Thus, the integer part estimation of the normalized CFO is not discussed in this paper.

3 Proposed Methods

3.1 CFO Estimation

When the length of the channel impulse response N_c is shorter than N_g, it is noted that more than one offset estimates can be obtained by exploiting the samples in the CP. Assume that the OFDM symbol synchronization is achieved at i. Then, these offset estimates can be denoted as $\hat{\delta}_j$, $j = -N_m, -N_m + 1, \cdots, -1 + i$ and $i = -N_m + 1, -N_m + 2, \cdots, 0$. Then, a new offset estimate $\delta_{i,j}$ is given by

$$\delta_{i,j} = \frac{1}{2}(\hat{\delta}_i + \hat{\delta}_j). \qquad (7)$$

Let the estimation error $\epsilon_{i,j} = \delta_{i,j} - \delta$. It can be easily shown that the estimation error has zero mean. Since $E[\epsilon_i^2] = E[\epsilon_j^2]$, the variance of the estimation error is given by

$$E[\epsilon_{i,j}^2] = \frac{1}{2}E[\epsilon_i^2] + \frac{1}{2}E[\epsilon_i\epsilon_j]. \qquad (8)$$

If ϵ_i and ϵ_j are uncorrelated, the error variance decreases to 1/2 of $E[\epsilon_i^2]$. However, ϵ_i and ϵ_j are not uncorrelated and have a positive correlation. We let $p = i - j$. After mathematical manipulations, we obtain

$$E[\epsilon_i \epsilon_j] = E[\epsilon_i^2]\left(1 - \frac{p}{L}\right). \tag{9}$$

Thus, the error variance is expressed as

$$E[\epsilon_{i,j}^2] = E[\epsilon_i^2]\left(1 - \frac{p}{2L}\right), \tag{10}$$

for $-N_m \leq i \leq 0$ and $-N_m \leq j \leq i$. It is noted that error variance $E[\epsilon_{i,j}^2]$ decreases as the distance between i and j increases.

Now, we use more than two estimates to form a new offset estimation. Namely, P estimates are used and an average is taken to get a new offset estimate Δ_{iP}.

$$\Delta_{iP} = \frac{1}{P} \sum_{p=0}^{P-1} \delta_{i,i-p}. \tag{11}$$

The estimation error $\epsilon_{iP} = \Delta_{iP} - \delta$ and it can be easily shown that $E[\epsilon_{iP}] = 0$. The variance of the estimation error is expressed as

$$E[\epsilon_{iP}^2] = \frac{1}{P} E[\epsilon_i^2] + \frac{2}{P^2} \sum_{k=0}^{P-2} \sum_{j=k+1}^{P-1} E[\epsilon_{i-k}\epsilon_{i-j}]. \tag{12}$$

Using (9) and after mathematical manipulations, (12) is simplified as

$$E[\epsilon_{iP}^2] = E[\epsilon_i^2]\left(1 - \frac{P^2 - 1}{3PL}\right), \tag{13}$$

for $-N_m \leq i \leq 0$ and $1 \leq P \leq N_m + i + 1$. The numerical values of the error variances are presented in Section 4.

3.2 RFO Estimation

Since the CFO estimation is not perfect, a frequency offset still remains in the demodulated samples, which is called an RFO. The RFO continuously rotates the signal constellation.

In a synchronization frame, the received OFDM symbols except the training symbols are demodulated with the FFT. Then, the CFO is corrected with the estimates $\delta_{i,j}$ or δ_{iP}. We denote the demodulated and corrected sequence as $\{Z_{l,k}\}$, where l and k represent the OFDM symbol and subcarrier index, respectively.

The samples $\{Z_{l,k}, k = 0, 1, \cdots, K-1\}$ are sent through a slicer to obtain $\{\overline{Z}_{l,k}, k = 0, 1, \cdots, K-1\}$. When QPSK data mapping is used, the sliced values have phases of $\{\pm\pi/4, \pm 3\pi/4\}$.

The RFO produces a small amount of rotation in the OFDM symbol next to the training symbol since the rotation angle is proportional to the time interval. Thus, in this OFDM symbol, the RFO hardly causes decision errors and there is a phase angle difference between $Z_{l,k}$ and $\overline{Z}_{l,k}$, which is proportional to the RFO. The phase angle is estimated as

$$\hat{\theta}_{l,k} = \angle \left(Z_{l,k} \overline{Z}_{l,k}^* \right), \qquad (14)$$

where $\angle(\cdot)$ stands for the phase angle of a complex number. However, due to the noise, there is a phase angle estimation error in (14). In order to reduce the estimation error, R estimates are computed and an average is taken to form an RFO estimate, where $1 \leq R \leq K$.

$$\hat{\theta}_l = \frac{1}{R} \sum_{k=0}^{R-1} \hat{\theta}_{l,k}. \qquad (15)$$

Then, the normalized RFO estimate is given by

$$\overline{\delta} = \frac{\hat{\theta}_l}{2\pi (N + N_g) T_s}, \qquad (16)$$

where T_s is the sampling period. Let us assume that the RFO is compensated by ψ_l at the lth OFDM symbol. Then, $(\psi_l - \psi_{l-1}) \cdot (2\pi (N + N_g) T_s)^{-1}$ should be equal to $\overline{\delta}$, making the following relation.

$$\psi_l = \hat{\theta}_l + \psi_{l-1}. \qquad (17)$$

At the beginning of a synchronization frame, the phase rotation due to an RFO is negligible. Thus, we let $\psi_0 = 0$.

At the $(l+1)$th OFDM symbol, the sequences $\{Z_{l+1,k}, k = 0, 1, \cdots, K-1\}$ are replaced by the phase corrected sequences, $\{Z_{l+1,k} \cdot e^{-j\psi_l}, k = 0, 1, \cdots, K-1\}$. And then, equations (14)–(17) are repeated, yielding a new estimate ψ_{l+1} of the phase angle arising from the RFO. These steps are iterated to the end of a synchronization frame.

4 Results and Discussions

When N_m samples in the CP are given, the error variances of δ_{i,N_m} and $\Delta_{i(N_m+1)}$ are computed by (10) and (13), respectively. The results are shown in Fig. 1. In this figure, it is assumed that the OFDM symbol synchronization is achieved at $i = 0$.

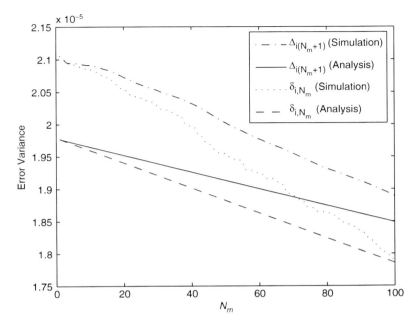

Fig. 1 Error variance comparison of carrier frequency offset estimations

From the results, it is found that δ_{i,N_m} show better performance than $\Delta_{i(N_m+1)}$. When $N_m = 100$, the variance of δ_{i,N_m} is 90% of that of $\hat{\delta}_0$ or $\hat{\delta}_{N_m}$ alone. It is also observed that the variance of $\Delta_{i(N_m+1)}$ is 93% of that of each $\hat{\delta}_j$.

In Figure 1, computer simulation results for the error variance are also given, where the number of subcarriers is 1,024 and the number of effective subchannels is 1,000. The length of the CP is set to 10% of the subcarriers, having one OFDM symbol composed of 1,126 samples. The period of the PN sequence for the first training symbol is 1,024 and QPSK is selected for a modulation scheme. The estimate error was observed over 50,000 training symbols. From Fig. 1, it is found that the difference between analysis and simulation results are within 7% of the analysis result, stating that the analysis is valid.

Next, the error variance of the RFO estimator is investigated by computer simulations. The SCA is used in an attempt to eliminate both the fractional and integer part of the normalized CFO. The length of a synchronization frame is 200 and 50 frames are used in the simulation. The results are shown in Fig. 2. It is observed that the error variance decreases significantly up to $R = 250$. However, beyond this point, reduction in the error variance is small. Namely, the variance at R=1000 is found to be only 1/4 of the error variance at $R = 250$ regardless of SNR. When $R = 500$, the error variance is only about 2.1 times larger than that at $R = 1000$. According to the equations (5) and (6), the error variance of the SCA estimator is 1.98×10^{-4} when $SNR = 0$dB. Thus, when $SNR = 0$dB, the performance

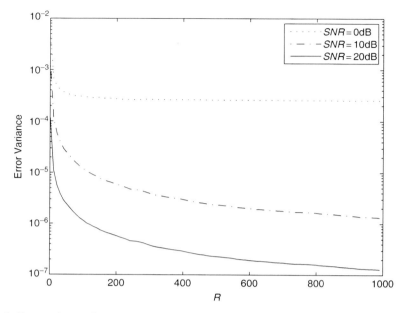

Fig. 2 Error variance of residual frequency offset estimations

improvement is not obtained with the RFO estimation since the error variance stays above 1.98×10^{-4}. However, when $SNR \geq 10$dB and $R \geq 100$, the performance improvement is prominent with the RFO estimation.

Now, the performances of the offset correction methods are investigated and compared through the bit error rate (BER). The length of the synchronization frame has been changed from 50 to 100 and the carrier frequency offset is set to 4.4 subcarrier spacings. Figure 3 shows the E_b/N_0 vs. BER result, where E_b/N_0 represents the received energy per bit to noise power spectral density ratio. When the frame length (FL) is 50, the SCA suffers a significant performance degradation due to the RFO. And the performance degradation is more prominent when the frame length is 100. The estimator $\delta_{i,j}$ gives better performance than the SCA. However, the estimator $\delta_{i,j}$ still produces an RFO, resulting in the BER performance degradation. When the frame length is 50, the estimator $\delta_{i,j}$ provides 3.6 dB gain over the SCA at the BER of 10^{-3}. However, when the frame length is 100, the estimator $\delta_{i,j}$ is observed to suffer a severe performance degradation due to the RFO and shows similar performance to the SCA.

When the RFO correction (RFOC) is applied, the BER performance improvement is significant. In Fig. 3, result for the perfect carrier synchronization is also provided for a comparison purpose. The OFDM receiver with RFO correction is found to give almost equivalent performance to the perfectly synchronized receiver. Especially, it is found that the RFO estimation with $R = 250$ possesses equivalent performance to that with $R = 1000$. Even when the frame length is increased to

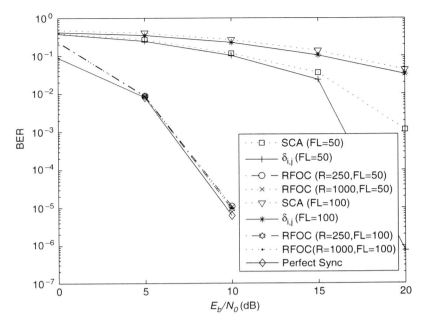

Fig. 3 BER performance comparison

100, it is also found that the performances of the RFO estimation with $R = 250$ and $R = 1000$ are equivalent and are slightly inferior to that of the perfect synchronization. Namely, the RFO estimator with $R = 250$ suffers only 0.32 dB loss compared with the perfect carrier synchronization when BER=10^{-5} and the frame length is 100. From the results, it is concluded that the RFO estimation with $R = 250$ provides almost equivalent performance to the perfect carrier synchronization when a constant carrier frequency offset is present.

5 Conclusions

In this paper, CFO estimations exploiting the samples in the CP are proposed and the performances of the proposed estimators are analyzed. And an RFO estimation with reduced complexity is proposed and the performance is investigated by computer simulations. The results show that the proposed CFO estimators provide better performance than the SCA in terms of the error variance. In addition, it is found that the RFO estimation obtained with 250 subcarrier symbols provides almost equivalent BER performance to the perfect carrier synchronization when a constant carrier frequency offset is present.

Acknowledgements This research was financially supported by Hansung University in the year of 2008.

References

1. T. M. Schmidl and D. C. Cox, "Robust frequency and timing synchronization for OFDM," *IEEE Transactions on Communications*, vol. 45, pp. 1613–1621, Dec. 1997.
2. V. S. Abhayawardhana and I. J. Wassell, "Residual frequency offset correction for coherently modulated OFDM systems in wireless communication," in *Proceedings of IEEE Vehicular Technology Conference*, Birmingham, USA, May 2002, pp. 777–781.
3. J. Jo, H.-W. Kim, and D.-S. Han, "Residual frequency offset compensation for IEEE 802.11a," in *Proceedings of IEEE Vehicular Technology Conference*, Los Angeles, USA, Sep. 2004, pp. 2201–2204.
4. M. Marey, M. Guenach, and H. Steendam, "Iterative residual frequency offset correction for OFDM system," in *Proceedings of IEEE Vehicular Technology Conference*, Singapore, May 2008, pp. 923–927.
5. P. H. Moose, "A technique for orthogonal frequency division multiplexing frequency offset correction," *IEEE Transactions on Communications*, vol. 42, pp. 2908–2914, Oct. 1994.
6. W. Kim and D. C. Cox, "Residual frequency offset and phase compensation for OFDM systems," in *Proceedings of IEEE Vehicular Technology Conference*, Baltimore, USA, Sep. 2007, pp. 2209–2213.
7. J.-H. Lee and S.-C. Kim, "Residual frequency offset compensation using the approximate SAGE algorithm for OFDM system," *IEEE Transactions on Communications*, vol. 54, pp. 765–769, May 2006.

A Joint Channel and Carrier Frequency Offset Estimation Based on Spread Pilot for Future Broadcasting Systems

Oudomsack Pierre Pasquero, Youssef Nasser, Matthieu Crussière, and Jean-François Hélard

Abstract In this paper, we propose a novel and simple joint channel frequency response (CFR) and carrier frequency offset (CFO) estimation based on 2D spread pilots for digital video broadcasting (DVB) systems. The advantage of the proposed algorithm compared to DVB-T/H standard, is to estimate the CFR and the CFO using the same pilot symbols. The performance evaluated over a realistic channel model, in a mobility scenario, in the presence of CFO, shows the efficiency of this technique which turns out to be a promising joint CFR and CFO estimation technique for the future DVB systems.

1 Introduction

Broadcasting digital TV is currently an area of intensive development and standardisation activities. Orthogonal frequency division multiplexing (OFDM) has been widely adopted in most of the digital video broadcasting standards for its potential for attaining high rate transmission over frequency selective channel. Nevertheless, the emergence of new consumer usages requires a better flow and quality transmission.

In order to increase the spectral efficiency of the system, we proposed in [1] to reduce the overhead part dedicated to CFR estimation by using a two-dimensional linear precoding (2D LP) function before OFDM modulation. The basic idea consists in dedicating one of the precoding sequences to transmit a so-called *spread pilot* information for CFR estimation. In this paper, we propose to use the same pilot symbols to jointly estimate the CFR and the CFO.

The article is organized as follows. In Section 2, we present the 2D LP OFDM transmitter structure. The CFR and CFO estimation based on spread pilots are

---

O.P. Pasquero (✉), Y. Nasser, M. Crussiere, and J.-F. Hélard
Institute of Electronics and Telecommunications of Rennes (IETR), UMR CNRS 6164,
INSA de Rennes, 20 Avenue de Buttes de Coësmes, 35043 Rennes, France
e-mail: {oudomsack.pasquero, youssef.nasser, matthieu.crussiere, jean-francois.helard}@insa-rennes.fr

described in Section 3. In Section 4, simulation results in terms of mean square error (MSE) and bit error rate (BER) are presented and compared with DVB-T standard. Finally, Section 5 concludes the paper.

2 System Model

Figure 1 exhibits the transmitter and receiver structures based on spread pilots. We consider an OFDM communication system using N subcarriers with a guard interval size of υ samples. The principle of 2D LP OFDM is to spread each complex symbol $x_{t,s}[i]$ over a subset of $L = L_t \times L_f = 2^n$ subcarriers, with L_t and L_f the time and frequency spreading factors respectively and $n \in \mathbb{N}$. To allow the receiver to estimate both the CFR and the CFO, we dedicate one of the L precoding sequences $\mathbf{c}_p$ to pilot information in each subset of subcarriers. Figure 2 illustrates the transmitted signal in three dimensions. In order to distinguish the different subsets of L subcarriers, we define t and s the indexes referring to the frame and the sub-band respectively, with $0 \leq s \leq S - 1$. The signal transmitted on a subset of subcarriers (t, s) writes:

$$\mathbf{y}_{t,s} = \mathbf{C} \mathbf{P} \mathbf{x}_{t,s} \qquad (1)$$

where $\mathbf{x}_{t,s} = [x_{t,s}[0] \ldots x_{t,s}[i] \ldots x_{t,s}[L-1]]^T$ is the complex symbol vector, $\mathbf{P} = diag\{\sqrt{P_0} \ldots \sqrt{P_i} \ldots \sqrt{P_{L-1}}\}$ is the diagonal matrix of the powers allocated to symbols, and $\mathbf{C} = [\mathbf{c}_0 \ldots \mathbf{c}_i \ldots \mathbf{c}_{L-1}]$ is the Walsh-Hadamard (WH) precoding matrix which ith column corresponds to ith precoding sequence $\mathbf{c}_i = [c_i[0] \ldots c_i[nL_t + q] \ldots c_i[L-1]]^T$. We assume normalized precoding sequences, i.e. $c_i[n, q] = \pm \frac{1}{\sqrt{L}}$. In each subset of subcarriers, each subcarrier is located by its

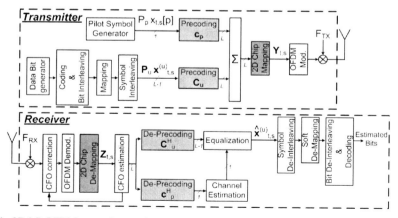

Fig. 1 2D LP OFDM transmitter and receiver based on spread pilot

A Joint Channel and Carrier Frequency Offset Estimation Based on Spread Pilot

Fig. 2 2D Spreading scheme

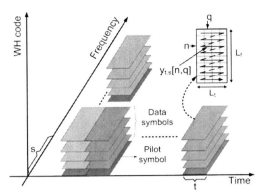

coordinates n and q corresponding to the subcarrier index and OFDM symbol index respectively. Hence, the symbol transmitted on the subcarrier (q, n) writes:

$$y_{t,s}[q,n] = \sum_{i=0}^{L-1} \sqrt{P_i} x_{t,s}[i] c_i [nL_t + q] \qquad (2)$$

2.1 Channel Estimation Principles

At the receiver side, if the transmitter and the receiver are perfectly synchronized, after OFDM demodulation, the signal received on a subset of subcarriers can be expressed as:

$$\mathbf{z}_{t,s} = \mathbf{H}_{t,s} \mathbf{y}_{t,s} + \mathbf{w}_{t,s} \qquad (3)$$

where $\mathbf{H}_{t,s}$ is the $[L \times L]$ diagonal matrix of the channel coefficients and $\mathbf{w}_{t,s}$ is the additive white Gaussian noise (AWGN) vector having zero mean and variance σ_w^2.

The basic idea of spread pilot CFR estimation technique is to estimate one average channel coefficient $\widehat{h}_{t,s}$ by subset of subcarriers. It is obtained by de-precoding the received signal $\mathbf{z}_{t,s}$ by the pilot sequence $\mathbf{c}_p^H$ and then dividing by the pilot symbol known by the receiver:

$$\widehat{h}_{t,s} = \frac{1}{\sqrt{P_p}\, x_{t,s}[p]} \mathbf{c}_p^H \mathbf{Z}_{t,s}$$

$$= \frac{1}{\sqrt{P_p}\, x_{t,s}[p]} \mathbf{c}_p^H [\mathbf{H}_{t,s} \mathbf{C P} \mathbf{x}_{t,s} + \mathbf{w}_{t,s}] \qquad (4)$$

Let us define $\mathbf{C}_u = [\mathbf{c}_0 \ldots \mathbf{c}_{i \neq p} \ldots \mathbf{c}_{L-1}]$ the $[L \times (L-1)]$ data precoding matrix, $\mathbf{P}_u = \mathrm{diag}\{\sqrt{P_0} \ldots \sqrt{P_{i \neq p}} \ldots \sqrt{P_{L-1}}\}$ the $[(L-1) \times (L-1)]$ diagonal matrix which entries are the powers assigned to the data symbols, and $\mathbf{x}_{t,s}^{(u)} = [x_{t,s}[0] \ldots x_{t,s}[i \neq p] \ldots x_{t,s}[L-1]]^T$ the $[(L-1) \times 1]$ data symbols vector. Given these notations, (4) can be rewritten as:

$$\begin{aligned}
\widehat{h}_{t,s} &= \frac{1}{\sqrt{P_p}\, x_{t,s}[p]} \left[\mathbf{c}_p^H \mathbf{H}_{t,s} \mathbf{c}_p \sqrt{P_p}\, x_{t,s}[p] + \mathbf{c}_p^H \mathbf{H}_{t,s} \mathbf{C}_u \mathbf{P}_u \mathbf{x}_{t,s}^{(u)} + \mathbf{c}_p^H \mathbf{w}_{t,s} \right] \\
&= \frac{1}{L} tr\{\mathbf{H}_{t,s}\} + \frac{1}{\sqrt{P_p}\, x_{t,s}[p]} \left[\mathbf{c}_p^H \mathbf{H}_{t,s} \mathbf{C}_u \mathbf{P}_u \mathbf{x}_{t,s}^{(u)} + \mathbf{c}_p^H \mathbf{w}_{t,s} \right] \\
&= \overline{h}_{t,s} + \mathrm{MCI}_{t,s} + \mathrm{WGN}_{t,s}
\end{aligned} \quad (5)$$

The first term $\overline{h}_{t,s}$ is the average channel response globally experienced by the subset of subcarriers (t,s). The second term represents the multiple code interference (MCI). It has been demonstrated in [1] that it results from the loss of orthogonality between the precoding sequences caused by the variance of the channel coefficients over the subset of subcarriers. Its variance is function of the autocorrelation of the channel $R_{HH}(n'-n, q'-q)$. It writes:

$$E\left\{ |\mathrm{MCI}|^2 \right\} = \frac{1}{P_p} \left(1 - \frac{1}{L^2} \sum_{n=0}^{L_f-1} \sum_{q=0}^{L_t-1} \sum_{n'=0}^{L_f-1} \sum_{q'=0}^{L_t-1} R_{HH}(n'-n, q'-q) \right) \quad (6)$$

One can actually check that if the channel is flat over a subset of subcarriers, then the MCI (6) is null. Therefore, it is important to optimize the time and frequency spreading lengths, L_t and L_f, according to the transmission scenario. Finally, the estimated channel coefficient (5) is used to equalize the data symbols spread over the same subset of subcarriers.

2.2 CFO Estimation Principles

In DVB-T standard [2], in addition of the scattered pilot subcarriers used for CFR estimation, continuous pilot subcarriers are defined for synchronization [3]. In our study, we propose to exploit the same spread pilot symbols to estimate the CFO. Contrary to CFR estimation, the CFO estimation is however processed before de-precoding function. Thus, let us express the received symbol during the qth OFDM symbol on the n'th subcarrier of the subset of subcarriers (t, s'):

$$\begin{aligned}
z_{t,s'}[q,n'] = e^{j2\pi q(N+v)\Delta F T_s} \sum_{s=0}^{S-1} \sum_{n=0}^{L_f-1} y_{t,s}[q,n] h_{t,s}[q,n] \\
\times \varphi(q,s',s,n',n) + w_{t,s'}[q,n']
\end{aligned} \quad (7)$$

where T_s is the sampling period and $\Delta F = F_{TX} - F_{RX}$ is the CFO. The phase rotation at left-hand side of (7) is due to the CFO increment in time. In our study, we assume that the CFO increments during a frame. $\varphi(q, s', s, n', n)$ is an equivalent transfer function describing the effect of the CFO, including the intercarrier interference (ICI). It is equal to:

$$\varphi(q, s', s, n', n) = \psi_N \left(\Delta F T_s + \frac{(s'-s) L_f + (n'-n)}{N} \right)$$
$$\times \exp\left(j\pi (N-1) \left(\Delta F T_s + \frac{(s'-s) L_f + (n'-n)}{N} \right) \right) \quad (8)$$

where $\psi_N(x)$ is the Dirichlet function defined by: $\psi_N(x) = \frac{\sin(\pi N x)}{N \sin(\pi x)}$.

It is clear from (7) that, due to the CFO, the phase rotation increments with q. Thus, we will benefit from this increment to define the CFO estimation metric $\Gamma_t(\Delta F T_s)$ computed from the pilot symbols by:

$$\Gamma_t(\Delta F T_s)$$
$$= \sum_{s'=0}^{S-1} \sum_{n'=0}^{L_f-1} \sum_{q=0}^{L_t-2} c_p^*[n' L_t + q + 1] z_{t,s'}[q+1, n'] c_p[n' L_t + q] z_{t,s'}^*[q, n']$$
$$= e^{j2\pi(N+v)\Delta F T_s} \sum_{s'=0}^{S-1} \sum_{n'=0}^{L_f-1} \sum_{q=0}^{L_t-2} |c_p[n' L_t + q + 1]|^2 \sqrt{P_p} \, x_{t,s'}[p] h_{t,s'}[q+1, n']$$
$$\times |c_p[n' L_t + q]|^2 \sqrt{P_p} \, x_{t,s'}^*[p] h_{t,s'}^*[q, n'] + \Xi_{t,s'}[q, n'] \quad (9)$$

where $\Xi_{t,s'}[q, n']$ results from the contributions of the interferences caused by the data chips superposed to the pilot chip, the ICI due to the CFO and the AWGN. Assuming that the channel is invariant during two consecutive OFDM symbols, (9) can be rewritten as:

$$\Gamma_t(\Delta F T_s) = \frac{P_p}{L} e^{j2\pi(N+v)\Delta F T_s}$$
$$\times \left(\sum_{s'=0}^{S-1} |x_{t,s'}[p]|^2 \sum_{n'=0}^{L_f-1} \sum_{q=0}^{L_t-2} |h_{t,s'}[q, n']|^2 + \Xi_{t,s'}[q, n'] \right) \quad (10)$$

All the interferences are assumed to have Gaussian distribution with zero mean [4, 5]. Consequently, if the product $S \times L_f \times (L_t - 1)$ is large enough, $\Xi_{t,s'}[q, n']$ tends to zero. Using (10), it is straightforward to say that the CFO is the measure of the phase of $\Gamma_t(\Delta F T_s)$. Let us note that to avoid any phase ambiguity, it is necessary that:

$$|2\pi(N+v)\Delta F T_s| < \pi \quad (11)$$

This constraint determinates the maximum estimable CFO value. Finally, a loop filter taking into account the estimated CFO of the previous and the current frames computes the final estimated CFO. It will be used to correct the signal in time domain in order to mitigate the ICI.

3 Simulation Results

In this section, we analyse our estimators in term of mean square error (MSE). Moreover, we compare the performance of the proposed system with the DVB-T standard under different channel models. Table 1 gives the simulation parameters and the useful bit rates of the DVB-T standard and the proposed system. In the proposed system, only one spread pilot symbol is used over L. Thus, the loss of useful bit rate due to spread pilot symbols is only 3.1% for $L = 32$ and 1.6% for $L = 64$, whereas it is 10.32% for DVB-T system. Therefore, a gain in terms of spectral efficiency and useful bit rate is obtained compared to DVB-T system. According to (11), the FFT size and the guard interval size values, the theoretical maximum value of CFO estimable is 40% of the intercarrier spacing.

Figures 3 and 4 depict the estimators performance in term of MSE for QPSK data symbols, different mobility scenarios and different spreading factors. We note that beyond a given ratio of energy per bit to the noise spectral density ($\frac{Eb}{No}$), the MSEs of both CFR and CFO estimators reach a floor. It is easily interpreted as being due to the MCI and the ICI.

Figures 5 and 6 give the BER measured at the output of the Viterbi decoder for both systems under F1 channel model and the COST207 Typical Urban 6 paths (TU6) channel model respectively. The F1 channel model defined in [2] corresponds to a fixed reception condition in line of sight. It is modelized without any Doppler effect. As well for fixed reception as mobile reception, we remark that the presence of CFO does not degrade significatively the performance of the proposed system. It proves the efficiency of the proposed CFO estimator. Moreover, it appears that

Table 1 Simulation parameters and useful bit rates

Bandwidth	8 MHz
FFT size (N)	2,048 samples
Guard interval size	512 samples (64 μs)
OFDM symbol duration (T_{OFDM})	280 μs
Rate of convolutional code	1/2 using $(133, 171)_o$
Constellations	QPSK and 16QAM
Carrier frequency	500 MHz
Channel models	F1 and TU6
Mobile speed for TU6	20 km/h, 60 km/h, 120 km/h, 350 km/h
Maximum doppler frequencies (f_D)	9.26 Hz, 27.78 Hz, 55.56 Hz, 162.04 Hz
$\beta = f_D \times T_{OFDM}$	0.003, 0.008, 0.012, 0.045
Normalized CFO: $\Delta F' = \Delta F.(NT_s)$	0 and 10%
Loop filter gain	1/16
Useful bit rates of DVB-T system	4.98 Mbits/s for QPSK
	9.95 Mbits/s for 16QAM
Useful bit rates of 2D LP OFDM for QPSK	5.51 Mbits/s for $L = 32$
	5.60 Mbits/s for $L = 64$
Useful bit rates of 2D LP OFDM for 16QAM	11.02 Mbits/s for $L = 32$
	11.20 Mbits/s for $L = 64$

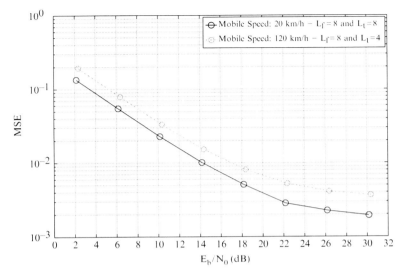

Fig. 3 MSE measured of the CFR estimation based on Spread Pilot under TU6 channel model; QPSK data symbols; Mobile Speeds: 20 and 120 km/h

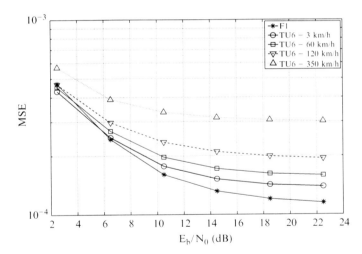

Fig. 4 MSE measured of the CFO estimation based on Spread Pilot; QPSK data symbols; $L_t = L_f = 4$

the performance of the proposed system is very close to that of DVB-T system with perfect CFR estimation. This can be explained by the efficient CFR estimation combined with the spectral efficiency improvement previously mentioned.

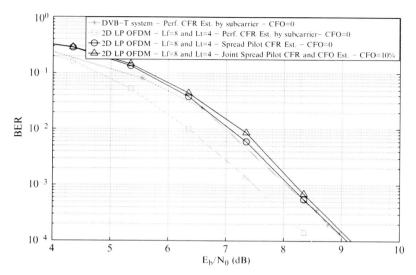

Fig. 5 BER performance under F1 channel; 16QAM; Rate of convolutional code: $R_c = 1/2$; $L_f = 8$ and $L_t = 4$

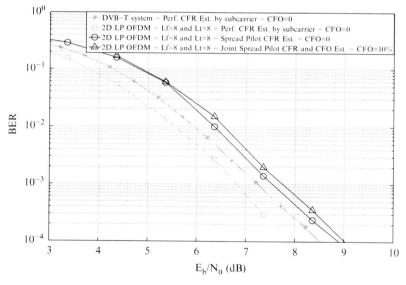

Fig. 6 BER performance under TU6 channel; Mobile speed: 60 km/h; QPSK; Rate of convolutional code: $R_c = 1/2$; $L_f = 8$ and $L_t = 8$

4 Conclusion

In this paper, we proposed an efficient and very simple joint CFR and CFO estimation for DVB systems. These algorithms based on spread pilots allow an improvement of the spectral efficiency and an increase of the useful bit rate of the system. Moreover, taking into account the power loss due to pilot symbols, we showed that the performance of the proposed system is equivalent to that of DVB-T with perfect CFR and CFO estimation.

Acknowledgements This work was supported by the European project CELTIC B21C ("Broadcast for the 21st Century") [6] and the French national project "Mobile TV World".

References

1. O. P. Pasquero, M. Crussière, Y. Nasser, J.-F. Hélard, "2D Linear Precoded OFDM for Future Mobile Digital Video Broadcasting," Signal Processing Advances in Wireless Communications, SPAWC 2008. IEEE 9th Workshop on, pp. 476–480, July 2008.
2. ETSI EN 300 744, "Digital Video Broadcasting (DVB); Framing structure channel coding and modulation for digital terrestrial television", Tech. Rep., Nov. 2004.
3. S. A. Fechtel, "OFDM carrier and sampling synchronization and its performance on stationary and mobile channels," IEEE Trans. Consum. Electron., vol. 46, no. 3, pp. 438–441, Aug 2000.
4. T. Pollet, M. Van Bladel, and M. Moeneclaey, "BER Sensitivity of OFDM Systems to Carrier Frequency Offset and Wiener Phase Noise," IEEE Trans. Commun., vol. 43, no. 234, pp. 191–193, Feb/Mar/Apr 1995.
5. Y. Nasser, M. Noes, L. Ros, and G. Jourdain, "Sensitivity of OFDM-CDMA Systems to Carrier Frequency Offset," Communications, ICC 2006. IEEE International Conference on, vol. 10, pp. 4577–4582, June 2006.
6. http://www.celtic initiative.org/Projects/B21C.

Channel Characteristics of Different Floors for Joint Communications and Positioning

Wei Wang, Thomas Jost, and Armin Dammann

Abstract Recently the fusion of positioning and wireless communication has gained many interests due to the merits of location information for the future communication systems. As one of the well known positioning techniques, Finger Printing (FP) based localization makes use of pre-collected wireless channel information to locate the target. Both positioning and communications are highly depending on the wireless channel. In this paper, we investigate and compare the channel characteristics of different floors based on an outdoor to indoor broadband channel sounder measurement campaign. The comparison between two floors are addressed in terms of delay spread, coherence time characteristic, Power Delay Profiles (PDPs), and Time of Arrival (ToA) bias from the Geometric Line-of-Sight (GLoS) to the first None Line-of-Sight (NLoS) path.

1 Introduction

Since the last decade, issues in wireless communication systems like the number of users and the demand for higher data rates are getting more and more critical. Therefore, it is essential to efficiently manage and utilize limited resources for future wireless network communication systems (e.g., Long Term Evolution (LTE)) [1], where Orthogonal Frequency Division Multiplexing (OFDM) is the most attractive modulation scheme. The cognitive radio can act as an allocator for frequency spectrum resources.

However, this aspect of how location information can benefit to communication systems has been rarely investigated. This has been demonstrated to be useful and necessary in [1]. As one well known positioning techniques, FP based positioning makes use of pre-collected information about the wireless channel saved in a

---

W. Wang (✉), T. Jost, and A. Dammann
German Aerospace Center (DLR), Institute of Communications and Navigation,
Oberpfaffenhofen, 82234 Wessling, Germany
e-mail: {Wei.Wang,Thomas.Jost,Armin.Dammann}@dlr.de

database, such as the Received Signal Strength (RSS), time difference of arrival, or PDP. The fusion of positioning and communications are gaining many interests.

Both FP based localization and wireless communication are strongly depending on the propagation channel. As one of the most challenging tasks, the indoor environment based propagation channel is focused in this paper. Modern residential houses, office buildings, and other constructions are usually multi-storey buildings. Therefore, it is necessary to investigate into channel characteristics of different floors in the same building. Several studies on the indoor to indoor multifloors haven been described before [3]. This paper discusses the outdoor to indoor multifloor channel characteristics for positioning and communication based on a broadband wireless channel measurement campaign at German Aerospace Center (DLR) premise in Oberpfaffenhofen.

In Section 2, the setup of the channel measurement campaign is addressed. Thereafter, Section 3 discusses the data processing methods and the evaluations. The corresponding results, like the Root Mean Square (RMS) delay spread and PDPs are shown in Section 4.

2 Channel Measurement Campaign

To gain insight into the differences of electromagnetic waves emitted from an outdoor transmitter, received at two different floors of the same building, DLR performed broadband channel measurements in April 2008. The measurements were accomplished using the Medav RUSK-DLR channel sounder at operating center frequency 5.2 GHz. The transmitter antenna was located on the rooftop of the office building TE02 of the Institute of Communications and Navigation (ICN) of DLR in a height of 11 m above ground. There the transmitter was emitting a 5 W chirp signal with a rectangular spectral shape of $B = 120$ MHz bandwidth. The transmitted periodic signal was vertical polarized with a period of 12.8 μs giving a maximum impulse response time length of 12.8 μs, which leads to a maximum propagation distance of 3.84 km. The measured i-th snapshot of the Channel Impulse Response (CIR), $h(i, n), n = 0, \ldots, N-1$ consists of $N = 1537$ samples at delays $\tau_n = n \Delta \tau$, with $\Delta \tau = 1/B$. The discrete transfer function $H(i, m)$ contains the same number of samples at frequencies $m \Delta f$, $m = 0, \ldots, N-1$ with spacing $\Delta f = 78$ kHz.

The channel sounder records one CIR every $T_g = 1.024$ ms providing a measurement rate of 976 CIRs per second (CIRs/s). The transmitter and receiver were synchronized by adjusting the 10 MHz Rubidium frequency normal of the receiver to the drifting of the transmitter frequency normal. Table 1 summaries the channel sounder setup for the measurements.

The receiver was located inside the office building TE01 of ICN with a spacing of 20 m between both buildings. The building itself can be characterized as a standard three story office building of concrete with metallic window glasses. As our primary goal was to simulate a moving receiver instead of point measurements, the receiving antenna was mounted on a model train, as displayed in Fig. 1 which was moving

Table 1 Channel sounder settings for the SISO measurement

RF center frequency	5.2 GHz
Bandwidth B	120 MHz
A/D converter at receiver	8 bits, 320 MHz
Transmit power	5 W $\cong$ 37 dBm
Signal period	12.8 µs
Number of sub-carriers	1536
Measurement time grid T_g	1.024 ms
Antennas	Omni-directional with vertical polarisation
Receiver speed v	0.1115 m/s
Transmit height	11 m

Fig. 1 The model train used as mobile receiver platform shown without receiving antenna

with a speed of $v = 0.1115$ m/s. For the same transmitter position, the model train was running on the tracks $R1$ and $R2$, where $R1$ was fixed on the ground of the first floor (ground floor), and $R2$ on the second floor. As shown in Figs. 2 and 3. Both tracks $R1$ and $R2$ were located at the same positions relative to the building walls in the corridor, only differing in height above ground, to compare the channel characteristics for both floors. The position of the transmitter was precisely determined using a Leica tachymeter giving a nominal accuracy in the sub-cm domain. To get a similar accuracy for the receiver antenna mounted on the model train, the train is equipped with a rotary encoder giving 500 impulses per motor turn as described in [2]. To prevent wheel slipping the model train runs by a cogwheel. Together with each CIR snapshot measurement done in a periodic cycle of T_g the number of impulses given by the rotary encoder since start of the model train is saved, which results in a precise traveled distance measure for each captured impulse response snapshot.

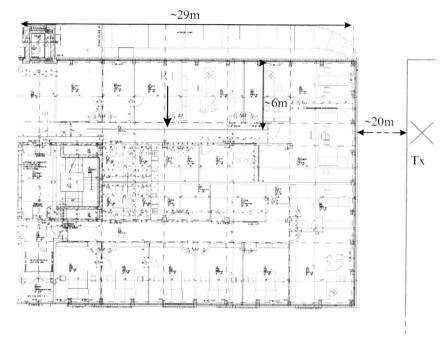

Fig. 2 Track $R1$ on the first floor (ground floor) which runs through the corridor

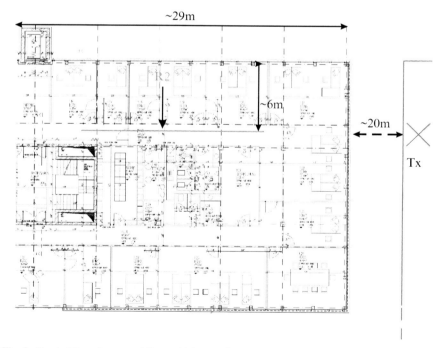

Fig. 3 Track $R2$ on the second floor which runs through the corridor

3 Data Processing and Evaluation

For wireless communication as well as for the FP based positioning, one of the most important measures of a wireless multipath channel is the PDP. The most common FP positioning algorithms make use of the offline-measured received signal power or PDPs for an online pattern recognition [6, 7]. The PDP in this paper is denoted as $P(l, n)$, where l denotes the number of a segment of 0.2 m length on the model railway track and n the delay bin of the sample at delay τ_n. $P(l, n)$ is calculated as

$$P(l,n) = \frac{1}{i_{l+1} - i_l} \sum_{i=i_l}^{i_{l+1}-1} |h_n(i,n)|^2 \qquad (1)$$

where i_l denotes the number of the CIR snapshot where the segment l starts and $h_n(i, n)$ is the CIR snapshot $h(i, n)$ normalized in power by the maximum peak power and in delay as the CIR is shifted such that the first detected path is at delay bin $n = 0$.

The RMS delay spread $\sigma_\tau(i)$ is calculated as

$$\sigma_\tau(i) = \sqrt{\frac{\sum_{n=0}^{N-1}(\tau_n - m_\tau(i))^2 \cdot |h_n(i,n)|^2}{\sum_{n=0}^{N-1}|h_n(i,n)|^2}} \quad \text{with} \quad m_\tau(i) = \frac{\sum_{n=0}^{N-1}\tau_n \cdot |h_n(i,n)|^2}{\sum_{n=0}^{N-1}|h_n(i,n)|^2}. \qquad (2)$$

A threshold calculated as the 99.9% quantile of a known noise region in $P(l, n)$ has been used. In this paper, the RMS delay spread is not calculated segment by segment like PDP, but snapshot by snapshot.

Frequency dispersion is usually measured by the coherence time. Since the model train was moving with a constant speed of $v = 0.1115$ m/s, the coherence time $C_T(\rho_t)$ can be equivalently expressed as the coherence distance $C_d(\rho)$. This provides more information to analyze the spatial correlation properties of the channel for FP based positioning which is highly depending on the spatial correlations of its features (as PDPs as example). Similar as the coherence time, the coherence distance is calculated by

$$C_d(\rho) = \frac{1}{W_T}\sum_{i=0}^{W_T-1} \frac{\sum_{n=0}^{N-1} h_n(i,n) \cdot h_n^*(i+\frac{\rho}{v},n)}{\sqrt{\sum_{n=0}^{N-1}|h_n(i,n)|^2 \cdot \sum_{n=0}^{N-1}|h_n(i+\frac{\rho}{v},n)|^2}} \qquad (3)$$

where $\rho = k \cdot v \cdot T_g$, with $k = 1, 2, \ldots$ and $C_d(\rho) = C_T(\rho_t \cdot v)$. A window length of 2.56 s containing $W_T = 2500$ CIR snapshots was utilized for the correlation. The "accelerating" parts of the track at the beginning and the end, where the model train traveling speed is not constant were excluded. Such equally spaced CIR snapshots were obtained for the coherence distance $C_d(\rho)$ calculation.

To get a insight into the channel time dispersion characteristics, we calculate the coherence bandwidth $C_B(i, w)$ as a correlation between two vectors of $H(i, m)$ shifted in frequency by w with a window length of $W_L = 768$ frequency samples with spacing of $\Delta f = 78$ kHz as shown in (4). As a result, the correlation for a frequency spacing w is possible up to 60 MHz. Similar as [4], the coherence bandwidth $C_B(i, w)$ for each CIR snapshot is calculated by

$$C_B(i, w) = \frac{\sum_{m=0}^{W_L-1} H(i, m) \cdot H^*(i, m+w)}{\sqrt{\sum_{m=0}^{WL-1} |H(i, m)|^2 \cdot \sum_{m=0}^{W_L-1} |H(i, m+w)|^2}}. \quad (4)$$

For coherence distance $C_d(\rho)$, we define the value of CD_x as the distance ρ where $C_d(\rho)$ falls to $x\%$ of the maximum. CB_x is defined in the same way for coherence bandwidth $CB_{i,w}$.

The ToA bias to the GLoS is an important fact in the positioning systems [2, 5]. It is defined as the absolute propagation delay difference between the first incoming path and the GLoS distance divided by the speed of light. To study the ToA bias characteristic for different floors, the ESPRIT super resolution algorithm has been utilized to the measured raw data to estimate the path delays, amplitudes, and phases. distance of GLoS.

4 Results

The sample Probability Density Functions (PDFs) of the PDPs $P(l, n)$ for tracks $R1$ on the first floor and tracks $R2$ on the second floor are presented in Figs. 4 and 5. As clearly to see the PDPs for both floor have a "two-cluster" structure, starting at 0 µs and approximately at 0.15 µs. The delay gap between both results in an equivalent propagation distance of approximately 45 m. They are most probably caused by reflections between both buildings which are 20 m spaced as shown in Figs. 2 and 3. The windows of the buildings are metallized and do have therefore good electromagnetic reflection properties.

For the first floor, the electromagnetic waves are most probably propagating through the big lab as shown in Fig. 2 before arriving at the antenna placed in the corridor. However, for the second floor, the amplitude of the waves suffer from additional transmission loss compared to the first floor which might be due to the office rooms located at the wall towards building TE02 on the second floor. Therefore,

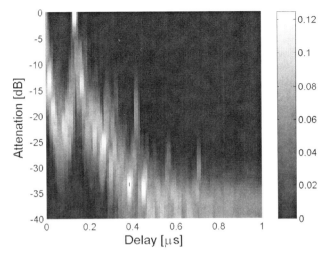

Fig. 4 PDF estimate for measurements taken while the model train is running on track $R1$

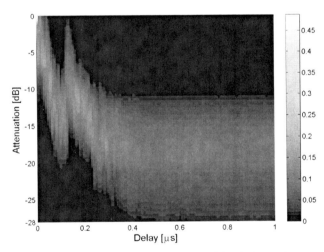

Fig. 5 PDF estimate for measurements taken while the model train is running on track $R2$

the power received when the model train runs on $R1$ is larger than the power obtained when running on $R2$. Although the PDPs of both tracks share the same cluster structure, in the first floor the second cluster has more power than the first cluster, whereas in the second floor the first cluster has more power than the second cluster. The PDPs of the different floors can explain the RMS delay spread values listed in Table 2. For the track $R1$, the power is more focused on the second cluster. For the track $R2$, the power is mostly focused on the beginning of the first cluster. As a result, the RMS delay spread when running on $R1$ is larger than the delay spread when running on $R2$.

Table 2 Statistical parameters of RMS delay spread, μ denotes the mean value and σ stands for the standard deviation

	$R1$	$R2$
μ [ns]	64.7438	39.6660
σ [ns]	12.4343	21.3655

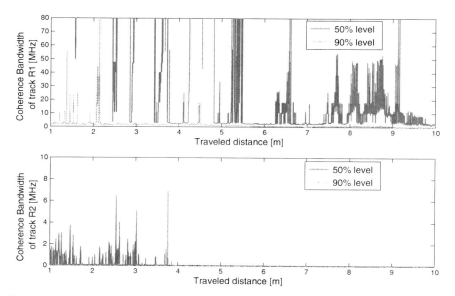

Fig. 6 Coherence bandwidth referring to a correlation of 50% and 90% when the receiver was running on tracks

The coherence distance computed according to the Section 3 is small for both tracks. The CD_{50} of track $R1$ and $R2$ are 1.8 cm and 0.088 cm respectively. And the CD_{90} are 0.0784 cm and 0.0014 cm respectively. The coherence distance calculated when running on $R1$ is higher compared to the value associated with $R2$. As the coherence distance along the corridor is rather small a FP based position algorithms could benefit as the CIRs for different locations are less correlated. Figure 6 shows the computed coherence bandwidth for the received signal while the model train was running on the tracks. For $R1$, the CB_{50} is always bigger than 60 MHz which means the correlation does not fall below a 50% level. After a traveled distance of 6 m, the CB_{50} gets smaller. For $R2$, the CB_{50} becomes quite small after traveling a distance of 4 m. This can be explained by the fact that, the received power in the parts after 4 m of $R2$ are small so that the correlations are more effected by noise. The statistical parameters of the coherence bandwidth are listed in Table 3. While calculating the CB_{50} and CB_{90}, the samples which do not fall below the corresponding correlation levels (50% and 90% respectively) are not taken into account in the calculations. The coherence bandwidth information is essential for communication

Table 3 Statistical parameters of coherence bandwidth, μ denotes the mean value and σ stands for the standard deviation

	μ [MHz]	σ [MHz]
$R1. CB_{50}$	8.6784	11.4623
$R1. CB_{90}$	1.3231	2.6828
$R2. CB_{50}$	0.3223	1.0298
$R2. CB_{90}$	0.0117	0.0041

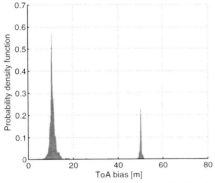
a Probability density function of ToA bias for track R1

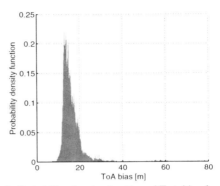
b Probability density function of ToA bias for track R2

Fig. 7 Probability density function of ToA bias for both tracks

systems. For instance, the typical cases for OFDM systems, the frequency spacing should be less than the coherence bandwidth and the symbol time is less than the coherence time [8].

Figure 7 shows the first results obtained for the ToA bias for track $R1$ and $R2$. It can be seen that the ToA biases when running on $R2$ are larger than the biases when running on $R1$. The sample PDF of the ToA bias of $R1$ has a narrow peak shape compared to the sample PDF of $R2$ which indicates the ToA biases do not vary much on the first floor. There is also a "tail" in Fig. 7a, which is 40 m away from the main part of the PDF. This is a result of the ESPRIT algorithm used for cases where the power of the first cluster falls under a threshold and the algorithm only recognizes the second cluster.

5 Conclusion

In this paper, based on an outdoor to indoor broadband wireless channel measurement campaign, we addressed the channel characteristics and a comparison between signals received on two different floors to study the wireless channel for joint communications and FP based positioning. The propagation characteristics are PDPs, coherence distance, coherence bandwidth, RMS delay spread, and the ToA bias. The evaluation shows for these measurements that CIRs measured on both floors have PDPs with a two-clusters structure. The obtained coherence distances are small for both floors which benefits the FP database collections. However, the RMS delay spread, the coherence bandwidth, and the ToA bias are different from floor to floor due to the different power allocations in the two clusters.

References

1. R. Raulefs and S. Plass, "Combining Wireless Communications And Navigation — The WHERE Project", *In Proceedings of 68th IEEE Vehicular Technology Conference (VTC 2008-Fall), Calgary, Sept. 2008*.
2. T. Jost, W. Wang, "Satellite-to-Indoor Broadband Channel Measurements at 1.51 GHz", *In Proceedings of the ION International Technical Meeting (ITM), Anaheim, California, Jan. 2009*.
3. M. Lott and I. Forkel, "A Multi-wall-and-floor Model for Indoor Radio Propagation", *In Proceedings of the 53rd IEEE Vehicular Technology Conference (VTC 2001-Spring), Rhodes, Greece, Mar. 2001*.
4. S. Gligorevic, R. Zierhut, T. Jost, W. Wang, "Airport Channel Measurements at 5.2 GHz", *In Proceedings of the 3rd European Conference on Antennas and Propagation (Eucap), Berlin, Germany, Mar. 2009*.
5. W. Wang, C. Mensing, S. Gligorevic, T. Jost, and A. Dammann, "Short Term Statistical Analysis of Outdoor to Indoor Propagation Channel for Geolocations", *In Proceedings of the 13th International OFDM Workshop (InOWo), Hamburg, Germany, Aug. 2008*.
6. P. Bahl and V. Padmanabhan, "RADAR: An In-Building RF-based User Location and Tracking System", *INFOCOM 19th Annual Joint Conference of the IEEE Computer and Communications Societies. Proceedings, IEEE. Vol. 2, pp. 775–784, 2000*.
7. M. Triki, D. Slock, V. Rigal, and P. Francois, "Mobile Terminal Positioning via Power Delay Profile Fingerprinting: Reproducible Validation Simulations", *In Proceedings of 64th IEEE Vehicular Technology Conference (VTC 2006-Fall), Montreal, Canada, 2006*.
8. F. Tufvesson and T. Maseng, "Pilot Assisted Channel Estimation for OFDM in Mobile Cellular Systems", *In Proceedings of 47th IEEE Vehicular Technology Conference (VTC 1997-Spring), Phoenix, USA, May. 1997*.

Part IV
Long Term Evolution (LTE)

Uplink Power Control Performance in UTRAN LTE Networks

Robert Müllner, Carsten F. Ball, Kolio Ivanov, Johann Lienhart, and Peter Hric

Abstract Uplink power control in 3GPP UTRAN Long Term Evolution (LTE) networks consists of a closed-loop scheme around an open-loop point of operation. The uplink performance of the network is decisively influenced by power control. This paper provides insight into the power control procedure and its interworking with Adaptive Transmission Bandwidth (ATB) as well as Adaptive Modulation and Coding (AMC) presenting a detailed performance evaluation by system level simulations for a fully loaded network. The analysis starts for pure open-loop power control as reference, for which the impact of parameter settings on resource allocation, utilization of specific modulation and coding schemes, retransmission rate, and resulting throughput has been determined. A two-dimensional parameter optimization for full path-loss compensation and fractional power control has been performed to conclude the best strategy for the trade-off between network capacity and coverage. Finally on top of this optimized open-loop power control parameter set the closed-loop component has been enabled and proposals for optimum power control threshold settings are provided. The beneficial effects of closed-loop power control are presented highlighting its ability to adjust the transmission power according to the desired quality and level requirements.

1 Introduction

Mobile broadband access with high data rates and low latencies becomes reality with the deployment of UTRAN Long Term Evolution (LTE) currently being standardized in 3GPP [1–4]. To achieve optimal network performance the role of Power

---

R. Müllner (✉), C.F. Ball, and K. Ivanov
Nokia Siemens Networks GmbH & Co. KG, Radio Access, St.-Martinstr. 76, 81541 Munich, Germany
e-mail: {robert.muellner, carsten.ball, kolio.ivanov}@nsn.com

J. Lienhart and P. Hric
Siemens AG, Autokaderstr. 29, 1210 Vienna, Austria
e-mail: {johann.lienhart, peter.hric}@siemens.com

Control (PC) becomes decisive for maintaining the required Signal over Interference plus Noise Ratio (SINR) according to Quality of Service (QoS) requirements while controlling at the same time the inter-cell interference [5]. Especially uplink (UL) PC is a means to effectively reduce interference in the network and to improve cell edge performance. Recall that LTE networks are based on tight frequency reuse 1. This study gives an insight into the behavior of the different types of UL PC such as fractional PC and full path-loss (PL) compensation as well as the impact of specific parameter settings using a full-blown system level simulation environment. Focus has been set on the investigation of the difference between open-loop (OL) and closed-loop (CL) PC algorithm. The paper is structured as follows. Section 2 describes the UL PC algorithm. The simulation model is introduced in Section 3. Detailed simulation results are presented and discussed in Section 4. Conclusions are given in Section 5.

2 Uplink Power Control Algorithm

The LTE PC algorithm specified in [6] is based on a combination of an OL and CL scheme. The user equipment (UE) controls its output power to keep the transmitted power per Resource Block (RB) constant irrespective of the allocated transmission bandwidth. OLPC is exclusively based on PL estimates, broadcast system parameters and some dedicated signaling without utilizing any feedback. Feedback from eNodeB, however, is used by the UE to adaptively correct its transmission power upon receiving CLPC commands derived from UL measurements. The transmission power for the Physical Uplink Shared Channel (PUSCH) is set by the UE according to

$$P = \min\{P_{\max}, 10 \cdot \log_{10} M + P_0 + \alpha \cdot PL + \Delta_{TF} + f(\Delta_i)\}. \quad (1)$$

P_{max} is the maximum allowed UE transmission power set, e.g. to 23 dBm (200 mW) for UE power class 3 [7]. M is the size of the PUSCH resource assignment to a specific UE expressed in number of RB. P_0 is the power offset composed of a cell specific and UE specific component [6].

The task of traditional PC schemes is to achieve equal received signal level at eNodeB for all UEs (full PL compensation). 3GPP has additionally introduced the use of fractional PC (FPC) [8,9] allowing that users with high PL operate at a lower receive signal requirement. Consequently UEs located at the cell border generate less interference to neighboring cells. The degree of FPC is controlled by the cell-specific PL compensation factor α.

PL in (1) is the downlink PL estimated by the UE. Δ_{TF} is a Transport Format (TF) dependent offset [6]. The term $f(\Delta_i)$ defines the CLPC specific correction term with relative or absolute power increase/decrease depending on the function $f(\cdot)$ [6], whereas only relative correction (accumulation) shall be considered in the following study. The correction values Δ_i are generated by comparing filtered level and quality measurements in UL with specific targets defined by a two-dimensional decision matrix. This matrix is implemented in eNodeB and designed in a generic and future-proof way with possible extensions to use *QoS*, *PL estimates* and *interference on neighboring cells* as further criteria to trigger power increase or decrease.

3 Simulation Model

A hexagonal regular cell layout with three sectors per site in an urban deployment scenario with 500 m Inter Site Distance (ISD) and frequency spectrum of 10 MHz has been assumed (frequency re-use 1). The deployment area comprises 21 cells in a wrap-around model using a Typical Urban (TU) channel model. A PL model for small cells with PL slope of 37.6 dB per decade is used. Basic configuration parameters such as PL model and antenna diagram are selected in accordance to [10].

The number of users within the simulation area is kept constant at 210 resulting in an average load of ten users per cell. The distribution of the users is unbalanced, i.e. the number of 210 UEs is distributed over the whole simulation area leading to a different number of UEs per cell. Slow moving subscribers have been assumed. During the simulation run an UE can change its serving cell by power budget handover. Full (infinite) buffer traffic model proposed for LTE benchmark evaluation in [11] is assumed. The simulation model includes synchronous non-adaptive Hybrid Automatic Repeat Request (HARQ) with Chase Combining.

LTE supports in UL QPSK and 16QAM modulation, 64QAM is optional. Slow Adaptive Modulation and Coding (AMC) [12] with 30 ms periods has been assumed for switching between Modulation and Coding Schemes (MCS) depending on radio conditions. Out of the available 28 MCS according to 3GPP only three MCS per modulation type have been selected: MCS-0 to 2 for QPSK, MCS-3 to 5 for 16QAM and MCS-6 to 8 for 64QAM as listed in Table 1. Adaptive Transmission Bandwidth (ATB) based on Power Headroom Reports (PHR) is updated every Transmission Time Interval (TTI) of 1 ms and has been modeled ideally, i.e. assuming no measurement error. The essential simulation parameters are listed in Table 1.

4 Simulation Results

4.1 Open-Loop Power Control

The impact of the power offset P_0 on the UL performance has been analyzed assuming full PL compensation ($\alpha = 1$). Figure 1a shows the cumulative distribution function (CDF) of transmission power per RB for four different P_0 settings in the range from -120 to -60 dBm. For low P_0 values (-120 and -100 dBm) the total UL transmission power according to (1) is low and the transmission power per RB is not limited. For both P_0 settings the total UE transmission power (not shown here) is below P_{max} of 23 dBm and no UE power shortage has been observed. In

Table 1 Parameters of the system level simulation model

Parameter	Value
Layout	7 sites – 3 cells/site – wrap around
Propagation scenario	Macro 1 (ISD 500 m)
System carrier frequency	2 GHz (according to [10] for case 1)
System bandwidth	10 MHz
Frequency re-use scheme	Re-use 1, no fractional re-use or interference avoidance applied
Traffic model	Full buffer [11], geometric session lifetime distribution (mean value 30 s)
Number of UEs	Constant number of 210 (in 21 cells)
eNodeB receiver	2 RX (maximum ratio combining)
UE speed	3 km/h
Scheduler	Round robin, channel unaware scheduler
Adaptive Transmission Bandwidth (ATB)	Ideal (based on power headroom report without measurement error)
Link adaptation	Slow Adaptive Modulation and Coding (AMC)
BLER target	10%
Modulation and Coding Schemes (MCS)	QPSK [R = 1/3, 1/2, 2/3], 16QAM [R = 1/2, 2/3, 5/6], 64QAM [R = 2/3, 5/6, 9/10]
Maximum UE transmission power P_{max}	23 dBm (200 mW) according to [7] for UE power class 3
Uplink power control	Open and closed-loop
Closed-loop PC interval	50 ms
Δ_{TF}	0, i.e. transport format dependent offset disabled
Δ_i for closed-loop PC	$\{-1; 0; 1; 3\}$ dB, accumulation enabled

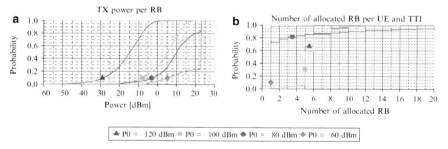

Fig. 1 CDF of (**a**) transmission power per resource block and (**b**) number of allocated resource blocks per UE and TTI for $\alpha = 1$ and various P0 settings

contrast, at higher power offset 17% of the RBs reach P_{max} for $P_0 = -80$ dBm, i.e. the requested transmission power according to (1) cannot be provided even for the allocation of a single RB only. For $P_0 = -60$ dBm limitation of transmission power per RB has been observed for 76% of the RBs. Hence proper cell-specific setting of P_0 is essential, especially if pure OLPC is applied. If P_0 is set too high, not only UEs located at the cell border, but also those in the vicinity of the eNodeB use unnecessarily high transmission power reducing battery life-time and might get less bandwidth assigned by ATB.

The number of RBs assigned in UL to the UE depends on (a) availability of physical resources in the cell and (b) available power headroom. The CDF of the number of allocated RBs per UE and TTI is depicted in Fig. 1b. For $P_0 = -120$ dBm five or more RBs have been assigned to the UE in 99.7% of the TTIs. A restriction in the number of assigned RBs is purely related to the number of users simultaneously served in the cell and not to UL power constraints. Recall that the 50 RBs in 10 MHz are shared by 10 UEs per cell in average.

For $P_0 = -100$ dBm five or more RBs have been assigned to the UE in 94% of the TTIs. Selecting higher P_0 results in a higher transmission power per RB and consequently leads to a reduction of the RBs assigned to the UE by ATB, because the total UE transmission power defined in (1) must not exceed P_{max}. For $P_0 = -80$ dBm only a single RB has been assigned to the UE in 73% of the TTIs. For $P_0 = -60$ dBm the ratio of TTIs for which only a single RB has been assigned to the UE is even 98%. Note that 10 UEs per cell with a single RB allocated per UE results in a fractional load of 20% only. Figure 1b demonstrates the utilization of the deployed air interface resources depending on P_0 setting.

Transmission power per RB and number of assigned RBs have significant impact on the SINR as shown in Fig. 2a. In the fully loaded network using $P_0 = -120$ dBm only 13% of the bursts are received at a SINR of 1 dB or higher, corresponding to the operating point of the most robust MCS-0 used in this study (cf. 2b). The SINR distribution gradually improves as P_0 increases to -100 and -80 dBm. For $P_0 = -60$ dBm the high interference at a low number of allocated RBs (cf. 1b) causes flattening of the SINR distribution, i.e. a considerable number of users enjoy high SINR, while other users suffer from low SINR resulting in a poor cell border throughput performance. Note that for $P_0 = -60$ dBm in most cases only one RB per UE has been assigned, i.e. the SINR distribution reflects the quality on the sparse number of allocated resources only. Figure 2b shows the throughput per RB for the first transmission vs. SINR obtained from link level simulations. The operating range of QPSK modulation is 1–7 dB. For 16QAM a SINR of 7–15 dB is required. The operating point of 64QAM is 15 dB and higher.

The optimum MCS per UE depends on radio conditions and is selected by AMC based on link quality measurements. For $P_0 = -120$ dBm the resulting SINR

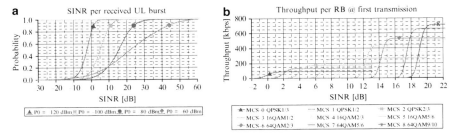

Fig. 2 (a) CDF of signal over interference plus noise ratio per received uplink burst for $\alpha = 1$ and various P_0 settings and (b) throughput per resource block vs. signal over interference plus noise ratio (first transmission) for various modulation and coding schemes

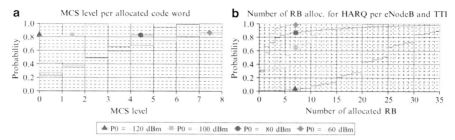

Fig. 3 CDF of (**a**) modulation and coding scheme utilization and (**b**) number of resource blocks allocated for HARQ for $\alpha = 1$ and various P_0 settings

(cf. 2a) is typically too low for selecting a high MCS and hence MCS-0 is dominating. The CDF of MCS level per allocated code word in Fig. 3a shows that for $P_0 = -120$ dBm almost 100% of the UEs use MCS-0. For $P_0 = -100$ dBm the ratio of MCS-0 utilization is 43%. For this P_0 setting still almost 100% QPSK utilization has been observed. For $P_0 = -80$ dBm the SINR is substantially higher and hence the percentage of QPSK modulation is only 50%, while further 45% of the code words have been transmitted in 16QAM. The ratio of 64QAM utilization is limited to 5%. In contrast for $P_0 = -60$ dBm a 64QAM utilization of 25% has been determined.

UE throughput is also affected by the percentage of cell resources utilized for re-transmissions. Figure 3b shows the CDF of RBs per TTI used for re-transmission of erroneous code words. For 50% of the TTIs 24 out of 50 available RBs in the cell have been occupied for re-transmissions using $P_0 = -120$ dBm. Note that the block error rate (BLER) target of 10% is not maintained. For $P_0 = -100$ dBm only 10% of the RBs (i.e. 5 RBs) are used with 50% probability for re-transmissions. This number amounts 1 RB for $P_0 = -80$ and -60 dBm, i.e. re-transmission gets more or less negligible.

The CDF of the resulting cell throughput is shown in Fig. 4a. Note that cell throughput is an indicator for spectrum efficiency and system capacity. Poor throughput (low capacity) has been obtained in the two extreme cases, $P_0 = -120$ dBm with a mean value of 1,969 kbps and $P_0 = -60$ dBm with 3,309 kbps, respectively. The major reason in the first case is the low transmission power resulting in a high number of assigned RBs but low SINR. Consequently only the most robust MCS has been applied combined with a high re-transmission rate. The reason for the low throughput obtained for $P_0 = -60$ dBm is the high transmission power per RB resulting for full PL compensation in a resource assignment mostly restricted to a single RB. Although high MCS have been selected due to the high SINR the resulting cell throughput is poor due to fractional loading. Choosing $P_0 = -80$ dBm the mean cell throughput was improved to 7,830 kbps.

The performance of individual connections has been applied as further criterion. The active session throughput is defined as the average throughput per session during active data transfer, shown in Fig. 4b.

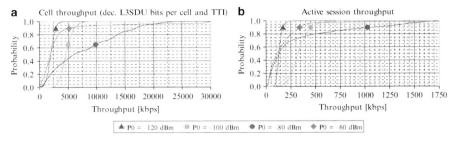

Fig. 4 CDF of (**a**) cell throughput and (**b**) end-to-end active session throughput for $\alpha = 1$ and various P_0 settings

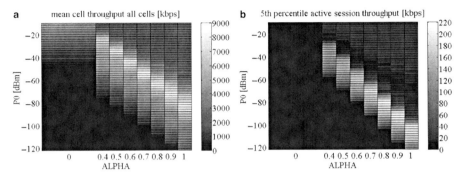

Fig. 5 (**a**) Mean cell throughput and (**b**) fifth percentile active session throughput depending on α and P_0

This analysis has shown the impact of OLPC P_0 settings on UE and cell throughput. A strong conjunction between PC, RB assignment by ATB, MCS selection handled by AMC and re-transmission rate has been observed.

4.2 Joint P_0 and α Optimization

For analyzing the impact of fractional PC ($\alpha < 1$) a two-dimensional optimization with respect to UE and cell throughput has been performed for P_0 ranging from -120 to -10 dBm and $\alpha \in \{0; 0.4; 0.5; 0.6; 0.7; 0.8; 0.9; 1\}$. Figure 5a shows the mean cell throughput, which is a measure for the network capacity. The fifth percentile active session throughput, which is used as measure for coverage, is shown in Fig. 5b.

For full PL compensation ($\alpha = 1$) the highest mean cell throughput of 8,213 kbps has been achieved for $P_0 = -82$ dBm. However, the fifth percentile active session throughput shows poor 19 kbps. The latter value reaches its maximum of 206 kbps at $P_0 = -106$ dBm. In case of FPC a maximum cell throughput of 8,738 kbps has been achieved for $\alpha = 0.7$ and $P_0 = -52$ dBm. For this combination the fifth percentile

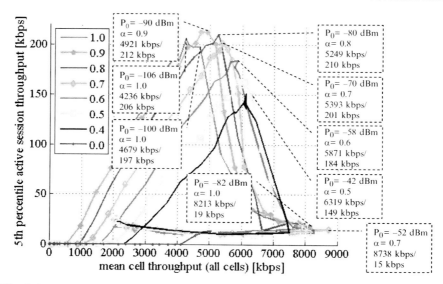

Fig. 6 Mean cell throughput vs. fifth percentile active session throughput

active session throughput reveals unacceptable 15 kbps. At constant P_0 a further decrease of α below 0.7 results in improved coverage but reduced capacity. Note that only a small range of reasonable P_0 and α settings is feasible (small stripe), which are not fully correlated for cell (cf. 5a) and UE throughput (cf. 5b).

Targeting at higher cell throughput always leads to user throughput degradation. Figure 6 shows the trade-off between capacity and coverage. For full PL compensation ($\alpha = 1$) a good compromise is achieved for $P_0 = [-106; -100]$ dBm, for FPC $P_0 = -80$ dBm and $\alpha = 0.8$ provide a good compromise.

4.3 Combined Open- and Closed-Loop Power Control

The proposed combination of OLPC settings of $P_0 = -80$ dBm and $\alpha = 0.8$ for FPC has been selected and the CL component has been enabled. The optimum CLPC target window for full load has been determined by simulations resulting in Received Signal Strength Indicator (RSSI) thresholds of -105 and -100 dBm and SINR thresholds of 2 and 5 dB. The procedure of the CL component is demonstrated in the RSSI/SINR plot in Fig. 7a. Each point characterizes a tuple of filtered RSSI and SINR values used for the decision on CLPC corrections $\{-1; 0; 1; 3\}$ dB. $\Delta_i = 0$ dB is commanded if both filtered RSSI and SINR are within the PC target window defined by the blue rectangle. Power increase by 1 dB is commanded if RSSI and/or SINR are at maximum 1 dB below the lower thresholds of the PC window shown in green color. The red cloud represents power increase by 3 dB, the

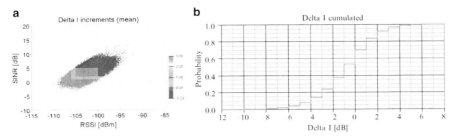

Fig. 7 (a) Closed-loop PC corrections in the RSSI/SINR plane, (b) CDF of accumulated Δ_i closed-loop transmission power corrections

violet cloud power decrease by 1 dB. The relative amount of power increase and decrease is shown in the CDF for accumulated Δ_i (i.e. sum of CLPC correction values determined at every CLPC interval) in Fig. 7b.

5 Conclusions

In this paper a detailed study of UL PC performance in LTE networks has been presented. Insight is given into the interworking between ATB, AMC, and PC. The impact of P_0 on OLPC assuming full PL compensation ($\alpha = 1$) has been analyzed thoroughly. For P_0 ranging from -106 to -100 dBm a good performance compromise in terms of capacity and coverage has been found. For lower P_0 settings both capacity and coverage degrade due to low transmission power causing poor SINR, robust MCS selection with low throughput and high re-transmission rate. An increase of P_0 beyond -100 dBm leads to increase in capacity but at the same time to drastic coverage degradations. Using even higher P_0 results in fractional loading since ATB restricts the UE RB allocation. For FPC ($\alpha < 1$) the combination of $P_0 = -80$ dBm and $\alpha = 0.8$ is recommended as a good compromise between capacity and coverage. Higher capacity has been achieved by decreasing α, however, implying poor cell border performance at the same time. The effect of the optional CL component has been investigated on top of OLPC with parameter settings ($P_0 = -80$ dBm and $\alpha = 0.8$). CLPC was based on two-dimensional optimization for RSSI/level and SINR/quality thresholds. In a fully loaded network the PC window set to RSSI $= [-105; -100$ dBm$]$ and SINR $= [2; 5$ dB$]$ provides a good compromise between capacity and coverage.

Both OLPC and CLPC components offer high capabilities to tune the UL performance to achieve highest capacity, optimized cell border performance or a trade-off of both. In combination OL and CL provide a powerful means for UL PC in upcoming UTRAN LTE networks.

References

1. 3GPP TS 36.300. Evolved Universal Terrestrial Radio Access (E-UTRA) and Evolved Universal Terrestrial Radio Access Network (E-UTRAN); Overall description (Release 8), December 2008.
2. 3GPP TS 36.211. Evolved Universal Terrestrial Radio Access (E-UTRA); Physical Channels and Modulation (Release 8), December 2008.
3. H. Holma, A. Toskala. *WCDMA for UMTS – HSPA Evolution and LTE*. Wiley, 2007.
4. C. F. Ball, T. Hindelang, I. Kambourov, S. Eder. Spectral Efficiency Assessment and Radio Performance Comparison between LTE and WiMAX. In *Proceedings of 19th Annual IEEE International Symposium on Personal, Indoor and Mobile Radio Communications (PIMRC 2008)*, Cannes, 2008.
5. C. Úbeda Castellanos, D. López Villa, C. Rosa, K. I. Pedersen, F. D. Calabrese, P.-H. Michaelsen, J. Michel. Performance of Uplink Fractional Power Control in UTRAN LTE. In *Proceedings of IEEE VTC Spring 2008*, Beijing, 2008.
6. 3GPP TS 36.213. Evolved Universal Terrestrial Radio Access (E-UTRA); Physical layer procedures (Release 8), December 2008.
7. 3GPP TS 36.101. Evolved Universal Terrestrial Radio Access (E-UTRA); User Equipment (UE) radio transmission and reception (Release 8), December 2008.
8. R1-073224. Way Forward on Power Control of PUSCH. 3GPP TSG RAN WG1 49bis, June 2007.
9. M. Boussif, N. Quintero, F. D. Calabrese, C. Rosa, J. Wigard. Interference Based Power Control Performance in LTE Uplink. In *Proceedings of International Symposium on Wireless Communication Systems 2008*, Reykjavik, 2008.
10. 3GPP TR 25.814 Annex A. Physical layer aspects for evolved Universal Terrestrial Radio Access (UTRA) (Release 7), September 2006.
11. R1-070674. LTE physical layer framework for performance verification. 3GPP TSG RAN1 48, February 2007.
12. C. F. Ball, K. Ivanov, R. Müllner. A BLER Based UL LTE AMC Featuring Fast Upgrade and Emergency Downgrade. In *Proceedings of ICT–MobileSummit 2008*, Stockholm, 2008.

Improving UMTS LTE Performance by UEP in High Order Modulation*

Helge Lüders, Andreas Minwegen, and Peter Vary

Abstract In this contribution we investigate the performance of the UMTS *Long Term Evolution* (LTE) physical layer using turbo coding and 64QAM with Gray mapping. We show how the mapping of systematic and parity bits to the six different bit positions defining one complex 64QAM symbol influences the convergence of the turbo decoder and thereby the bit error rate (BER) performance as well as number of necessary decoding iterations. Exploiting the *unequal error protection* (UEP) property of Gray mapped 64QAM results in an SNR performance gain of approximately 2 dB for the non-iterative system and in addition leads to a significant reduction of the necessary decoding iterations when iterative decoding is performed.

1 Introduction

The latest release of the UMTS LTE standard [1, 2] features a flexible physical layer employing turbo channel coding, rate-matching by adaptive puncturing or repetition of encoded bits, *hybrid automatic repeat-request* (HARQ), a choice of complex signal constellations (BPSK, QPSK, 16QAM, and 64QAM all with Gray mapping), and cyclic prefix *Orthogonal Frequency Division Multiplexing* (OFDM) with a bandwidth dependent number of subcarriers. The choice of the modulation and coding scheme for each individual user, i.e., the code rate and complex signal constellation, is left to the scheduler based on the instantaneous channel conditions and the current load of the radio cell.

In this contribution we show how significant gains in terms of necessary turbo decoding iterations can be achieved by a simple reordering of the encoded bits of

---

H. Lüders (✉), A. Minwegen, and P. Vary
Institute of Communication Systems and Data Processing
RWTH Aachen University, Germany
e-mail: {lueders|minwegen|vary}@ind.rwth-aachen.de

* This work has been supported by the Research Centre, RWTH Aachen University.

the LTE turbo coder taking into account the unequal error protection (UEP) property of Gray mapped 64QAM. Therefore, following this introduction the system model for the considered UMTS LTE physical layer and its parameters are sketched in Section 2. Section 3 analyzes the employed UEP property and introduce the proposed reordering. Simulation results demonstrating the convergence behavior and the bit error rate performance are given in Section 4.

2 System Model

The model of the considered transmission system is depicted in Fig. 1. According to the LTE standard, a block of data bits $\underline{x}$ of a given size is encoded by a systematic rate-$\frac{1}{3}$ turbo coder consisting of two parallel concatenated convolutional codes (PCCC) with octal generator polynomial $G = \{1, 15/13\}_8$ each generating one parity bit per data bit. The encoded bits are then separated into three streams: The first contains the systematic, i.e., the uncoded data bits $\underline{x}$, whereas the second and third contain the parity bits of the two constituent encoders $\underline{p}_\mathrm{I}$ and $\underline{p}_\mathrm{II}$, respectively. For an efficient and easy to implement rate matching the three streams are individually interleaved and written to a ring buffer [3]. For the sake of simplicity the subinterleavers of these streams are omitted as they do not influence the results for the selected channel model.

For a given number n of data bits a block of m encoded bits is selected for transmission resulting in an effective code rate $r = \frac{n}{m}$. The user's requested throughput and the size m of the block of encoded bits is determined by the scheduler according to the user's instantaneous channel quality, maximum delay, target BER and the current load of the radio cell. Thereby the scheduler implicitly influences the code

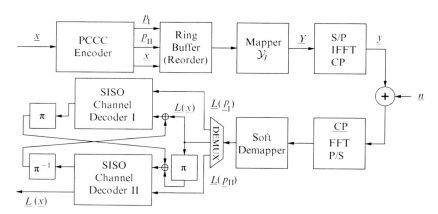

Fig. 1 Transmission system model

rate r of the user. A block size $m < 3 \cdot n$ results in a code rate $r > \frac{1}{3}$, whereas if m is sufficiently large, the code rate r can take values $r < \frac{1}{3}$ by repetition of systematic and parity bits.

Furthermore, the LTE HARQ scheme allows for up to four transmissions of different combinations of systematic and parity bits, the so-called *redundancy versions*. For the initial transmission, first, the systematic bits are selected and the remaining space of the code block of size m is then filled up with parity bits. Each following retransmission which may be requested by the receiver starts at a different position within the ring buffer, i.e., each redundancy version consists of a different combination of systematic and parity bits. Obviously, each retransmission of a code block implicitly results in a decrease of the effective code rate and directly leads to losses in throughput and latency. It is therefore highly desirable to achieve a certain target BER already after decoding of the initial transmission of a code block. For details of the rate matching algorithm the reader is referred to [2]. In the following investigations we exemplarily choose $r = \frac{1}{3}$ coding, i.e., the transmission of all systematic and parity bits obtained from the turbo coder for a fixed data block size of 6,144 bits. This corresponds to the maximum data block size in LTE systems [2]. Furthermore, only the first redundancy version, i.e., the initial transmission is considered.

The bits selected for transmission are grouped to vectors of I bits with $I \in \{1, 2, 4, 6\}$ which then are assigned to complex modulation symbols $Y \in \mathcal{Y}_I$ out of a of signal constellation symbols $\mathcal{Y}_I$, i.e. BPSK, QPSK, 16QAM, or 64QAM. In this work we consider $I = 6$ bits per modulation symbol and $\mathcal{Y}_6$ corresponding to a Gray mapped 64QAM. The complex modulation symbols $\underline{Y}$ are then OFDM modulated and a cyclic prefix (CP) is added to form the transmit signal $\underline{y}$ with averaged unit power. OFDM modulation is realized using an Inverse Fast Fourier Transform (IFFT) of size 2048 (20 MHz system bandwidth). To prove the potential of our approach we start with a complex additive white Gaussian noise (CAWGN) channel model: The signal is disturbed by noise samples $\underline{n}$ with zero mean and one-sided noise power $N_0/2$.

On the receiving side the CP is removed and OFDM demodulation is performed employing the Fast Fourier Transform (FFT). The demodulated complex symbols are fed to a soft demapper (SDM) which delivers reliability information in form of *log-likelihood ratios* (LLR) $\underline{L}(\underline{x})$, $\underline{L}(\underline{p}_I)$, and $\underline{L}(\underline{p}_{II})$ on the systematic (data) bits $\underline{x}$ and the parity bits of the two constituent encoders $\underline{p}_I$, $\underline{p}_{II}$, respectively. The LLRs are then passed on to a parallel turbo decoding structure consisting of two soft input soft output (SISO) channel decoders (CD) using the MaxLogMAP algorithm [4] for soft channel decoding. For the initial decoding step after soft demapping the LLRs are fed into CD I without using any a priori information. The obtained extrinsic information is properly interleaved and fed into CD II as a priori information. From this a priori information together with the LLRs from the soft demapper CD II generates extrinsic information for the next decoding iteration in CD I etc. After a fixed number of decoding iterations the sequence of data bits are estimated from the resulting LLRs.

3 UEP and Proposed Bit Reordering

Considering a high order modulation scheme like 64QAM with Gray mapping as given in [1] it can be observed that each bit position or bit level in the 6 bit symbol exhibits a different BER and therefore a different level of protection in an AWGN environment. This is illustrated by Fig. 2 where the mutual information $\mathcal{I} \hat{=} \mathcal{I}(X;L)$ between transmitted bit X and received LLR L as obtained from the SDM for each bit position of the 64QAM mapping is plotted versus the channel quality given as signal-to-noise ratio (SNR) E_S/N_0. The mutual information

$$\mathcal{I}(X;L) = \frac{1}{2} \cdot \sum_{x=0,1} \int_{-\infty}^{+\infty} p_L(\xi|X=x) \cdot \mathrm{ld} \frac{2 \cdot p_L(\xi|X=x)}{p_L(\xi|X=0) + p_L(\xi|X=1)} \mathrm{d}\xi$$

with $\quad 0 \leq \mathcal{I} \leq 1$

as defined in [5] uses distributions p_L determined in simulations and serves here as a measure of the protection of each bit level.

It is observed that for the six bits (Bit $0, \ldots, 5$) of each Gray mapped 64QAM symbol there exist four different levels of protection, i.e., two groups consisting of two bits (first group: Bits 2 and 3, second group: Bits 4 and 5, cf. Fig. 2) exhibit the same error protection properties while the best protection against bit errors at low channel qualities ($E_S/N_0 < 7$ dB) is provided by Bits 0 and 1. This behavior can be interpreted as an unequal error protection property inherent in 64QAM with Gray mapping.

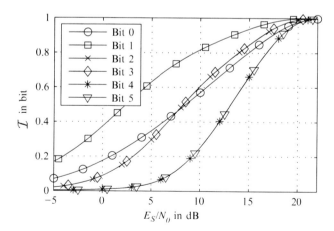

Fig. 2 Mutual information $\mathcal{I}$ between transmitted bits and received LLRs versus channel quality for the different bit levels of 64QAM with Gray mapping

There are various ways of making use of UEP in communication systems. In this contribution we exploit it for unequally protecting systematic and parity bits. The employment of UEP for systematic and parity bits of a turbo coded system has been shown, e.g., in [6] where a faster convergence of a joint turbo decoding and channel estimation process is observed if systematic bits get stronger protection against errors than parity bits. This directly leads to gains in the BER performance in case of a fixed number of turbo decoding and channel estimation iterations. In LTE, the unequal error protection property present in the complex signal constellations is not taken into account. This motivates us to propose the following approach to exploit this UEP property: By a simple reordering of the encoded bits, systematic bits are placed in the well protected positions of the complex signal constellation symbol while the parity bits are asymmetrically placed in the less protected positions.

This scheme is depicted in Fig. 3: With rate-$\frac{1}{3}$ coding the data bits are encoded to become the systematic bits $\underline{x}$ with their corresponding parity bits $\underline{p}_\mathrm{I}$ and $\underline{p}_\mathrm{II}$ (cf. Fig. 1). In the standard LTE system these bits are sequentially read out of the ring buffer and mapped to the 6 bit positions of the 64QAM symbol resulting in complex modulation symbols that either consist of 6 systematic bits x or of 6 parity bits p_I and p_II (with the possible exception of one symbol consisting of both types of bits). In our proposed reordering, the bits are mapped according to the levels of protection illustrated by Fig. 2: The systematic bits $\underline{x}$ are mapped to Bits 0 and 1 of the constellation symbol, the parity bits $\underline{p}_\mathrm{I}$ of the first constituent encoder of the PCCC are mapped to Bits 2 and 3 respectively, leaving Bits 4 and 5 for the parity bits $\underline{p}_\mathrm{II}$ of the second constituent encoder. In that way the systematic bits obtain the highest level of protection while the parity bits $\underline{p}_\mathrm{II}$ are most error-prone. The implications of this UEP scheme are explained in the following section.

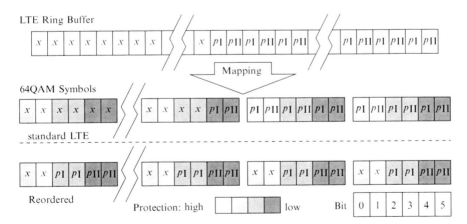

Fig. 3 Standard LTE and reordered mapping of encoded bits to complex 64QAM symbols

4 Simulation Results

To illustrate the impact of our modified mapping scheme we first depict in Fig. 4 the average mutual information $\mathcal{I}$ obtained by the systematic and parity bits of both constituent encoders and compare it to the average mutual information of all coded bits obtained for the standard LTE system. Obviously, the average of the mutual information obtained by systematic and parity bits of the proposed scheme equals the average mutual information obtained for standard LTE. Furthermore, the average mutual information of p_I corresponds to the curves for Bit 2 and 3 of the 64QAM symbol while the average mutual information of p_{II} corresponds to the curves for Bit 4 and 5 (cf. Fig. 2). The quintessence in Fig. 4, however, is the significance of the unequal error protection of systematic and the two different types of parity bits in the proposed scheme.

This fact is illustrated even more vividly by the Extrinsic Information Transfer (EXIT) characteristics [5] which describe the amount of extrinsic information $\mathcal{I}^{[ext]}$ obtained from a SISO decoding unit given a certain amount of a priori information $\mathcal{I}^{[apri]}$. The EXIT characteristics of the two component SISO channel decoders exchanging extrinsic information in a turbo decoding process in one EXIT chart give a distinct impression of the convergence behavior of this iterative decoding process. For a detailed description of EXIT analysis the reader is referred to [5].

Figure 5 compares the characteristics of SISO CD I and SISO CD II obtained from simulations of the proposed system to those obtained from simulations of standard LTE under different channel conditions. The difference is evident: While in the standard LTE case in Fig. 5a the EXIT characteristics of both SISO CDs show the expected symmetric behavior, the proposed reordering of bits leads to highly asymmetric characteristics, i.e., SISO CD I and SISO CD II yield different

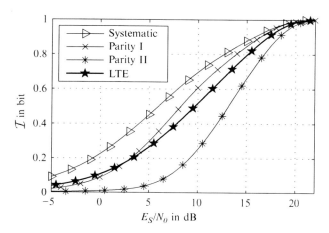

Fig. 4 Average mutual information $\mathcal{I}$ for transmitted systematic and parity bits using the proposed scheme compared to the average mutual information for all transmitted bits in standard LTE versus channel quality

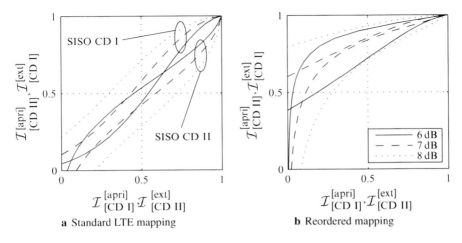

Fig. 5 EXIT characteristics of the SISO component decoders for different channel qualities

transfer functions leading to a different convergence behavior (cf. Fig. 5b). The higher protection of its parity bits $\underline{p}_\mathrm{I}$ enables SISO CD I to produce a high amount of extrinsic information $\mathcal{I}^{[\mathrm{ext}]}_{[\mathrm{CD\ I}]}$ in the initial decoding step, i.e., for $\mathcal{I}^{[\mathrm{apri}]}_{[\mathrm{CD\ I}]} = 0$. Contrary, SISO CD II suffers from its less protected parity bits $\underline{p}_\mathrm{II}$ which results in a poor performance, i.e., $\mathcal{I}^{[\mathrm{ext}]}_{[\mathrm{CD\ II}]}(\mathcal{I}^{[\mathrm{apri}]}_{[\mathrm{CD\ II}]} = 0)$ is very low for the initial decoding step. However, such an initial decoding step without a priori information is never executed by the second decoder in a parallel turbo decoding structure: SISO CD II will always perform its first decoding step with a priori knowledge obtained from the initial decoding of SISO CD I.

Recording and averaging the mutual information during simulations of both systems for a SNR of 7.5 dB enables the plotting of the staircase-like decoding trajectories depicted in Fig. 6. These trajectories confirm the above statements for the depicted case: As the trajectory of the system with reordered bit mapping in Fig. 6b reaches further to the right after the first complete turbo decoding iteration, i.e., the first stair step, compared to the standard LTE system in Fig. 6a, a better BER performance for the non-iterative case can be predicted. Furthermore, while in case of standard LTE approximately five turbo decoding iterations are necessary to reach a residual bit error rate of 10^{-3} depicted by the dashed line in both subplots, only three such iterations suffice in the modified system. This behavior indicates a faster convergence speed of the system with reordered mapping.

Both predictions, the better performance in the non-iterative case, as well as the faster convergence are backed by the simulation results given in Fig. 7: For the non-iterative case depicted in Fig. 7a the proposed mapping yields an SNR gain of approximately 2 dB over the standard LTE system. For the iterative system on the other hand, the BER performance of standard LTE with nine turbo decoding iterations is already reached or even outperformed after five iterations with the proposed

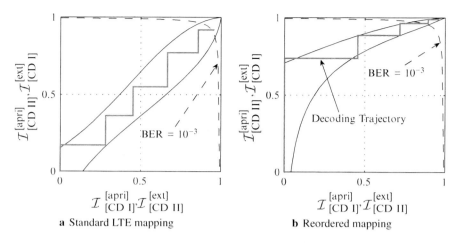

Fig. 6 EXIT characteristics of the SISO component decoders and decoding trajectory for $E_S/N_0 = 7.5\,\mathrm{dB}$

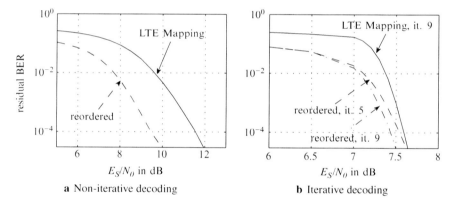

Fig. 7 Residual BER for standard LTE mapping and the proposed reordered mapping with non-iterative and iterative decoding, block length: 6,144 data bits

reordering, cf. Fig. 7b. Note that we do not claim a superiority in terms of residual BER of either one of the systems after convergence.

So far we used the maximum allowed block length of 6,144 data bits for all simulations. However, it is a well known fact, that the code block length influences the performance of a turbo coded system: For small code block lengths the system converges earlier than predicted by the intersection of the decoder characteristics in the EXIT chart which leads to a degradation in BER performance. That is why we give an example for the BER performance of a system with a block length of 280 data bits in Fig. 8. For the non-iterative case depicted in Fig. 8a the same SNR gain

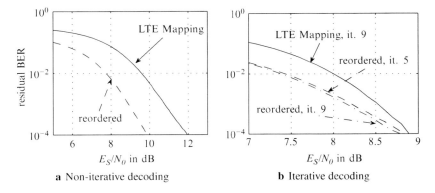

Fig. 8 Residual BER for standard LTE mapping and the proposed reordered mapping with non-iterative and iterative decoding, block length: 280 data bits

between standard LTE mapping and the proposed reordered mapping is observed although the absolute performance decreases slightly. As expected, the BER performance of both systems with iterative decoding is reduced due to the smaller code block length. Therefore, the advantage in the convergence behavior of the modified system appears smaller, but is nevertheless still visible.

5 Conclusion

In this contribution we illustrated the employment of UEP inherent in 64QAM Gray signalling by slightly modifying the mapping of coded bits to complex 64QAM symbols in a UMTS LTE system. It is shown that for rate-$\frac{1}{3}$ coding SNR gains of approximately 2 dB can be achieved for non-iterative decoding while a significant reduction of turbo decoding iterations (44 % in the given example) can be obtained for iterative systems. Basis is the unequal error protection of the parity bits of the two constituent encoders employed in LTE leading to an advantageous convergence behavior which has been analyzed using EXIT charts. Simulations have been conducted using an CAWGN channel model. Complex channel models like flat fading Rayleigh channels or time-variant frequency selective fading channels are currently considered.

Acknowledgment The authors wish to thank Laurent Schmalen for providing the software basis of the simulation tool used for this work.

References

1. 3GPP TS 36.211. Evolved Terrestrial Radio Access (E-UTRA); Physical Channels and Modulation. Version 8.2.0, 3GPP Technical Specification, 2008.
2. 3GPP TS 36.212. Evolved Terrestrial Radio Access (E-UTRA); Multiplexing and Channel Coding. Version 8.2.0, 3GPP Technical Specification, 2008.
3. J. Cheng, A. Nimbalker, Y. Blankenship, B. Classon, and T. Blankenship. Analysis of Circular Buffer Rate Matching for LTE Turbo Code. In *Proceedings of IEEE Vehicular Technology Conference (VTC-Fall)*, Calgary, Canada, Sept. 2008.
4. P. Robertson, E. Villebrun, and P. Hoeher. A Comparison of Optimal and Sub-Optimal MAP Decoding Algorithms Operating in the Log Domain. In *Proc. IEEE Int. Conf. Commun. (ICC)*, pages 1009–1013, Seattle, USA, 1995.
5. S. ten Brink. Convergence Behaviour of Iteratively Decoded Parallel Concatenated Codes. *IEEE Transactions on Communications*, October 2001.
6. S. Godtmann, H. Lüders, G. Ascheid, and P. Vary. A Bit-Mapping Strategy for Joint Iterative Channel Estimation and Turbo-Decoding. In *Proceedings of IEEE Vehicular Technology Conference (VTC-Fall)*, Calgary, Canada, Sept. 2008.

Performance Evaluation of a Low-Complexity LTE Base Station Receiver

Xiaoyin Zhao and Andreas Ibing

Abstract We describe algorithms for the LTE uplink which implement the functionalities of random access preamble detection, time and frequency synchronisation, channel estimation and equalization (including MIMO) and error correction decoding. Evaluation is done by means of preamble detection and false detection rates, error vector magnitude and bit error rates as well as packet error rates after turbo decoding.

1 Introduction

LTE uplink uses DFT-spread OFDMA transmission, which in comparison to OFDMA reduces peak to average power ratio and allows for more efficient power amplifiers at the terminal [1]. Multi-user MIMO (MU-MIMO) as well as single-user MIMO (SU-MIMO) are supported [2]. For entering the system by non-synchronized transmission, a random access channel (RACH) is available. Resource scheduling is done with a granularity of one resource block, which spans 12 subcarriers times 7 OFDM symbols [3]. Forward error correction is done using a rate 1/3 turbo code.

We present an evaluation of a low-complexity uplink signal processing and decoding chain by describing and evaluating individual processing blocks as well as the performance of the complete chain in different scenarios. The focus of the evaluation is on MU-MIMO, which compared to SU-MIMO has the additional task of dealing with different carrier frequency offsets (CFOs) on the same subcarrier (each terminal has its own CFO relative to the base station).

X. Zhao (✉) and A. Ibing
Fraunhofer Institute for Telecommunications, Heinrich-Hertz-Institut, Einsinufer 37,
10587 Berlin, Germany
e-mail: {xiaoyin.zhao,andreas.ibing}@hhi.fraunhofer.de

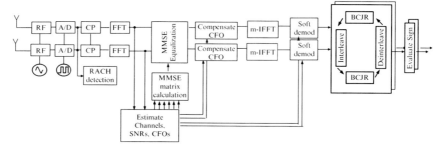

Fig. 1 Receiver overview

2 Receiver Overview

Figure 1 gives an overview of the receiver. Detection of RACH preambles is done on the time domain sample stream before OFDM cyclic prefix removal and the receiver FFTs. After the FFTs, the known pilot symbol sequences are used to estimate channels, SNRs, carrier frequency offsets (CFOs) and fine timing offsets. Estimated channels and SNRs are used to compute MIMO equalization matrices, CFOs are compensated after the separation of data streams to support MU-MIMO (different terminals have different CFOs). After DFT despreading with adequate IFFT size, symbols are soft demodulated taking into account the noise variance after equalization (output are log-likelihood ratios, LLRs). LLRs are iteratively transformed into LLRs with lower error rate by turbo decoding, before the information bits are output by evaluating the signs of the LLRs. Simulations are done for FFT size 2048 (20 MHz, 1,200 subcarriers).

3 Preamble Detection and Timing Advance

For detection of RACH preambles we use a large FFT of the time domain samples in the RACH window, followed by elementwise multiplication with the stored FFT of a preamble sequence [1]. The resulting vector is IFFT transformed, followed by peak to average detection (comparison to a threshold value) [4]. If a peak is detected, the round-trip delay estimate is given by the peak position and phase. This computation of correlation in frequency domain is enabled by the random access burst cyclic prefix (which is much longer than the OFDM cyclic prefix). The structure of the RACH in time and frequency is illustrated in Fig. 2, the timing of an unsynchronized random access burst at the base station is depicted in Fig. 3. Evaluation of preamble miss rate and false detection rate show that a threshold value of the peak-to-average ratio of around 20 enables both no preamble misses and no false detections in the receiver operating SNR region (Figs. 4 and 5). After detection of an initial ranging preamble, the base station protocol stack instructs the terminal to adjust its clock to compensate the round-trip delay (timing advance).

Fig. 2 Position of random access channel in time and frequency. A random access burst is much longer than an OFDM symbol, to allow for several kilometers of round-trip delay. A 1 ms subframe contains 14 OFDM symbols

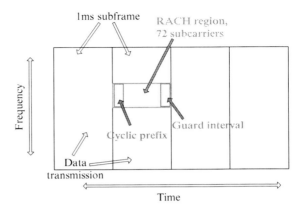

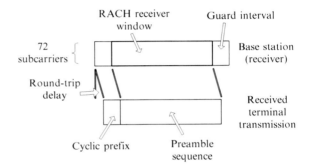

Fig. 3 Unsynchronized random access burst

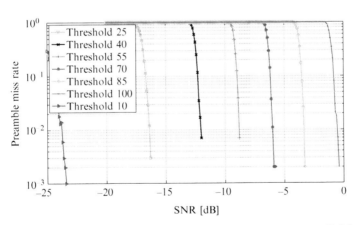

Fig. 4 Preamble miss rate in dependence on threshold (AWGN channel). For negligible miss rate in the operating SNR region the threshold should be smaller than around 40

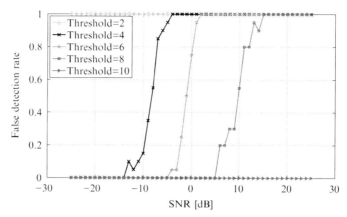

Fig. 5 Preamble false detection rate in dependence on threshold (AWGN channel). For negligible false detection rate the threshold should be larger than 10

4 Channel Estimation

Channel coefficients as well as noise and interference level are estimated to enable (MIMO) equalization. Noise and interference level are also used to compute reliability information in soft demodulation (computation of LLRs).

We estimate channels per physical resource block. The received pilot symbols in frequency domain are first divided through the known transmitted pilot symbols to obtain initial raw channel estimates (noisy).

The pilot structure in frequency domain which we assume for MIMO is illustrated if Fig. 6. To enable independent channel estimation for different transmit-receive antenna pairs in frequency domain, the pilot positions are alternating over subcarriers (similar to the downlink MIMO pilot structure). Note that the inserted zeros differ from the specification in [3].

Signal to noise ratio (SNR) is estimated using a subspace method as in [5]. Interference is implicitly treated as noise. Initial channel estimates are projected onto the noise subspace using a precomputed singular value decomposition [6].

Channel coefficients per transmit-receive antenna pair and subcarrier are estimated by filtering the noisy initial estimates using a one-dimensional Wiener Filter [7] in frequency direction. Filter matrices are precomputed for different SNR levels, using robust channel reference statistics [8]. The Wiener Filter makes use of channel power delay profile statistics and SNR level chosen at design time. We use a different precomputed filter matrix for each 5 dB SNR range.

Unlike in the downlink, tracking loops for frequency and time as described e.g. in [9] are not possible, because different terminals have different timing errors and different frequency offsets (which we treat in Sections 5 and 6). For the same reason we also do not apply 2D channel estimation (frequency and time direction) over neighbouring resource blocks: the correlation of pilots in different slots also depends on the frequency offset.

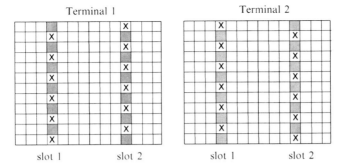

Fig. 6 Assumed pilot structure for MIMO transmission. Note: this structure differs from [3]

Fig. 7 Estimation of carrier frequency offset

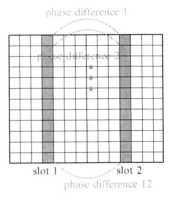

The estimated channel coefficients (after Wiener Filtering) are used to estimate carrier frequency offset (individually per terminal). CFO leads to (apart from inter-carrier interference) common phase error of symbols over different subcarriers. We compute CFO as average phase increase between the pilots of neighbouring slots (depicted in Fig. 7). This is applicable when intra-subframe frequency hopping [3] is not used.

5 Discontinuous Timing Advance Tracking

For only one transmitter like in downlink, timing errors could be tracked and compensated at the receiver by estimating FFT window position offset (using the estimated channel) and following adjustment of FFT window position [9].

But in the uplink, different terminals transmit at the same time while having individual timing errors. In addition, individual terminals are not scheduled in all subframes (and they may move of course, changing the round-trip delay). Initial timing advance is performed after RACH preamble detection. After a certain time (synchronisation timer), a terminal switches to unsynchronised mode. In case of transmission, it then has to transmit a RACH preamble again. To reduce RACH overhead (number of transmitted preambles), we estimate the remaining (uncompensated) round-trip delay at the base station (using the estimated channel frequency response) at each uplink transmission as in [10]. The terminal synchronisation timer can be reset by a MAC information element as part of a (downlink) MAC layer packet, timing advance adjustment is possible with a granularity of 16 samples [3]. This discontinuous timing advance tracking is applied for terminals with more frequent transmission. The required accuracy of timing error estimation using the pilot symbols can be achieved for terminals transmitting more than around 10 PRB in 1 s [10].

6 MIMO Equalization and Frequency Offset Compensation

For only one transmitter like in downlink, the frequency offset could be compensated at the receiver in time domain, reducing inter-carrier interference [9]. But different terminals (transmitting at the same time in uplink) have different frequency offsets, so that we compensate CFOs in frequency domain after the spatial separation (MIMO equalization) of the data streams.

For MIMO equalization we use the unbiased MMSE filter, which is given by

$$G_{unb} = S(H^H H + \sigma^2 I)^{-1} H^H,$$

where S is a diagonal scaling matrix which removes the MMSE bias (scaling does not change SNR). The equalization matrix is used for one subcarrier for the duration of one slot (seven symbols, six of them being data symbols to be equalized).

The error vector magnitude (average squared distance of symbols from their ideal locations in the complex plane) after equalization is illustrated in Fig. 8 for different noise levels and velocities (Doppler spread, Jakes Spectrum).

CFO compensation is performed individually per data stream after spatial separation, to stop modulation set rotation. Data symbol phases are rotated according to estimated CFO, so that the reference phase is that of the middle symbol in the slot (where the equalizer matrix was computed). Uncoded bit error rates without CFO and with two different CFOs (two terminals) and compensation are illustrated in Fig. 9 for different bandwidths.

Data symbols are then soft demodulated to generated LLRs for the decoder. The max-log approximation is used, which only considers the Euclidean distances to the closest two symbol candidates:

$$L = -\frac{1}{\sigma_{eq}^2} \left(min_{s \in S_1} ||\mathbf{y} - \mathbf{s}||^2 - min_{s \in S_0} ||\mathbf{y} - \mathbf{s}||^2 \right)$$

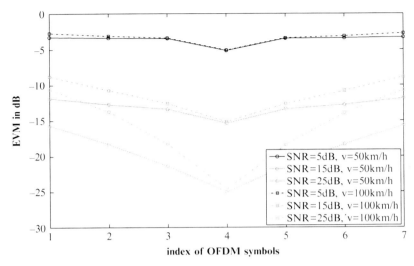

Fig. 8 Error vector magnitude for equalization of one resource block for different noise levels and terminal velocities (at 2.6 GHz). Pilot symbols are transmitted in the fourth of 7 OFDM symbols (middle OFDM symbol)

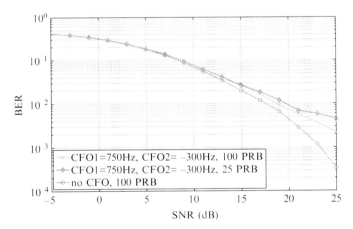

Fig. 9 Uncoded bit error rates without CFO and with compensation of different CFOs (two terminals) for different transmission bandwidths

Considering that the LTE modulation sets are separable, implementation reduces to independent one-dimensional table lookups for inphase and quadrature bits. For LLR scaling, the noise variance after equalization (per symbol) is used.

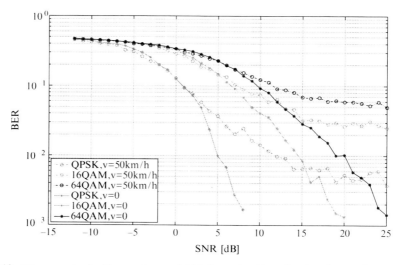

Fig. 10 Bit error rates for the worst case of 40 information bit packet (smallest packet size), for different terminal velocities and urban micro channel

7 Turbo Decoding

The turbo decoder consists of two (identical) 8-state soft-in soft-out convolutional decoders (we use the max-log BCJR implementation) and turbo (de)interleaver. The decoders iteratively recompute the LLRs based on the known code structure and extrinsic information exchange with the respective other decoder. The pseudo-random (de)interleaver between the two decoders is based on quadratic polynomial permutation (QPP) [11]. For our evaluation we run eight iterations.

Figure 10 shows the resulting bit error rates (BER) after turbo decoding for QPSK, 16QAM and 64QAM for the worst case of 40 bit packets (smallest packet size) for different terminal velocities, using an urban micro channel model [12]. Figure 11 shows frame error rates (FER) for packet size of 512 information bits for all modulations and different velocities. Scheduling normally targets a specific frame error rate (like e.g. 10%) by choosing modulation level and transmission mode, the remaining packet errors are dealt with by (hybrid) automatic repeat request (HARQ).

8 Discussion

We presented the evaluation of an LTE base station receiver, which despite its low complexity achieves reasonable performance. The pilot symbol pattern we assumed differs from [3], to enable low-complexity channel estimation in frequency domain. The evaluation focused on MU-MIMO and showed that MU-MIMO reception with

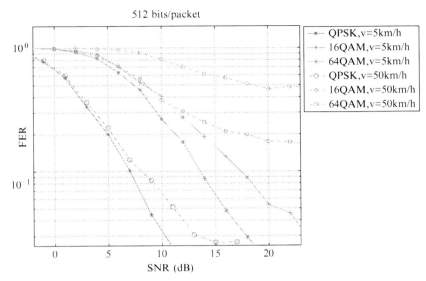

Fig. 11 Frame error rates for different modulations and terminal velocities, packet size 512 information bits

this receiver is very sensitive to Doppler spread caused by terminal velocity. Performance improvement opportunities include more sophisticated channel estimation and/or MIMO detection (e.g. MMSE-SIC as applied in [13]). For higher terminal velocity the transmission/reception mode falls back to SIMO transmission (scheduler decision).

References

1. E. Dahlman, S. Parkvall, J. Sköld, and P. Beming. *3G Evolution: HSPA and LTE for Mobile Broadband*. Academic Press, 2007.
2. 3GPP. *3GPP TSG RAN: TS36.300v8.7.0 E-UTRA and E-UTRAN; Overall Description; Stage 2*, Jan 2009.
3. 3GPP. *3GPP TSG RAN: TS36.211v8.5.0 Physical Channels and Modulation*, Dec. 2008.
4. Y. Wen, W. Huang, and Z. Zhang. Cazac sequence and its application in LTE random access. In *Information Theory Workshop*, 2006.
5. Stefan Schiffermuller and Volker Jungnickel. Practical channel interpolation for OFDMA. In *Proceedings IEEE Global Telecommunications Conference (GLOBECOM 2006)*, San Francisco, California, November 2006.
6. O. Edfors, M. Sandell, S. van de Beek, S. Wilson, and P. Börjesson. OFDM channel estimation by singular value decomposition. *IEEE Transactions on Communications*, 46(7):931–939, July 1998.
7. K. Fazel and S. Kaiser. *Multi-Carrier and Spread Spectrum Systems: From OFDM and MC-CDMA to LTE and WiMAX*. Wiley, Nov. 2008.
8. Y. Li, L. Cimini, and N. Sollenberger. Robust channel estimation for OFDMsystems with rapid dispersive fading channels. *IEEE Transactions on Communications*, 46(7):902–915, July 1998.

9. Konstantinos Manolakis, Andreas Ibing, and Volker Jungnickel. Performance evaluation of a 3GPP LTE terminal receiver. In *European Wireless Conference (EW2008)*, June 2008.
10. A. Ibing, X. Zhao, and D. Kühling. Discontinuous timing advance tracking for OFDMA uplink. In *International OFDM Workshop*, 2008.
11. 3GPP. *3GPP TSG RAN: TS36.212v8.4.0 E-UTRA; Multiplexing and Channel Coding*, Sept. 2008.
12. 3GPP. *3GPP TR25.814v7.1.0 Physical Layer Aspects for Evolved UTRA (Release 7)*, September 2006.
13. M. Ohm. SIC receiver in a mobile MIMO-OFDM system with optimization for HARQ operation. In *International OFDM Workshop*, 2008.

Part V
Peak-to-Average Power Ratio (PAPR)

On DT-CWT Based OFDM: PAPR Analysis

Mohamed Hussien Mohamed Nerma, Nidal S. Kamel,
and Varun Jeoti Jagadish

Abstract High peak - to - average power ratio (PAPR) of transmitted signals is a major shortcoming for multicarrier modulation system. This paper introduces a novel OFDM system based on dual–tree complex wavelet transform (DT-CWT). In the proposed scheme, DT-CWT is used in place of fast Fourier transform (FFT). The complementary cumulative distribution function (CCDF) of PAPR for the proposed scheme signal achieves about 3 dB improvement in PAPR over the traditional OFDM and wavelet packet modulation (WPM) signals at 0.1% of CCDF at the cost of acceptable increase in computational complexity without using any pruning techniques.

1 Introduction

OFDM and WPM have emerged as efficient multicarrier modulation schemes over wireless multipath communication channels. However, a major drawback in the signals of these two systems is their large envelope function. This problem is quantified by the peak - to - average power ratio (PAPR). From the central limit theorem (CLT) [1], this causes the OFDM and WPM signals to have complex Gaussian process behavior and the instantaneous power is chi-square distributed. Various schemes have been developed to reduce the PAPR [2, 3], and WPM [4, 5]. In this paper, we develop a DT-CWT based OFDM system and investigate its PAPR.

1.1 Wavelet Modulation

Wavelet transform (WT) is used as a modulation technique in many communication fields including multicarrier modulation (MCM) and wireless communication [6].

M.H.M. Nerma (✉), N.S. Kamel, and V.J. Jagadish
Electrical and Electronic Engineering Department, University Technology PETRONAS, Bandar Seri Iskandar, Tronho, 31750 Perak, Malaysia
e-mail: mohamed_hussien@utp.edu.my, {nidal_kamel, varun_jetoi}@petronas.com.my

The conventional multicarrier systems are DFT based systems [7]. However, DFT based MCM systems suffer from the high side lobes due to rectangular shaped DFT window and also these systems waste precious bandwidth due to the redundant cyclic prefix (CP). WT based MCM systems, can help mitigate these problems. The wavelet filters possess the advantages of having greater side-lobe attenuation and requires no CP [8]. Using wavelet packet transform (WPT) as basis functions, the sensitivity to multipath channel distortion and synchronization can be reduced as compared to OFDM and the system performance of WOFDM with reference to ISI, ICI and SNR is shown to be far better than the conventional DFT-OFDM [9,10].

1.2 Peak-to-Average Power Ratio (PAPR)

The PAPR of the baseband transmitted signal x[n] is expressed as [11]:

$$PAPR = \frac{\max |x[n]^2|}{E|x[n]^2|} \quad (1)$$

where E{.} denotes the average. The PAPR performance of OFDM systems is demonstrated using what is called the complementary cumulative distribution function (CCDF) of PAPR. Given the reference level $PAPR_0 > 0$, the probability of a PAPR being higher than the reference value is the CCDF expressed as follows [12]:

$$CCDF(PAPR_0) = PrPAPR > PAPR_0 \quad (2)$$

To reduce the PAPR in OFDM and WPM systems, several techniques have been proposed [13]. In this work, however, we propose a novel DTCWT based OFDM system and analyze the PAPR performance of the same. It is shown that the PAPR performance of the proposed system is superior to other systems without having to use any PAPR reduction technique.

2 The Dual-Tree Complex Wavelet Transform (DT-CWT)

Since the early 1990s the WT and WPT have been widely used in wireless communication [14]. A number of modulation schemes based on wavelets have been proposed [10, 15–22]. Kingsbury [23–27] introduced and made a complete description of DT-CWT. The DT-CWT employs two real discrete WT (DWT). This transform uses the pair of the filters ($h_0(n)$, $h_1(n)$ the low-pass/high-pass filter pair for the upper FB respectively) and ($g_0(n)$, $g_1(n)$ the low-pass/high-pass filter pair for the lower FB respectively) that are used to define the sequence of wavelet function $\psi(t)$ and scaling function $\phi(t)$ as follows

Fig. 1 The dual tree discrete CWT (DT-DCWT) Analysis (demodulation) FB

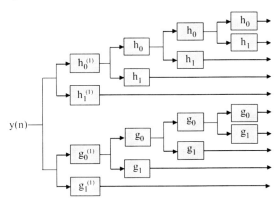

Fig. 2 The Inverse dual tree discrete CWT (IDT-DCWT) Synthesis (modulation) FB

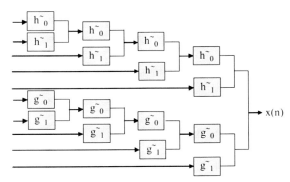

$$\psi_h(t) = \sqrt{2} \sum_n h_1(n)\phi_h(2t) \quad (3)$$

$$\phi_h(t) = \sqrt{2} \sum_n h_0(n)\phi_h(2t) \quad (4)$$

To satisfy the perfect reconstruction (PR) conditions, the filters are designed so that $\psi_g(t)$ is approximately the Hilbert transform of $\psi_h(t)$.

$$\psi_g(t) = H\psi_h(t) \quad (5)$$

The analysis and the synthesis FBs used to implement the DT-CWT and their inverses are illustrated in Figs. 1 and 2 respectively.

3 OFDM and WPM Systems

The WPM system has a higher spectral efficiency and provides robustness with regard to inter-channel interference than the conventional OFDM system. Moreover WPM is able to decompose time-frequency plane flexibly by arranging FB constructions [16]. WPM system does not require CP. According to the IEEE broadband wireless standard 802.16.3, avoiding CP gives wavelet OFDM an advantage of roughly 20% in bandwidth efficiency. Moreover as pilot tones are not necessary for wavelet based OFDM system, this gives wavelet based OFDM system another 8% advantage over typical OFDM implementations [30]. We expect the OFDM system based on DT-CWT will take all the advantages of WPM system. However, a major problem of the common discrete WPT (DWPT) is its lack of shift invariance [15]. To overcome the problem of shift dependence, one possible approach is to simply omit the sub-sampling causing the shift dependence. Techniques that omit or partially omit sub-sampling are also known as cycle spinning, oversampled FBs or undecimated WTs. However, these transforms are redundant [28], which is not desirable in multicarrier modulation. As an alternative, we used a non-redundant WT called DT-CWT that achieves approximate shift invariance [29]. This transform yields to complex wavelet coefficients that modulate the data stream in the same way that WPM do.

3.1 OFDM Based on DT-CWT

Similar to the conventional OFDM and WPM systems, a functional block diagram of OFDM based on DT-CWT is shown in Fig. 3. The inverse DT-CWT (IDT-CWT) and DT-CWT blocks are used in place of inverse FFT (IFFT) and FFT blocks respectively. Data to be transmitted are typically in the form of a serial data stream. PSK or QAM modulations can be implemented in the proposed system the choice depends on various factors like the bit rate and sensitivity to errors. The transmitter accepts modulated data (in this paper we use 16 QAM). This stream is passed through a serial to parallel (S/P) converter, giving N lower bit rate data stream, and then this stream is modulated through an IDT-CWT matrix realized by an N-band

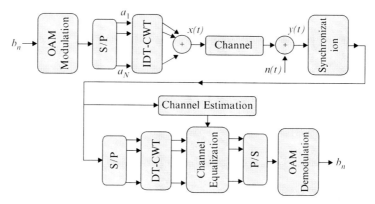

Fig. 3 DT-CWT modulation (DT-CWTM) functional block diagram

synthesis FB. Before the receiver can demodulate the subcarriers, it has to perform the synchronization. For the proposed system, known data interleaved among unknown data are used for channel estimation. Then, the signal is down sampled by N and demodulated using elements of the DT-CWT matrix realized by an N-band analysis FB. The signal is equalized after DT-CWT stage. IDT-CWT works in a similar fashion as IFFT or IDWPT. It takes as the input QAM symbols and outputs them in parallel time-frequency "subcarriers". In Fig. 2 as the synthesis process, it can be shown that the transmitted signal, x[n] is constructed as the sum of N waveform $\psi[k]$ individually modulated with the QAM or PSK symbols [29] as follows:

$$x[n] = \sum_{i} \sum_{j=0}^{N-1} a_{i,j} \psi_j [n - iN] \qquad (6)$$

where $a_{i,j}$ is a constellation encoded i-th data symbol modulating the j-th DT-CWT function. The IDT-CWT synthesis is discrete representation of the transmitted signal as sum of N waveforms shifted in time that embed information about data symbols. Those waveforms are built by j successive iterations of $\hat{h}_0$, $\hat{h}_1$ and $\hat{g}_0$, $\hat{g}_1$. The DT-CWT at the receiver recovers the transmitted symbols $a_{i,j}$ through the analysis formula exploiting orthogonality properties of DT-CWT and schematically represented in Fig. 1. In the baseband equivalent OFDM transmitter with m frame of N QAM symbols, a_k^m, k = 0, 1, ..., N − 1, the OFDM frame is given by

$$x^m[n] = \sum_{k=0}^{N-1} a_k^m e^{j2\pi nk/N} \qquad (7)$$

While for WPM system, the transmitted signal x[n] is constructed as the sum of N wavelet packet function $\varphi_j[n]$ individually modulated with the QAM symbols as the case in this paper.

$$x[n] = \sum_{i} \sum_{j=0}^{M-1} a_{i,j} \varphi_j [n - iM] \qquad (8)$$

In order to achieve fair comparisons, same simulation parameters are used. The simulations were carried out for conventional OFDM with a 64 subcarriers using a 16 QAM modulation. The Daubechies-1 (Daub-1) wavelet packet bases were used to construct the wavelet packet trees in WPM system. For the proposed scheme, near symmetric 13,19 tap filters and quarter sample shift orthogonal 14 tap filters were used to construct the real and the imaginary part of DT-CWT respectively. From the CLT, x[n] is complex Gaussian distributed, and the sequence x[n] has high PAPR. Furthermore, to demonstrate the similarities between power spectrum density (PSD) characteristic of conventional OFDM, WPM and the proposed scheme, the simulated PSD characteristics are presented in Fig. 4. This figure shows that the proposed scheme performs better than the other two systems.

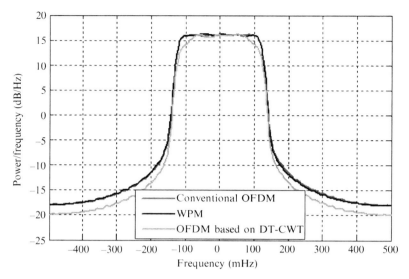

Fig. 4 PSD of the Conventional OFDM, WPM and OFDM based on DT-CWT

4 PAPR in OFDM Based on DT-CWT

In order to analyze PAPR, we generate the transmitted waveforms using 16 QAM modulation with 64 subcarriers for all these systems. Figure 5 shows, in the TD, the envelope of the proposed system. For the comparison, we also plotted the envelope of the conventional OFDM and WPM waveforms corresponding to the same information symbol pattern. The transmitted envelopes for the conventional OFDM and WPM systems illustrate approximately similar behavior, where the peak is about 2.25, while the transmitted envelope for the proposed system demonstrates better behavior than the other two systems, where the peak is only about 1.25 and this is the reason for that the proposed system gives better result for PAPR than the other two systems. The results for PAPR are best quantified using CCDF. In Fig. 6, for 64 subcarriers with 16 QAM modulation, the CCDF plots for the proposed system, conventional OFDM and WPM system are shown. This figure shows that the OFDM based on DT-CWT offers the best PAPR performance without using any reduction techniques. The proposed scheme signal achieves about 3 dB improvement in PAPR over the traditional OFDM and WPM signals at 0.1% of CCDF while the other two systems, are approximately given same results of PAPR. The simulation results for PAPR were also repeated with 16 QAM modulation and 64 subcarriers (Fig. 7) using different set of filters. OFDM DT-CWT$_1$ illustrate the PAPR when using near-symmetric (n-sym) 13, 19 tap filters in the first stage of the FB and quarter sample shift orthogonal (q-sh) 14 tap filters in the succeeding stages, OFDM DT-CWT$_2$ using (n-sym 13, 19 with q-sha 10 (10 non zero taps) filters), OFDM DT-CWT$_3$ using (antonini (anto) 9, 7 tap filters with q-sha 10 (only 6 non zero taps) filters), OFDM

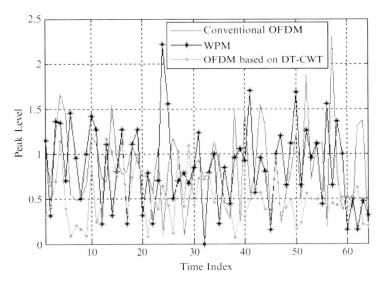

Fig. 5 The Envelope of the Conventional OFDM, WPM and OFDM based on DT-CWT

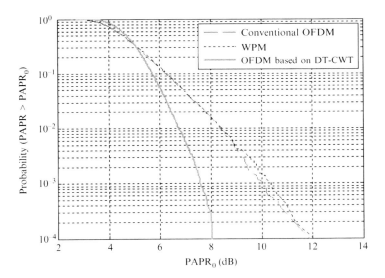

Fig. 6 CCDF the Conventional OFDM, WPM and OFDM based on DT-CWT

DT-CWT_4 using (anto 9,7 with q-sh 14 filters), OFDM DT-CWT_5 using (n-sym 5, 7 with q-sh 14 filters), OFDM DT-CWT_6 using (LeGall (leg) 5, 3 tap filters with q-sh 14 filters), OFDM DT-CWT_7 using (n-sym 5, 7 with q-sh 16 filters) and OFDM DT-CWT_8 using (leg 5,3 with q-sh 18 filters). The results in Figure 7 show that there is no observed degradation as a result of using different set of mismatching filters in the design of the proposed scheme.

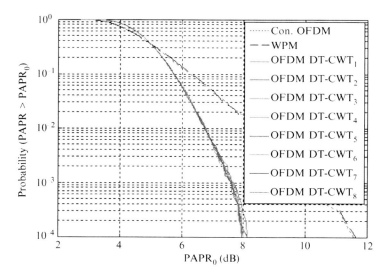

Fig. 7 The effect of using different set of filters in design of the OFDM based on DT-CWT

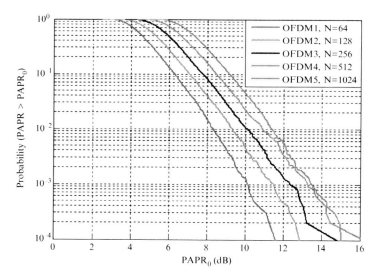

Fig. 8 CCDF of PAPR for conventional OFDM symbol with various values of subcarriers

Again, the results for PAPR were repeated in both Fig. 8 (for the conventional OFDM system) and Fig. 9 (for the proposed system) using different numbers of subcarriers (64, 128, 256, 512, and 1024) with 16QAM modulation. We observe from these figures that the PAPR increases as the number of subcarriers numbers (N) increases. As shown in Fig. 6–9 the PAPR performance of DT-CWT based OFDM systems is better than the conventional OFDM system or even WPM systems. As

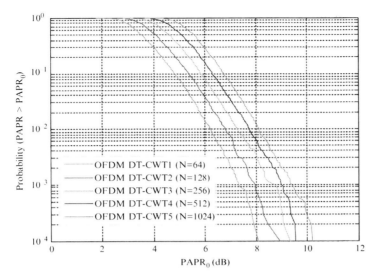

Fig. 9 CCDF of PAPR for OFDM based on DT-CWT symbol with various values of subcarriers

we saw in Fig. 5 showing the time-domain signal envelop of all the systems, this improvement in PAPR performance is explained from lower peaks of DTCWT systems observed in the Fig. 5.

5 Conclusions

This paper a new OFDM scheme that is based on DT-CWT is proposed. Comparing the proposed scheme in terms of PAPR with the traditional OFDM and WPM systems we see that the proposed scheme offers 3dB better PAPR performance over the conventional OFDM and WPM systems at 0.1% of CCDF. While the conventional OFDM and WPM systems shows similar behavior. Simulation results shows that there is no observed PAPR degradation as a result of using different set of mismatching filters in DT-CWT based system.

References

1. Papoulis A., Pillai S. U., *Probability Random Variables and Stovhastic Processes*. New York: McGraw-Hill, 2002.
2. M. Breiling, S. H. muller, and J. B. Huber, *SLM peak power reduction without explicit side information*. IEEE Commun. Lett., vol. 5, no. 6, pp. 239–241, 2001.
3. J. Tellado, Ed., *Multicarrier Modulation with low PAPR Application to DSL and wireless*. New York: Kluwer, 2002.

4. M. Baro and J. Ilow, *PAPR reduction in wavelet packet modulation using tree pruning.* in 2007 IEEE 65th Vehicular Technology Conference VTC 2007 – spring, Dubin, Ireland, Apr. 2007.
5. M. Baro and J. Ilow, *PAPR reduction in OFDM using wavelet packet pre-processing.* Consumer Communications and Networking Conference, 2008. CCNC 2008. 5th IEEE.
6. M. K Lakshmanan, H. Nikookar, *A review of Wavelets for Digital Communication.* Wireless Personal Communication (2006) 37: 387- OFDM* 420, Springer 2006.
7. J. A. C. Bingham, *Multicarrier modulation for data transmission: An Idea Whose Time Has Come.* IEEE Communications Magazine, vol. 28, no. 5, pp. 5–14, 1990.
8. H. M. Newlin, *Developments in the use of wavelets in communication systems.* TRW Systems and Information Technology Group, Sunnyvale, California.
9. A. Jamin, P. Mahonen, *Wavelet packet modulation for wireless communications.* Wireless Communications and Mobile Computing vol. 5, no. 2, pp. 123–137, 2005.
10. C. V. Bouwel, J. Potemans, S. Schepers, B. Nauwelaers and A.V. de Capelle, *Wavelet Packet Based Multicarrier Modulation.* Proc. IEEE Benelux Symposium on Communications and Vehicular Technology, Leuven, Belgium, 19 October 2000.
11. C. Schurgers and M. B. Srivastava, *approach to peak-to-average power ratio in OFDM.* in SPIE's 47th Annual Meeting, San Diego, CA, 2001, pp. 454–464.
12. S. H. Han and J. H. Lee, *An Overview of to peak-to-average power ratio reduction techniques for multimedia transmission.* IEEE Wireless Communications, vol. 12, no. 2, pp. 56–65, 2005.
13. Richard Van Nee and Ramjee Prasal, *OFDM for Wireless Multimedia Communications.* P. cm Artech House Universal Personal Communications Series, Inc. Boston. London, 2000.
14. Panchamkumar D. Shukla, *Complex wavelet Transforms and Their Applications.* Master Thesis 2003. Signal Processing Division. University of Strathclyde Department of Electronic and Electrical Engineering.
15. M. Gautier, J. Lienard, and M. Arndt, *Efficient Wavelet Packet Modulation for Wireless Communication.* AICT'07 IEEE Computer Society, 2007.
16. A. Jamin and P. Mahonen, *Wavelet Packet Modulation for Wireless Communications.* Wiley Wireless Communications and networking, Journal, vol. 5, no. 2, pp. 123–137, Mar. 2005.
17. Xiaodong Zhang and Guangguo Bi, *OFDM Scheme Based on Complex Orthogonal Wavelet Packet.* http://ieeexplore.ieee.org/iel5/7636/20844/00965270.pdf.
18. M. Gautier and J. Lienard, *Performance of Complex Wavelet packet Based Multicarrier Transmission through Double Dispersive Channel.* NORSIG 06, IEEE Nordic Signal Processing Symposium (Iceland), June 2006.
19. C. J. Mtika and R. Nunna, *A wavelet-based multicarrier modulation scheme.* in Proceedings of the 40th Midwest Symposium on Circuits and Systems, vol. 2, August 1997, pp. 869–872.
20. N. Erdol, F. Bao, and Z. Chen, *Wavelet modulation: a prototype for digital communication systems.* in IEEE Southcon Conference, 1995, pp. 168–171.
21. A. R. Lindsey and J. C. Dill, *Wavelet packet modulation: a generalized method for orthogonally multiplexed communications.* in IEEE 27th Southeastern Symposium on System Theory, 1995, pp. 392–396.
22. A. R. Lindsey, *Wavelet packet modulation for orthogonally multiplexed communication.* IEEE Transaction on Signal Processing, vol. 45, no. 5, pp. 1336–1339, May 1997.
23. Ivan W. Selesnick, Richard G. Baraniuk, and Nick G. Kingsbury, *The Dual-Tree Complex Wavelet Transform.* IEEE Signal Processing Mag, pp. 1053–5888, Nov. 2005.
24. N. G. Kingsbury, *The dual-tree complex wavelet transform: A new technique for shift invariance and directional filters.* in Proc. 8th IEEE DSP Workshop, Utah, Aug. 9–12, 1998, paper no. 86.
25. N. G. Kingsbury, *Image processing with complex wavelets.* Philos. Trans. R. Soc. London A, Math. Phys. Sci., vol. 357, no. 1760, pp. 2543–2560, Sept. 1999.
26. N. G. Kingsbury, *A dual-tree complex wavelet transform with improved orthogonality and symmetry properties.* in Proc. IEEE Int. Conf. Image Processing, Vancouver, BC, Canada, Sept. 10–13, 2000, vol. 2, pp. 375–378.
27. N. G. Kingsbury, *Complex wavelets for shift invariant analysis and filtering of signals.* Appl. Comput. Harmon. Anal., vol. 10, no. 3, pp. 234–253, May 2001.

28. I. W. Selesnick, *the Double Density Dual-Tree DWT*. IEEE Transactions on Signal Processing, 52(5): 1304–1315, May 2004.
29. J. M. Lina, *Complex Daubechies Wavelets: Filter Design and Applications*. ISAAC Conference, June 1997.
30. M. K. Lakshmanan and H. Nikookar, *A Review of Wavelets for Digital Wireless Communication*. Wireless Personal Communications Springer, 37: 387–420, Jan. 2006.

SOCP Approach for PAPR Reduction Using Tone Reservation for the Future DVB-T/H Standards

Irène Mahafeno, Jean-François Hélard, and Yves Louet

Abstract High PAPR of OFDM signals is an important problem for DVB transmission. Due to the nonlinearity of the HPA, it can result in low power efficiency and performance degradation of the system. This paper presents the relevance of a tone reservation method with SOCP optimization approach for PAPR reduction, for the future DVB-T/H standards. A study of dedicated subcarriers power control is first presented. Simulation results show that without increasing the BER for a given SNR, this method reduces the PAPR significantly when only a small set of subcarriers is dedicated for PAPR reduction. Moreover, it presents a good trade-off between the PAPR reduction gain and the increase in mean transmitted power. Performance evaluation in the presence of a nonlinear HPA also demonstrates its relevance, when the gain in terms of IBO is evaluated assuming system constraints on spectral mask and MER performance respectively.

1 Introduction

Multicarrier orthogonal frequency division multiplexing (OFDM) transmission has received interest, especially in wireless applications [1, 2]. It has been adopted in several standards, including the current terrestrial digital video broadcasting (DVB-T). High peak-to-average power ratio (PAPR) is known to be one of the main drawbacks of OFDM systems. Many methods have been proposed to reduce the OFDM PAPR signal [3]. Some use clipping and filtering techniques, others are based on coding or multiple signal representation techniques. However, these

---

I. Mahafeno (✉) and J.-F. Hélard
Institute of Electronics and Telecommunications of Rennes, UMR CNRS 6164, INSA de Rennes, 20 Avenue de Buttes de Coësmes, 35043 Rennes, France
e-mail: {masinjara-irene.mahafeno, jean-francois.helard}@insa-rennes.fr

Y. Louet
Institute of Electronics and Telecommunications of Rennes, UMR CNRS 6164, SUPELEC, Avenue de la Boulaie, BP 81127, 35511 Cesson-Sévigné, France
e-mail: yves.louet@supelec.fr

methods may lead to an increase in the bit error rate (BER) or require side information (SI) transmission to recover the original data. Recent methods use allocated subcarriers to generate some additional information for PAPR reduction [4,5]. These methods are commonly known as tone reservation (TR) based methods. TR method was first proposed by Tellado in [4] and has been extended using second order cone programming (SOCP) formulation in [6, 7]. In [7], the additional information is added to all data subcarriers which degrades the BER performance. In [6], the authors proposed to use only out-of-band (OOB) subcarriers, which avoid BER performance degradation. However, since the power of the OOB subcarriers is limited by the standard spectrum mask, a significant number of these subcarriers have to be allocated for PAPR reduction, in order to achieve a significant PAPR reduction gain. In this paper, we propose a TR based method with SOCP formulation as a candidate for the future generation of the DVB transmission. A very small set of subcarriers in the useful bandwidth is allocated for PAPR reduction. This method does not need any SI and does not lead to any BER increase for a given signal to noise ratio (SNR). To avoid high peak power on dedicated subcarrier, a study of its power control is carried out. A performance evaluation of the system in the presence of a nonlinear HPA is also performed. The gain of the method is evaluated assuming system constraints on spectral mask and on modulation error ratio (MER) performance respectively.

This paper is organized as follows. Section 2 describes the proposed TR based PAPR reduction method. A study of dedicated subcarriers power control is presented in Section 3. Section 4 assesses the benefit of the proposed method in the presence of a nonlinear HPA. Finally, conclusions are drawn in Section 5.

2 System Description

Figure 1 presents the structure of an OFDM transmitter. Let $\mathbf{X} = [X_0 \ldots X_{N-1}]$ be a sequence of complex symbols and $\mathbf{x} = [x_0 \ldots x_{N-1}]$ be its inverse Fourier transform. We denote $x(t)$ the OFDM baseband signal. Assuming NL equidistant samples of $x(t)$, the oversampled version of $x(t)$ is given by

$$x_n = \frac{1}{\sqrt{NL}} \sum_{k=0}^{N-1} X_k e^{\frac{j2\pi nk}{NL}}, \, n \in [0 \ldots NL - 1] \qquad (1)$$

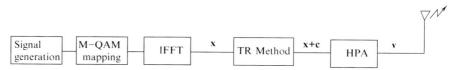

Fig. 1 Structure of an OFDM transmitter with tone reservation method

where $j = \sqrt{-1}$, L represents the oversampling factor and N denotes the number of subcarriers. The PAPR of **x** is defined by

$$PAPR\{\mathbf{x}_L, L \geq 4\} = \frac{\max_{0 \leq n \leq NL-1} |x_n|^2}{E\{|\mathbf{x}_L|^2\}}, \text{ with } \mathbf{x}_L = \mathcal{Q}_L \mathbf{X}_L. \quad (2)$$

where $\mathbf{X}_L$ is the zero-padded vector of **X** by factor L, $E[.]$ denotes the expectation operation and $\mathcal{Q}_L$ is the inverse discrete Fourier transform matrix of size NL scaled by $\sqrt{L}$.

TR method involves allocating a set of subcarriers which are optimized for PAPR reduction [4]. The idea is to add a time domain vector **c** to the original multicarrier vector **x** in order to reduce its PAPR, i.e., to verify $PAPR(\mathbf{x}+\mathbf{c}) < PAPR(\mathbf{x})$. Vector **c** is computed at the transmitter and stripped off at the receiver. The problem can be formulated as follows: reducing PAPR leads to minimization of maximum peak value of the combining signal ($\mathbf{x}+\mathbf{c} = \mathbf{x} + \mathcal{Q}_L \mathbf{C}$) while keeping the average power constant. Mathematically, this problem can be written as

$$\min_{\mathbf{c}} \max_{n} |x_n + \mathbf{q}_{n.L}^{row} \mathbf{C}|, \quad (3)$$

where $\mathbf{q}_{n.L}^{row}$ is the n-th row of $\mathcal{Q}_L$. TR techniques using SOCP formulation is applied. SOCP is a convex optimization problem class that minimizes a linear function over the intersection of an affine set and the product of second-order (quadratic) cones [8]. The PAPR minimization problem given in (3) can then be rewritten as

$$\text{Minimize } \beta$$
$$\text{subject to } \|x_n + \mathbf{q}_{n.L}^{row} \mathbf{C}\| \leq \beta,$$
$$0 \leq n \leq NL - 1. \quad (4)$$

In this study, the dedicated subcarriers are located in the useful bandwidth. Since these dedicated subcarriers are not used for data transmission, spectral efficiency will naturally slightly decrease but the advantage of the method is that no SI is required. Both transmitter and receiver have the knowledge of the location of reserved subcarriers. Thus, the receiver only decodes data information and ignores the additional information. Furthermore, it does not increase the BER for a given SNR.

3 Dedicated Subcarriers Power Control

The performance of the proposed method has been evaluated using complementary cumulative distribution function (CCDF) [9]. Simulation results using the current DVB-T parameters showed that this method provides a PAPR reduction gain of 3.65 dB when only 0.47% of subcarriers is used. It also offers an interesting trade-off between the PAPR reduction gain and the mean transmitted power increase. The dedicated subcarriers location does not have any influence on the PAPR reduction gain. Furthermore, the gain does not depend on the constellation type. Indeed, similar gain is obtained for 4-QAM, 16-QAM and 64-QAM constellations. However, one drawback of this method is that, to achieve a significant value of PAPR reduction gain, the dedicated subcarriers power level can greatly increase (see Fig. 2a). So, we propose to control each dedicated subcarrier power in the way that the power level difference between the dedicated subcarriers and the data subcarriers cannot exceed a given value λ [dB]. To do so, upper bound condition is added as a constraint in the optimization problem. Equation (4) becomes

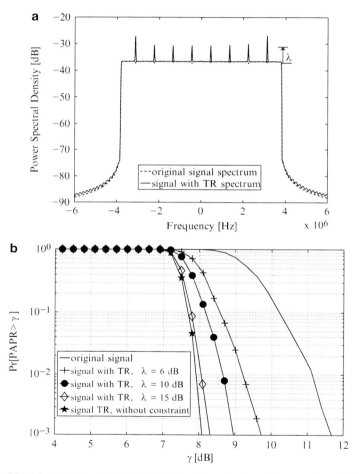

Fig. 2 (a) PSD of the signal - (b) PAPR CCDF for different values of the gap ($\lambda = 6, 10, 15$ dB), $N_d = 8, L = 2$, 16-QAM constellation

$$\text{Minimize } \beta$$
$$\text{subject to } \|x_n + \mathbf{q}_{n,L}^{row}\mathbf{C}\| \leq \beta, 0 \leq n \leq NL-1$$
$$\|C_n\| \leq \sqrt{\Gamma}, \quad (5)$$

where $\Gamma = 10^{\frac{\lambda}{10}}$ is the power level gap and $C_n = [C_1 \ldots C_{N_d}]$.

The dedicated subcarriers power limitation leads to a slight degradation of the PAPR reduction gain.

Figure 2b presents the PAPR CCDF of the signal for different values of λ and for $N_d = 8$ dedicated subcarriers. Current DVB-T parameters are used, $N = 2048$ (2K mode) with $N_u = 1705$ useful subcarriers is assumed. Data are randomly generated and 16-QAM constellation is used.

The PAPR decreases as λ increases. Indeed, optimal solutions are achieved when dedicated subcarriers power is not limited. With the power control, although the PAPR reduction gain has slightly decreased compared to the optimal solutions, it is still significant. For example, with $\lambda = 10$ dB, for an exceed probability of 10^{-3}, the PAPR reduction gain is equal to 2.75 dB while only 0.47% of useful subcarriers is used. The corresponding mean transmitted power increase is equal to 0.50 dB, which corresponds to a good trade-off between the PAPR reduction gain and the power increase.

4 Performance Evaluation of the Method

Performance of the method is evaluated in the presence of a nonlinear HPA. A Rapp's solid state power amplifier (SSPA) model is assumed [10]. Current DVB-T parameters are used, but extension to DVB-T2/H is straightforward. DVB-T2 standard specification has been published very recently [11]. Two different PAPR reduction methods have been adopted for the DVB-T2 transmission, one of which is a TR based method. The algorithm presented in [11] is based on a gradient method first proposed in [4]. It involves in iteratively cancelling out the peaks of the time domain signal, by a set of impulse-like kernels, using dedicated subcarriers. In this section, we assume the same number of subcarriers reserved for PAPR reduction and the same power level limitation proposed in DVB-T2 standard specification, i.e. $N_d = 17(1\%)$ for $2K$ mode and $\lambda = 10$ dB.

The PAPR reduction gain is evaluated in terms of input back-off (IBO) gain obtained for different system constraints and for different values of the SSPA smoothness factor p. Comparison results are presented in Table 1. Here, $p = 100$ defines a linearized SSPA while $p = 3$ corresponds to a typical nonlinear SSPA. When a constraint on the spectral mask is applied, the IBO gain is obtained by comparing the adjacent channel power ratio (ACPR) values of the signal with and without the method. In this case, the target ACPR value is equal to -52 dB. When a constraint on performance is set, the IBO gain is deducted by comparing the MER values of the signal with and without the method. In that case, the target value of MER is equal to 40 dB.

Table 1 IBO gain for different system constraints and $p = 3, 100$ ($N_d = 17(1\%)$ dedicated subcarriers)

System constraints	$p = 100$	$p = 3$
Spectral mask constraint (ACPR = -52 dB)	IBO gain of 3 dB	IBO gain of 2 dB
MER constraint (MER = 40 dB)	IBO gain of 1.45 dB	IBO gain of 0.8 dB

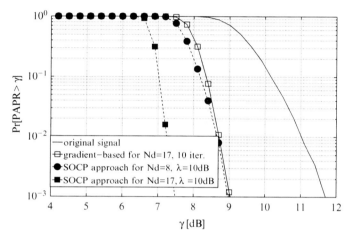

Fig. 3 PAPR CCDF for $N_d = 8, 17$, $L = 2$, 16-QAM constellation: SOCP approach with $\lambda = 10$ dB *versus* gradient-based method with 10 iterations

Assuming constraint on ACPR, the IBO gain is equal to 3 dB for $p = 100$ and to 2 for $p = 3$. When avoiding the MER performance degradation of the system, this gain is equal to 1.45 and 0.8 dB for $p = 100$ and $p = 3$ respectively. The proposed approach then achieves a significant reduction gain and thus allows the use of a small value of back-off. That advantage makes this approach an interesting technique for the future DVB transmission. Indeed, it can reserve less subcarrier for PAPR reduction in comparison with the gradient-based method to achieve the same gain, as presented in Fig. 3.

Figure 3 compares the PAPR CCDF of the signal obtained with the proposed approach with that of the gradient-based method (for a number of iteration equals 10 and a clipping amplitude equal to 3.5 dB [11]). The proposed approach outperforms the gradient-based method by 4.2 dB when comparing the CCDF curves with the same number of dedicated subcarriers (1%), for an exceed probability of 10^{-3}. With this approach, practically half of the number of dedicated subcarriers can be used to achieve the same performance as the gradient-based method. The proposed approach then provides a better spectral efficiency.

5 Conclusions

A PAPR reduction method based on TR technique has been studied for DVB-T systems. Only a small set of subcarriers located in the useful bandwidth is dedicated for PAPR reduction. This is done in the same manner as some subcarriers are reserved for channel estimation and frequency synchronisation in the current DVB-T standard. A SOCP optimization is applied to find the peak reduction signal that minimizes the PAPR. A study of dedicated subcarriers power control has then been carried out. By allocating 0.47% of useful subcarriers, a PAPR reduction gain of 2.75 dB is still achieved for 10 dB of power level limitation. In the presence of a SSPA, the proposed method enhances the amplifier power efficiency by allowing the use of a small value of IBO. IBO gains of about 3 and 2 dB (for $p = 100$ and $p = 3$ respectively) are achieved when assuming constraint on ACPR, i.e. when avoiding the increase of OOB radiation of the signal spectrum. These gains are equal to 1.45 and 0.8 dB respectively when avoiding the MER performance degradation of the system. In conclusion, the study presented in this paper demonstrates the relevance of the proposed TR based method for the future DVB transmission.

Acknowledgements The authors would like to thank the European CELTIC project "Broadcast for the twenty-first century (B21C)" for its support of this work.

References

1. U. Reimers, "Digital Video Broadcasting," *IEEE Commun. Mag.*, vol. 36, no. 10, pp. 104–110, June 1998.
2. B. R. Saltzberg, "Comparison of Single-Carrier and Multitone Digital Modulation for ADSL Applications," *IEEE Commun. Mag.*, vol. 36, no. 11, pp. 114–121, Nov. 1998.
3. Y. Louët and J. Palicot, "A Classification of Methods for Efficient Power Amplification of Signals," *Annals of Telecom.*, vol. 63, no. 7–8, pp. 351–368, July–Aug. 2008.
4. J. Tellado, "Peak-to-Average Power Ratio Reduction for Multicarrier Modulation," *PhD thesis*, Stanford University, Sept. 1999.
5. B. S. Krongold and D. L. Jones, "An Active-Set Approach for OFDM PAPR Reduction via Tone Reservation," *IEEE Trans. on Signal Processing*, vol. 52, no. 2, pp. 495–509, Feb. 2004.
6. S. Zabre, J. Palicot, Y. Louët and C. Lereau, "SOCP Approach for OFDM Peak-to-Average Power Ratio Reduction in the Signal Adding Context," *IEEE Symp. on Signal Processing and Information Technology*, Vancouver, Canada, Aug. 2006.
7. A. Aggarwal and T. H. Meng, "Minimising the Peak-to-Average Power Ratio of OFDM Signals via Convex Optimization," *Proc. IEEE Globecom Conference*, vol. 4, San Francisco, California, pp. 2385–2389, Dec. 2003.
8. M. Lobo, L. Vandenberghe, S. Boyd and H. Lebret, "Applications of Second Order Cone Programming," *Linear Algebra and its Applications*, 284:193–228, Nov. 1998. Special Issue on Linear Algebra in Control, Signals and Image Processing.
9. I. Mahafeno, Y. Louët, J.F. Hélard, "PAPR reduction method for OFDM systems using dedicated subcarriers: a proposal for the future DVB-T standard," *Proc. IEEE International Symposium on Broadband Multimedia System*, Las Vegas, Nevada, March–Apr. 2008.

10. C. Rapp, "Effects of HPA-nonlinearity on 4-DPSK/OFDM-signal for a digital sound broadcasting system," *2nd European conference on satellite communications*, Luettich, pp. 179–184, Oct. 1991.
11. "Frame structure channel coding and modulation for a second generation digital terrestrial television broadcasting system (DVB-T2)," *DVB Document A122*, June 2008.

Reduced-PAPR Code Allocation Strategy for MC-CDMA Transmissions

Filippo Giannetti, Vincenzo Lottici, Ivan Stupia, and Nunzio Aldo D'Andrea

Abstract First, an approximate expression for the peak-to-average power ratio (PAPR) of a multi-carrier code-division multiple access (MC-CDMA) signal is analytically derived. Then, it is demonstrated that the PAPR of a MC-CDMA signal can be suitably reduced by resorting to a judicious strategy for the allocation of the signature codes. Based on such a result, a low-complexity implementation of the proposed strategy is presented and its performance gain over the random allocation strategy is numerically assessed in terms of PAPR and out-of-band emissions.

1 Introduction

The large amplitude fluctuations of MC-CDMA signals, quantified by the PAPR parameter, entail a considerable vulnerability to nonlinear distortions induced by transmitter high-power amplifier (HPA). This detrimental effect causes significant performance degradation that requires adequate countermeasures. Some PAPR reduction techniques, originally devised for orthogonal frequency-division multiplexing (OFDM) systems can be effectively applied also to MC-CDMA [1], but they require additional processing at the transmitter and/or additional side information at the receiver. As suggested in [2], the PAPR of a MC-CDMA signal can be effectively reduced (with no, or minimal, complexity increase) also by resorting to a judicious policy for the selection of the signatures to be assigned to the active users.

In the present work we focus on the down-link transmission of a wireless MC-CDMA network, with a twofold aim: (i) to derive an (approximate) analytical characterization of the PAPR, for a generic code set, following a statistical approach, and (ii) based on the analytical findings, to derive a judicious code allocation strategy which reduces the PAPR, so as mitigate the impact of nonlinear distortions at the transmitter.

F. Giannetti (✉), V. Lottici, I. Stupia, and N.A. D'Andrea
University of Pisa, Department of Information Engineering, Via G. Caruso, 16 - I-56122 Pisa, Italy
e-mail: {filippo.giannetti, vincenzo.lottici, ivan.stupia, aldo.dandrea}@iet.unipi.it

After system description (Section 2), the analytical characterization of the PAPR is illustrated in Section 3 and the reduced PAPR code allocation strategy is derived in Section 4. Numerical results are presented in Section 5 and some conclusions are eventually drawn in Section 6.

2 System Description

The nth information-bearing symbol $a_n^{(k)}$ of the kth user ($1 \leq k \leq N_u$) runs at rate $1/T_s$, and belongs to a M-QAM constellation with $\mathrm{E}\{a_n^{(k)}\} = 0$, $\mathrm{E}\{|a_n^{(k)}|^2\} \triangleq A_2$, and $\mathrm{E}\{|a_n^{(k)}|^4\} \triangleq A_4$, where $\mathrm{E}\{\cdot\}$ represents statistical expectation. Each symbol is copied into N branches (equal to the number of subcarriers), and each of the N replicated symbols is then multiplied by a different chip of the binary spreading sequence, which acts as the user's channel signature, $c_m^{(i_k)} \in \{\pm 1\}$ ($0 \leq m \leq N-1$), with i_k an integer index such that $1 \leq i_k \leq N$, and $1 \leq k \leq N_u$. The resulting spreading factor, i.e., the number of replicas of each information symbol transmitted on an OFDM block, is thus coincident with the number of subcarriers N. The spread data $a_n^{(k)} c_m^{(i_k)}$ of the N_u active users are then summed, and mapped to the N available subcarriers by an OFDM modulator. To avoid interference between successive OFDM blocks and preserve the orthogonality among subcarriers, a cyclic prefix of L samples is inserted at the beginning of each block to produce a $(N+L)$-size block containing the samples

$$b_{l,n} = \frac{1}{\sqrt{N}} \sum_{k=1}^{N_u} \sum_{m=0}^{N-1} a_n^{(k)} c_m^{(i_k)} e^{j2\pi ml/N}, \quad -L \leq l \leq N-1. \quad (1)$$

Considering the T_s-long interval $I \triangleq (nT_s - LT_c \leq t < nT_s + NT_c)$ of the nth OFDM block, the signal at the output of the shaping filter is

$$s(t) = \begin{cases} \sum_{l=0}^{N-1} b_{l,n} g_T(t - nT_s - lT_c) & nT_s \leq t < nT_s + NT_c \\ s(t + NT_c) & nT_s - NT_c \leq t < 0 \end{cases}, \quad (2)$$

where $T_c \triangleq T_s/(N+L)$ is the chip interval, while $g_T(t)$ is a T_c-energy a root-raised-cosine (RRC) pulse with rolloff α. The HPA is modeled as a Rapp's nonlinear memoryless device with amplitude and phase characteristics $M(\rho) = \rho/(1+\rho^q)^{1/q}$, and $\Phi(\rho) = 0$, respectively, where $\rho(t) \triangleq \xi_0|s(t)|$ is the instantaneous amplitude of the signal at the HPA input, ξ_0 is the input back-off (IBO), and q is an integer value which controls the smoothness of the transition between the linear region and the saturation one.

3 PAPR Analysis

The PAPR of the signal which enters the HPA is defined as $PAPR \triangleq P_{\max}/\overline{P}_s$, where $P_{\max} \triangleq \max_{t \in I}\{|s(t)|^2\}$ is the peak power of the signal $s(t)$, and $\overline{P}_s \triangleq \frac{1}{T_s}\int_I P_s(t)\,dt$ is the time-average of the signal statistical power $P_s(t) \triangleq E\{|s(t)|^2\}$, both evaluated over the interval I. Assuming that the cyclic prefix is much shorter than the block interval ($L << N$), after some manipulations, we obtain $\overline{P}_s \simeq N_u A_2$ [3]. By replacing (1) into (2), by assuming that the pulse $g_T(t)$ has a small roll-off, ($\alpha << 1$), and by defining the following modified code sequence

$$\gamma_m^{(i_k)} \triangleq \begin{cases} c_m^{(i_k)} & 0 \leq m \leq N/2 - 1 \\ c_{m+N}^{(i_k)} & -N/2 \leq m \leq -1 \end{cases}, \tag{3}$$

after some manipulations we obtain the instantaneous signal power

$$|s(t)|^2 \simeq \sum_{k=1}^{N_u} |a_n^{(i_k)}|^2$$
$$+ \frac{2}{N}\Re\left\{\sum_{\mu=1}^{N-1}\left[\sum_{k=1}^{N_u}|a_n^{(i_k)}|^2 A_\mu^{(i_k)} + \sum_{k=1}^{N_u}\sum_{\substack{k'=1 \\ k' \neq k}}^{N_u} a_n^{(i_k)} a_n^{(i_{k'})*} X_\mu^{(i_k,i_{k'})}\right] e^{j 2\pi\mu \frac{t}{T_s}}\right\}, \tag{4}$$

where * denotes conjugation, and we defined the aperiodic auto-correlation of the generic i_k th user modified code, and the aperiodic cross-correlation between the generic i_k th and the $i_{k'}$ th user modified codes as

$$A_\mu^{(i_k)} \triangleq \sum_{\nu=-N/2}^{N/2-1-\mu} \gamma_\nu^{(i_k)} \gamma_{\nu+\mu}^{(i_k)}, \quad X_\mu^{(i_k,i_{k'})} \triangleq \sum_{\nu=-N/2}^{N/2-1-\mu} \gamma_\nu^{(i_k)} \gamma_{\nu+\mu}^{(i_{k'})}, \tag{5}$$

with the time shift $0 \leq \mu \leq N - 1$, the code indexes $1 \leq i_k \leq N_u$, and $1 \leq i_{k'} \leq N_u$. The peak power shall be then obtained by taking the maximum value of (4), over the interval I. Hereafter, the peak power, and hence the PAPR, is evaluated by resorting to a statistical approach. In the sequel, we will denote with $\eta_C \triangleq E\{C\}$, and $\sigma_C^2 \triangleq E\{|C|^2\} - |\eta_C|^2$, the mean and the variance, respectively, of the complex-valued random variable (RV) C. Assuming $N_u \gg 1$ and invoking the Central Limit Theorem, after some manipulations and a few approximations [3], the RV $P(t) \triangleq |s(t)|^2$ which represents the instantaneous power at a fixed time instant t, can be (approximately) modeled as a Gaussian RV having mean and variance

$$\eta_P = N_u A_2 + \frac{2}{N} A_2 \sum_{\mu=1}^{N-1}\left\{\left[\sum_{k=1}^{N_u} A_\mu^{(i_k)}\right]\cos\left(2\pi\mu\frac{t}{T_s}\right)\right\}, \tag{6}$$

$$\sigma_P^2 = N_u \left(A_4 - A_2^2 \right)$$
$$+ \frac{2}{N^2} \sum_{k=1}^{N_u} \left\{ A_4 \sum_{\mu=1}^{N-1} A_\mu^{(i_k)2} + A_2^2 \sum_{\substack{k'=1 \\ k' \neq k}}^{N_u} \sum_{\mu=1}^{N-1} \left[A_\mu^{(i_k)} A_\mu^{(i_{k'})} + X_\mu^{(i_k, i_{k'})2} \right] \right\}, \quad (7)$$

respectively. We evaluate now the statistics of the stochastic process $P(t)$ at the instants $t_h \triangleq nT_s + hT_c$, with $-L \leq h < N$ and n any integer, taken at sample rate $1/T_c$ over one interval I, and we get the set of RVs $P_h \triangleq P(t_h)$, having mean $\eta_P(t_h)$, variance σ_P^2 and cumulative distribution function (CDF) $\Pr\{P_h \leq p\}$. The peak signal power is thus expressed by the RV $P_{\max} \triangleq \max_h \{P_h\}$, and the CDF of the maximum of a set of RVs (assumed independent for the sake of simplicity) is

$$\Pr\{P_{\max} \leq p\} = \Pr\{P_{-L} \leq p, \ldots, P_{N-1} \leq p\} = \prod_{h=-L}^{N-1} \Pr\{P_h \leq p\}. \quad (8)$$

Now, for a given threshold value p_0 of the peak power, we obtain the relevant crossing probability, expressed by the complementary CDF (CCDF)

$$\Pr\{P_{\max} > p_0\} = 1 - \prod_{h=-L}^{N-1} \Pr\{P_h \leq p\}. \quad (9)$$

By resorting to a Gaussian model, the CDF of P_h is expressed by

$$\Pr\{P_h \leq p\} = 1 - Q\left(\frac{p - \eta_P(t_h)}{\sigma_P} \right), \quad (10)$$

where $Q(x) \triangleq \int_x^\infty \exp\{-y^2/2\}/\sqrt{2\pi} \, dy$. At a given time instant, the *PAPR* is a RV which can be expressed as $PAPR = P_{\max}/(N_u A_2)$, and therefore, for a given threshold value $PAPR_0$ the relevant crossing probability is expressed by the CCDF

$$\Pr\{PAPR > PAPR_0\} = \Pr\{P_{\max} > PAPR_0 \cdot (N_u A_2)\}$$
$$= 1 - \prod_{h=-L}^{N-1} \left[1 - Q\left(\frac{PAPR_0 N_u A_2 - \eta_P(t_h)}{\sigma_P} \right) \right]. \quad (11)$$

4 Reduced PAPR Code Allocation

Let define now the following metric

$$\Lambda(\mathbf{C}) \triangleq \sum_{k=1}^{N_u} \left\{ \sum_{\mu=1}^{N-1} A_\mu^{(i_k)2} + \sum_{\substack{k'=1 \\ k' \neq k}}^{N_u} \sum_{\mu=1}^{N-1} \left[A_\mu^{(i_k)} A_\mu^{(i_{k'})} + X_\mu^{(i_k, i_{k'})2} \right] \right\}, \quad (12)$$

where $\mathbf{C} \triangleq [\mathbf{c}^{(i_1)}, \mathbf{c}^{(i_2)}, \ldots, \mathbf{c}^{(i_{N_u})}]$ ($1 \leq i_k \leq N$, and $i_k \neq i_h$ if $k \neq h$) is an $N \times N_u$ array containing the subset of N_u user signatures selected among an N-size code set, and $\mathbf{c}^{(i_k)} \triangleq [c_0^{(i_k)}, c_1^{(i_k)}, \ldots, c_{N-1}^{(i_k)}]^T$ ($1 \leq i_k \leq N_u$) is the spreading code assigned to the generic kth user, and ^{T} denotes the transpose. Also, by defining

$$\mathcal{E}_A^{(i_k)} \triangleq \sum_{\mu=1}^{N-1} A_\mu^{(i_k)2}, \quad \mathcal{E}_X^{(i_k,i_{k'})} \triangleq \sum_{\mu=1}^{N-1} \left[A_\mu^{(i_k)} A_\mu^{(i_{k'})} + X_\mu^{(i_k,i_{k'})2} \right], \tag{13}$$

the metric (12) can be rearranged as

$$\Lambda(\mathbf{C}) = \sum_{k=1}^{N_u} \left[\mathcal{E}_A^{(i_k)} + \sum_{\substack{k'=1 \\ k' \neq k}}^{N_u} \mathcal{E}_X^{(i_k,i_{k'})} \right]. \tag{14}$$

Now, let notice that letting $A_4 \simeq A_2^2$ in (7), we get $\sigma_P^2 \simeq \frac{2}{N^2} A_4 \Lambda(\mathbf{C})$, therefore, the CCDF of the PAPR (11) turns out to be (approximately) a function of the metric $\Lambda(\mathbf{C})$. Such a dependence is clearly evidenced by Fig. 1, which plots the value $PAPR_0$ @ 99% (i.e., that threshold which is crossed by the $PAPR$ with probability 10^{-2}) vs. $\Lambda(\mathbf{C})$, for a 16-QAM with $N = 64$, $N_u = 32$, and eight different subsets of N_u codes taken from the WH set. The dot-dash line is a second order polynomial fitting in a least squares sense that reveals a (quasi) monotonic trend between metric and PAPR. Such a feature, suggests an efficient strategy for the allocation of the user signatures aimed at reducing the PAPR. Actually, the problem of choosing a subset of N_u codes that suitably reduces the PAPR of the aggregate MC-CDMA signal, can be reformulated, in a sub-optimal way, as finding the array $\hat{\mathbf{C}} \triangleq [\mathbf{c}^{(\hat{i}_1)}, \mathbf{c}^{(\hat{i}_2)}, \ldots, \mathbf{c}^{(\hat{i}_{N_u})}]$ containing the set of codes $\mathbf{c}^{(\hat{i}_k)}$ that minimizes the metric (12). In the sequel, such a procedure will be referred to as "Reduced PAPR

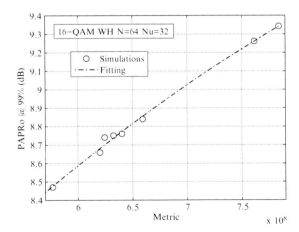

Fig. 1 PAPR @ 99% vs. Metric Λ (16-QAM, WH set, $N = 64$ and $N_u = 32$)

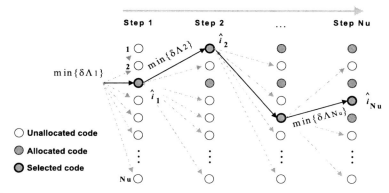

Fig. 2 Tree-Search method

Allocation" (RPA) method, whilst the conventional allocation strategy wherein the spreading codes of **C** are randomly selected will be labelled as "Random Allocation" (RA) method. For given N and N_u, solving the RPA through an exhaustive search method (ESM), would result in a prohibitively large number of tests. The RPA complexity can be effectively reduced by resorting to a search method, named "Tree-Search Method" (TSM), which is outlined step by step hereafter, and is pictorially illustrated in Fig. 2.

STEP 1: The signature of user #1 is chosen among the N available codes by computing the initial metric

$$\Lambda_1(i_1) \triangleq \delta\Lambda_1(i_1) \triangleq \mathcal{E}_A^{(i_1)}, \tag{15}$$

and we allocate to user #1 that code whose index $\hat{i}_1$ minimizes (15)

$$\Lambda_1(\hat{i}_1) = \min_{i_1} \{\Lambda_1(i_1)\}. \tag{16}$$

STEP k: The signature of user #k is chosen among the $N - k + 1$ available codes by computing the accumulated metric

$$\Lambda_k(\hat{i}_1, \hat{i}_2, \ldots, \hat{i}_{k-1}, i_k) \triangleq \Lambda_{k-1}(\hat{i}_1, \hat{i}_2, \ldots, \hat{i}_{k-1}) + \delta\Lambda_k(\hat{i}_1, \hat{i}_2, \ldots, \hat{i}_{k-1}, i_k), \tag{17}$$

where we defined the "branch metric" at step k as

$$\delta\Lambda_k(\hat{i}_1, \hat{i}_2, \ldots, \hat{i}_{k-1}, i_k) \triangleq \mathcal{E}_A^{(i_k)} + \sum_{h=1}^{k-1} \mathcal{E}_X^{(\hat{i}_h, i_k)} + \sum_{h=1}^{k-1} \mathcal{E}_X^{(i_k, \hat{i}_h)}, \tag{18}$$

and we allocate to user #k that code whose index $\hat{i}_k$ minimizes (17)

$$\Lambda_k(\hat{i}_1, \hat{i}_2, \ldots, \hat{i}_k) = \min_{i_k} \left\{\Lambda_k(\hat{i}_1, \hat{i}_2, \ldots, \hat{i}_{k-1}, i_k)\right\}. \tag{19}$$

Notice that, since Λ_{k-1} does not depend on i_k, the minimization of Λ_k is achieved by minimizing the branch metric $\delta\Lambda_k$, only. However, by this way we achieve a "local minimum" with respect to subset of the previously allocated codes $[\mathbf{c}^{(\hat{i}_1)}, \mathbf{c}^{(\hat{i}_2)}, \ldots, \mathbf{c}^{(\hat{i}_{k-1})}]$.

STEP N_u: The signature of user #N_u, is chosen among the $N - N_u + 1$ available codes by computing the total accumulated metric

$$\Lambda_{N_u}(\hat{i}_1, \ldots, \hat{i}_{N_u-1}, i_{N_u}) \triangleq \Lambda_{N_u-1}(\hat{i}_1, \ldots, \hat{i}_{N_u-1}) + \delta\Lambda_{N_u}(\hat{i}_1, \ldots, \hat{i}_{N_u-1}, i_{N_u}), \quad (20)$$

and we allocate to user #N_u that code whose index $\hat{i}_{N_u}$ minimizes (20)

$$\Lambda_{N_u}(\hat{i}_1, \hat{i}_2, \ldots, \hat{i}_{N_u}) = \min_{i_{N_u}} \left\{ \Lambda_{N_u}(\hat{i}_1, \hat{i}_2, \ldots, \hat{i}_{N_u-1}, i_{N_u}) \right\}. \quad (21)$$

By recalling (18) and by including all the selected code vectors into the array $\hat{\mathbf{C}}$, the total accumulated metric (21) can be rewritten as

$$\Lambda_{N_u}(\hat{\mathbf{C}}) = \sum_{k=1}^{N_u} \delta\Lambda_k(\hat{i}_1, \ldots, \hat{i}_k) = \sum_{k=1}^{N_u} \left[\varepsilon_A^{(\hat{i}_k)} + \sum_{\substack{k'=1 \\ k' \neq k}}^{N_u} \varepsilon_X^{(\hat{i}_k, \hat{i}_{k'})} \right], \quad (22)$$

and represents thus an (almost) minimized version of the target metric (14), i.e., $\Lambda(\hat{\mathbf{C}}) = \Lambda_{N_u}(\hat{\mathbf{C}})$. The computational burden of the TSM procedure outlined above is expressed by the total number of tests to be performed during the minimum metric computation, which is N at the first step, $N - 1$ at the second one, and so forth. The complexity of the TSM for $N = 64$ with $N_u = 32$ (worst case), amounts to 1,552 tests, while the ESM would require about 10^{18} tests. Also, when allocating a number of users $N_u > N/2$, the TSM algorithm reveals faster if used in the reverse order, i.e., starting from the whole code set and by decrementing the number of allocated codes from N down to N_u by discarding those having the maximum branch metrics.

5 Numerical Results

Numerical results have been derived for $N = 64$, 16-QAM, RRC pulses with $\alpha = 0.125$, and the following N-periodic sequence sets [4]: Walsh-Hadamard (WH), Golay (CG), Orthogonal Gold (OG), Small Kasami (SK), Shapiro-Rudin (SR). Figure 3 compares the analytical CCDF (solid line) and the simulated one (dashed line) for $N_u = 32$, and various spreading code sets. As apparent, despite the approximations, the analytical expression (11) provides a rather satisfactory accuracy. Figure 4 compares the CCDFs of the PAPR, for seven RA realizations (dashed lines) and for the RPA method (solid line), with WH set and $N_u = 8$. As apparent the

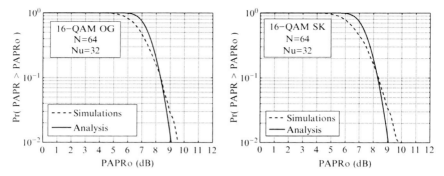

Fig. 3 CCDF of the PAPR (16-QAM, OG and SK sets, $N = 64$, $N_u = 32$)

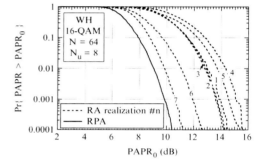

Fig. 4 CCDF of the PAPR (16-QAM, WH set, $N = 64$, $N_u = 8$)

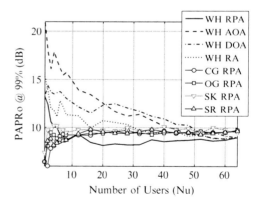

Fig. 5 PAPR @ 99% vs. N_u (16-QAM, $N = 64$)

RPA is shown to provide significant PAPR reductions with respect to the "blind" RA approach. Figure 5 compares the simulated values of the $PAPR_0$ @ 99% vs. N_u, for various code configurations. In Ascending Order Allocation (AOA) and Descending Order Allocation (DOA) strategies the WH codes are assigned in ascending and in descending row order from the WH matrix, respectively. As apparent, the WH set greatly benefits from the use of the RPA strategy and for loads greater than 20%

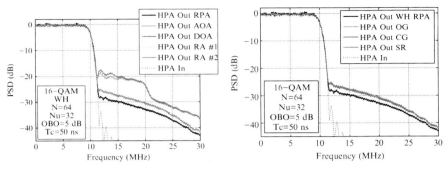

Fig. 6 Spectra (16-QAM, $N = 64$, $N_u = 32$, $OBO = 5$ dB, $T_c = 50$ ns)

it yields the best PAPR performance among all the considered code sets. The use of the RPA strategy reveals extremely beneficial in the presence of a transmit HPA which induces nonlinear distortions. Figure 6 presents the simulated power spectral density (PSD) for $N_u = 32$ and $T_c = 50$ ns, after the HPA, modeled by the Rapp model, with $q = 10$ and output back-off $OBO = 5$ dB. The chart on the left shows the spectrum at the HPA output for a WH set, with different allocation strategies. The chart on the right compares the spectrum at the HPA output for a WH set, obtained with RPA strategy, with the spectra of other sets (which have almost coincident plots). The graphs demonstrate that RPA is particularly effective in reducing out-of-band emissions of the WH set.

6 Conclusions

We have illustrated a simple method, named RPA, based on a proper allocation of the users' signatures to reduce a given PAPR metric figure of the transmitted signal and to improve the transmission system performance over the nonlinear channel. The proposed RPA strategy does not require changing the spreading code set, is advantageous for any traffic load and does not require any modification of the transmitter. A low-complexity tree-search implementation of the RPA has been devised, and simulation results demonstrated that RPA remarkably boosts link performance, in terms of PAPR distribution and out-of-band emissions. Finally, we remark that, the adoption of the proposed RPA scheme in combination with a data-predistortion scheme [5], can provide a further improvement of the system performance.

Acknowledgment The authors wish to acknowledge the activity of the Network of Excellence in Wireless COMmunications NEWCOM++ of the European Commission (contract n. 216715) that motivated this work.

References

1. N. Ohkubo, T. Ohtsuki, "A Peak to Average Power Ratio Reduction of Multicarrier CDMA Using Selected Mapping", *IEEE VTC 2002 – Fall*, pp. 2086–2090.
2. N. Hathi, M. Rodrigues, I. Darwazeh, J. O'Reilly, "Analysis of the Influence of Walsh-Hadamard Code Allocation Strategies on the Performance of Multi-Carrier CDMA Systems in the Presence of HPA Non-Linearities", *IEEE PIMRC 2002*, Sept. 2002, pp. 1305–1309.
3. F. Giannetti, V. Lottici, I. Stupia, N.A. D'Andrea, "PAPR Analytical Characterization and Reduced-PAPR Code Allocation Strategy for MC-CDMA Transmissions over Nonlinear Channels", *NEWCOM++ Project Technical Report*, Pisa, Italy, February, 2009.
4. E.H. Dinan, B. Jabbari, "Spreading Codes for Direct Sequence CDMA and Wideband CDMA Cellular Networks", *IEEE Commun. Mag.*, Sept. 1998, pp. 48–54.
5. V. Lottici, F. Giannetti, "Iterative Nonlinear Channel Compensation in MC-CDMA Systems", *Fifth International Workshop on Multi-Carrier Spread Spectrum (MC-SS 2005)*, Oberpfaffenhofen, Germany, Sept. 14–16, 2005, pp. 461–472.

Part VI
Adaptive Transmission

Adaptive Multiuser OFDMA Systems with High Priority Users in the Presence of Imperfect CQI

Alexander Kühne and Anja Klein

Abstract In this paper, an adaptive multiuser Orthogonal Frequency Division Multiple Access (OFDMA) system in the downlink is investigated which serves two sets of users differing in their priority regarding channel access. A Weighted Proportional Fair Scheduling (WPFS) approach is applied using instantaneous Channel Quality Information (CQI) and user priorities to allocate the different subcarriers to the different users. These CQI values are assumed to be imperfect due to time delays and estimation errors. The joint impact of imperfect CQI and user priority on the performance of the system is investigated analytically and assessed by numerical results. It appears that serving users with different priorities comes at the expense of reduced system data rate and less robustness against imperfect CQI.

1 Introduction

The Orthogonal Frequency Division Multiple Access (OFDMA) transmission scheme is a promising candidate for future mobile networks [1]. It allows an efficient adaptation to the channel conditions by performing time-frequency scheduling of the different subcarriers to the different users. In systems where users experience different channel conditions, Proportional Fair Scheduling (PFS) approaches provide a good trade-off between system throughput and fairness. OFDMA systems applying PFS are well discussed in the literature [2, 3]. If furthermore different user priorities shall be considered, Weighted Proportional Fair Scheduling (WPFS) approaches can be applied, which are discussed, e.g., in [4–6]. These WPFS algorithms favour high priority users to get channel access even if their channel gain is low which leads to a degradation of the system throughput compared to PFS approaches. Both PFS and WPFS algorithms require channel knowledge at the transmitter. However, in a realistic scenario, the channels are not perfectly known at the transmitter which also results in performance degradations compared to the

A. Kühne (✉) and A. Klein
Technische Universität Darmstadt, Communications Engineering Lab, 64283 Darmstadt, Germany
e-mail: {a.kuehne,a.klein}@nt.tu-darmstadt.de

case of perfect channel knowledge. The joint impact of imperfect channel knowledge and different user priorities on the performance of an OFDMA system has rarely been mentioned in the literature and the present paper will contribute to this aspect. Assuming outdated and estimated Channel Quality Information (CQI) at the transmitter, we analytically investigate this joint impact on the performance of an OFDMA system applying WPFS. The remainder of this paper is organised as follows. In Section 2, the considered system model is presented. In Section 3, the assumptions on the CQI are discussed. Section 4 introduces the adaptive OFDMA scheme applying WPFS. In Section 5, closed form expressions for the data rate and Bit Error Rate (BER) are derived analytically taking into account the joint impact of imperfect CQI and user priorities. In Section 6, numerical results illustrate the impact of both user priorities and imperfect CQI on the achievable system data rate.

2 System Model

In this work, we consider a one cell OFDMA downlink scenario with N subcarriers with index $n = 1, \ldots, N$. One Base Station (BS) and U Mobile Stations (MSs) with user index $u = 1, \cdots, U$ are located in the cell. The BS and the MSs are equipped with one antenna each. Each user u experiences a different average Signal-to-Noise-Ratio (SNR) $\bar{\gamma}_u$ depending on the pathloss. The fast fading on the n-th subcarrier of user u in time slot k with $k \in \mathbb{Z}$ is expressed by the channel transfer function $H_u(n,k)$ which is modeled as a complex normal distributed random variable with zero mean and variance one. Thus, the instantaneous SNR $\gamma_u(n,k)$ in time slot k of subcarrier n of user u is calculated according to

$$\gamma_u(n,k) = \bar{\gamma}_u \cdot |H_u(n,k)|^2. \tag{1}$$

Further on, we assume that there are two disjoint sets of users. The first set $\mathcal{S}_H$ contains U_H high priority users and the second set $\mathcal{S}_L$ contains U_L low priority users, with $U_H + U_L = U$.

3 Channel Quality Information

In order to perform an adaptive transmission, channel knowledge at the BS is required. In this work, we use the instantaneous SNR values of (1) as CQI which are fed back from the MSs to the BS in a Frequency Division Duplex system or measured at the BS in a Time Division Duplex system. In a realistic scenario, the CQI available at the BS suffers from different sources of error and, thus, cannot be assumed to be perfectly known. In the following, two sources of error together with the error modelling are introduced. To ease the comprehensibility, the user, subcarrier and time indices u, n and k are omitted in the notation of the channel transfer function.

3.1 Estimated CQI

The CQI values are assumed to be noisy estimates. The actual channel transfer factor H is modeled as a superposition of the estimated channel transfer factor $\hat{H}$ and an additional error term E leading to $H = \hat{H} + E$, where E is modeled as a complex normal distributed random variable with zero mean and variance σ_E^2. $\hat{H}$ is also modeled as zero-mean complex normal distributed random variable, but with variance $1 - \sigma_E^2$. The error variance $\sigma_E^2 \in [0, 1]$ depends on the conditions of the channel and the applied estimation scheme and, according to [7], is given by $\sigma_E^2 = \frac{1}{1 + T_\tau P_\tau}$, where T_τ is the number of training symbols per coherence time and P_τ the SNR during the training phase. In the following, we assume $T_\tau = 1$ and $P_\tau = \bar{\gamma}_u$, i.e., the error variance $\sigma_{E,u}^2$ of user u is given by

$$\sigma_{E,u}^2 = (1 + \bar{\gamma}_u)^{-1}. \tag{2}$$

3.2 Outdated CQI

Since there exists a time delay T between the time instance when measuring the SNR and the actual time of data transmission, the CQI available at the BS is outdated. Assuming that the channel follows Jakes' model, the actual channel and the outdated channel are correlated with a correlation coefficient of $\rho = J_0(2\pi f_D T)$, with $J_0(x)$ denoting the 0th-order Bessel function of the first kind and f_D the Doppler frequency. Hence, the correlation coefficient ρ_u of user u is given by

$$\rho_u = J_0 \left(2\pi f_{D,u} T\right), \tag{3}$$

where $f_{D,u}$ designates the Doppler frequency of user u.

4 Adaptive Transmission Applying WPFS

In the following, the CQI values are applied to perform WPFS in order to allocate the different subcarriers to the different users according to their priority and channel quality. Let p_u be the priority factor which is $p_u = 1$ for all users of set $\mathcal{S}_L$ and $p_u = p$, $p \geq 1$, for all users of set $\mathcal{S}_H$. Subcarrier n in time slot k is allocated to user $u^*(n, k)$ with the highest ratio between the weighted instantaneous SNR and the average SNR given by

$$u^*(n, k) = \arg\max_u \left\{ \frac{p_u \cdot \gamma_u(n, k)}{\bar{\gamma}_u} \right\}. \tag{4}$$

Integrating over the joint probability density function of the weighted and normalised SNR values of (4), the probability $F_H(p)$ that a subcarrier is allocated to a high priority user as a function of the priority factor p is calculated by

$$F_H(p) = \int_{y_1=0}^{\infty} \int_{y_2=0}^{y_1} \cdots \int_{y_{U_H}=0}^{y_1} \int_{z_1=0}^{y_1} \cdots \int_{z_{U_L}=0}^{y_1} \left(\frac{1}{p} \cdot e^{-\frac{y_1}{p}}\right) \cdot \left(\frac{1}{p} \cdot e^{-\frac{y_2}{p}}\right)$$
$$\cdots \left(\frac{1}{p} \cdot e^{-\frac{y_{U_H}}{p}}\right) \cdot e^{-z_1} \cdot \ldots \cdot e^{-z_{U_L}} \, dy_1 dy_2 \ldots dy_{U_H} dz_1 \ldots dz_{U_L}$$
$$= \int_0^{\infty} \left(1 - e^{-\frac{y_1}{p}}\right)^{U_H - 1} \cdot (1 - e^{-y_1})^{U_L} \cdot \frac{1}{p} \cdot e^{-\frac{y_1}{p}} \, dy_1. \tag{5}$$

Applying the binomial theorem, the integral in (5) can be solved resulting in

$$F_H(p) = \sum_{m=0}^{U_H-1} \binom{U_H-1}{m} \sum_{l=0}^{U_L} \binom{U_L}{l} \frac{(-1)^{m+l}}{1+m+p \cdot l}. \tag{6}$$

The probability $F_L(p)$ that a subcarrier is allocated to a low priority user can be calculated directly from (6) resulting in

$$F_L(p) = \frac{1}{U_L} \cdot (1 - F_H(p) \cdot U_H) \text{ with } U_L \geq 1. \tag{7}$$

In the following, we introduce the priority gain g, which denotes the increase of channel access probability for high priority users compared to PFS, where all users have the same priority and the channel access probability is $1/U$ for each user. Hence, we have to determine the priority factor p in such a way that

$$F_H(p) = \frac{g}{U}, \tag{8}$$

which can be done numerically using for example the *fzero* function in MATLAB™. From (7) it can be seen that g is upper bounded by $g \leq \frac{U}{U_H}$ since $F_L(p)$ has to be non-negative. Furthermore, $g \geq 1$, since for the priority factor $p \geq 1$ holds true, with $g = 1$, i.e. $p = 1$, corresponding to PFS where each user has the same channel access probability. Thus,

$$1 \leq g \leq \frac{U}{U_H}. \tag{9}$$

After all subcarriers are allocated to the different users, the modulation scheme is selected for each allocated subcarrier based on the SNR value, i.e., the modulation is adapted to the pathloss and the fast fading. In this work, uncoded M-QAM and M-PSK modulation are considered.

5 Joint Impact of Imperfect CQI and User Priority

In the following, the joint impact of imperfect CQI and user priority on the performance of an OFDMA system applying WPFS is considered. In order to do so, the distribution of the SNR values of the selected users has to be derived, i.e. the Probability Density Function (PDF) and the Cumulative Density Function (CDF). Subsequently, closed form expressions for the average data rate and BER are derived analytically taking into account imperfect CQI and user priority. Finally, the data rate is maximised subject to a target BER.

5.1 SNR Distribution Considering User Priority

The PDF $p_{H,\hat{\gamma}}^{(u)}(\hat{\gamma})$ of the outdated and estimated SNR $\hat{\gamma}$ of a scheduled high priority user that successfully competed against $U_H - 1$ other high priority users and U_L low priority users is calculated according to

$$p_{H,\hat{\gamma}}^{(u)}(\hat{\gamma}) = a_H \underbrace{\int_0^{\hat{\gamma}} \cdots \int_0^{\hat{\gamma}}}_{U_H-1 \text{ times}} \underbrace{\int_0^{p \cdot \hat{\gamma}} \cdots \int_0^{p \cdot \hat{\gamma}}}_{U_L \text{ times}} \left(\frac{1}{\bar{\gamma}_{E,u}} \cdot e^{\frac{-y_1}{\bar{\gamma}_{E,u}}} \right) \cdots \left(\frac{1}{\bar{\gamma}_{E,u}} \cdot e^{\frac{-y_{U_H-1}}{\bar{\gamma}_{E,u}}} \right)$$

$$\cdot \left(\frac{1}{\bar{\gamma}_{E,u}} e^{\frac{-z_1}{\bar{\gamma}_{E,u}}} \right) \cdots \left(\frac{1}{\bar{\gamma}_{E,u}} \cdot e^{\frac{-z_{U_L}}{\bar{\gamma}_{E,u}}} \right) \cdot \left(\frac{1}{\bar{\gamma}_{E,u}} \cdot e^{\frac{-\hat{\gamma}}{\bar{\gamma}_{E,u}}} \right)$$

$$\cdot dy_1 \ldots dy_{U_H-1} dz_1 \ldots dz_{U_L}$$

$$= \frac{a_H}{\bar{\gamma}_{E,u}} \cdot e^{\frac{-\hat{\gamma}}{\bar{\gamma}_{E,u}}} \cdot \left(1 - e^{\frac{-p \cdot \hat{\gamma}}{\bar{\gamma}_{E,u}}} \right)^{U_L} \cdot \left(1 - e^{\frac{-\hat{\gamma}}{\bar{\gamma}_{E,u}}} \right)^{U_H-1}, \quad (10)$$

with $\bar{\gamma}_{E,u} = \bar{\gamma}_u \cdot (1 - \sigma_{E,u}^2)$. The factor a_H ensures that $\int_0^{\infty} p_{H,\hat{\gamma}}^{(u)}(\hat{\gamma}) d\hat{\gamma} = 1$, leading to

$$a_H = \left[\sum_{v=0}^{U_H-1} \binom{U_H-1}{v} \sum_{w=0}^{U_L} \binom{U_L}{w} \frac{(-1)^{v+w}}{1+v+p \cdot w} \right]^{-1}. \quad (11)$$

The PDF $p_{L,\hat{\gamma}}^{(u)}(\hat{\gamma})$ of the outdated and estimated SNR $\hat{\gamma}$ of a scheduled low priority user is calculated by exchanging U_L with U_H and $p \cdot \hat{\gamma}$ with $\hat{\gamma}/p$ in (10), respectively, resulting in

$$p_{L,\hat{\gamma}}^{(u)}(\hat{\gamma}) = \frac{a_L}{\bar{\gamma}_{E,u}} \cdot e^{\frac{-\hat{\gamma}}{\bar{\gamma}_{E,u}}} \cdot \left(1 - e^{\frac{-\hat{\gamma}}{p \cdot \bar{\gamma}_{E,u}}} \right)^{U_H} \cdot \left(1 - e^{\frac{-\hat{\gamma}}{\bar{\gamma}_{E,u}}} \right)^{U_L-1} \quad (12)$$

with

$$a_L = \left[\sum_{v=0}^{U_L-1} \binom{U_L-1}{v} \sum_{w=0}^{U_H} \binom{U_H}{w} \frac{p \cdot (-1)^{v+w}}{w + p \cdot (1+v)}\right]^{-1}. \quad (13)$$

Integrating (10) and (12), the CDF $F_{\hat{\gamma}}^{(u)}(\hat{\gamma})$ of the outdated and estimated SNR of a scheduled high and low priority user is given by

$$F_{\hat{\gamma}}^{(u)}(\hat{\gamma}) = \alpha \cdot \sum_{v=0}^{V-1} \binom{V-1}{v} \sum_{w=0}^{W} \frac{\binom{W}{w}(-1)^{v+w}}{1+v+\varphi \cdot w} \cdot \left(1 - e^{-\frac{\hat{\gamma}(1+v+\varphi \cdot w)}{\bar{\gamma}_{E,u}}}\right) \quad (14)$$

with $\alpha = a_H$, $\varphi = p$, $V = U_H$ and $W = U_L$ for high priority users and $\alpha = a_L$, $\varphi = 1/p$, $V = U_L$ and $W = U_H$ for a low priority users.

5.2 Average Data Rate

The average sum bit per symbol rate is formulated as the sum rate of the different modulation constellations weighted by their probability. Assuming that there are M modulation schemes available, $\gamma^{(u)} = [\gamma_0^{(u)}, \gamma_1^{(u)}, ..., \gamma_M^{(u)}]^T$, with $\gamma_0^{(u)} = 0$ and $\gamma_M^{(u)} = \infty$, denotes the threshold vector of user u which contains the SNR threshold values determining the interval in which a particular modulation scheme is applied. Thus, the average data rate $\bar{R}_{H/L}^{(u)}$ of user u for high and low priority users can be formulated as

$$\bar{R}_{H/L}^{(u)} = \sum_{m=1}^{M} \int_{\gamma_{m-1}^{(u)}}^{\gamma_m^{(u)}} b_m \cdot p_{H/L,\hat{\gamma}}^{(u)}(\hat{\gamma}) \, d\hat{\gamma} \quad (15)$$

with b_m denoting the number of bits per symbol corresponding to the applied modulation scheme. Using (14), (15) can be written as

$$\bar{R}_{H/L}^{(u)} = \sum_{m=1}^{M} b_m \cdot \left(F_{\hat{\gamma}}^{(u)}(\gamma_m^{(u)}) - F_{\hat{\gamma}}^{(u)}(\gamma_{m-1}^{(u)})\right). \quad (16)$$

5.3 Average BER

In the following, we use the approximation of the instantaneous BER for M-QAM and M-PSK modulation introduced in [8] given by

$$BER_m(\gamma) = 0.2 \cdot \exp(-\beta_m \gamma) \quad (17)$$

with $m = 1, ..., M$, where $\beta_m = \frac{1.6}{2^{b_m}-1}$ using M-QAM modulation and $\beta_m = \frac{7}{2^{1.9b_m}+1}$ using M-PSK modulation, respectively. The average BER is then defined as the sum of the average bit errors of the different modulation constellations divided by the average bit rate [9]. To determine the average BER, we introduce the conditional PDF $p_{\gamma|\hat{\gamma}}^{(u)}(\gamma|\hat{\gamma})$ of the actual SNR γ and the outdated and estimated SNR $\hat{\gamma}$ of user u given by

$$p_{\gamma|\hat{\gamma}}^{(u)}(\gamma|\hat{\gamma}) = \frac{1}{\bar{\gamma}_u \sigma_{r,u}^2} \cdot \exp\left(-\frac{\rho_u^2 \cdot \hat{\gamma} + \gamma}{\bar{\gamma}_u \sigma_{r,u}^2}\right) \cdot I_0\left(\frac{2\rho\sqrt{\gamma \cdot \hat{\gamma}}}{\bar{\gamma}_u \sigma_{r,u}^2}\right), \quad (18)$$

with $\sigma_{r,u}^2 = 1 - \rho_u^2(1 - \sigma_{E,u}^2)$ and $I_0(x)$ denoting the 0th-order modified Bessel function of the first kind. The average BER $\overline{BER}_{H/L}^{(u)}$ of user u for high and low priority users is then given by

$$\overline{BER}_{H/L}^{(u)} = \frac{1}{\bar{R}_{H/L}^{(u)}} \sum_{m=1}^{M} b_m \int_{\gamma_{m-1}^{(u)}}^{\gamma_m^{(u)}} p_{H/L,\hat{\gamma}}^{(u)}(\hat{\gamma}) \cdot \left[\int_0^\infty BER_m(\gamma) \cdot p_{\gamma|\hat{\gamma}}^{(u)}(\gamma|\hat{\gamma}) d\gamma\right] d\hat{\gamma}. \quad (19)$$

Inserting (10), (12), (17) and (18) in (19), (19) can be rewritten to

$$\overline{BER}_{H/L}^{(u)} = \frac{\alpha \cdot U}{5 \cdot \bar{R}_{H/L}^{(u)}} \sum_{m=1}^{M} b_m \sum_{v=0}^{V-1} \binom{V-1}{v} \sum_{w=0}^{W} \binom{W}{w} \quad (20)$$

$$\cdot \frac{(-1)^{v+w}}{A(m,v,w)} \cdot \left[e^{\frac{-\gamma_{m-1}^{(u)} \cdot A(m,v,w)}{\bar{\gamma}_{E,u} \cdot (1+\beta_m \bar{\gamma}_u \sigma_{r,u}^2)}} - e^{\frac{-\gamma_m^{(u)} \cdot A(m,v,w)}{\bar{\gamma}_{E,u} \cdot (1+\beta_m \bar{\gamma}_u \sigma_{r,u}^2)}} \right]$$

with $A(m,v,w) = (1 + v + \varphi \cdot w) \cdot (1 + \beta_m \bar{\gamma}_u \sigma_{r,u}^2) + \bar{\gamma}_{E,u} \beta_m \rho_u^2$. Note that $\alpha = a_H$, $\varphi = p$, $V = U_H$ and $W = U_L$ for high priority users and $\alpha = a_L$, $\varphi = 1/p$, $V = U_L$ and $W = U_H$ for low priority users.

5.4 Optimising Data Rate

In the following, we are looking for the optimal modulation scheme threshold vector $\gamma^{(u)}$ of user u which maximises the average data rate under the constraint of a target BER BER_T, i.e., we have to solve the following optimization problem:

$$\bar{R}_{H/L,opt}^{(u)} = \max_{\gamma^{(u)}} \left(\bar{R}_{H/L}^{(u)}(\gamma^{(u)}) \right) \quad (21)$$

$$\text{subject to} \quad \overline{BER}_{H/L}^{(u)}(\gamma^{(u)}) \leq BER_T.$$

To solve (21), we perform a Lagrange multiplier approach similar to [9] where the objective function $\Phi^{(u)}(\gamma)$ is given by

$$\Phi^{(u)}(\gamma^{(u)}) = \bar{R}_{H/L}^{(u)}(\gamma^{(u)}) + \lambda$$
$$\cdot \left(\bar{R}_{H/L}^{(u)}(\gamma^{(u)}) \overline{BER}_{H/L}^{(u)}(\gamma^{(u)}) - \bar{R}_{H/L}^{(u)}(\gamma^{(u)}) BER_T \right) \qquad (22)$$

with λ denoting the Lagrange multiplier. In order to determine the optimal threshold vector $\gamma_{opt}^{(u)}$, we have to differentiate $\Phi^{(u)}(\gamma^{(u)})$ with respect to the elements of $\gamma^{(u)}$, where $\frac{\partial \Phi^{(u)}(\gamma_{opt}^{(u)})}{\partial \gamma_m^{(u)}} = 0$ must hold for all $m = 1, ..., M-1$. Let $\zeta^{(u)}(\hat{\gamma}, m, \sigma_{E,u}^2, \rho_u)$ denote the solution of the inner integral of (19) given by

$$\zeta^{(u)}(\hat{\gamma}, m, \sigma_{E,u}^2, \rho_u) = \frac{0.2}{1 + \beta_m \bar{\gamma}_u \sigma_{r,u}^2} \cdot \exp\left(-\frac{\hat{\gamma} \rho_u^2 \beta_m}{1 + \beta_m \bar{\gamma}_u \sigma_{r,u}^2} \right). \qquad (23)$$

Inserting (15) and (19) in (22) and using (23), the derivation results in $M - 1$ equations given by

$$\frac{(1 - \lambda BER_T)}{\lambda} = \frac{1}{b_{m+1} - b_m} \left(\zeta^{(u)}(m, \gamma_m^{(u)}, \sigma_{E,u}^2, \rho_u) \right. \qquad (24)$$
$$\left. \cdot b_m - \zeta^{(u)}(m+1, \gamma_m^{(u)}, \sigma_{E,u}^2, \rho_u) \cdot b_{m+1} \right), \quad m = 1, \ldots, M - 1.$$

From (24) it can be seen that each element $\gamma_m^{(u)}$ of the optimal threshold vector $\gamma_{opt}^{(u)}$ can be calculated using an initial value $\gamma_1^{(u)}$. Thus, each threshold vector $\gamma^{(u)}$ is a function of the initial value $\gamma_1^{(u)}$, i.e., $\gamma^{(u)} = f(\gamma_1^{(u)})$. Determining the maximum average data rate subject to the target BER, we have to find the optimal initial value $\gamma_{1,opt}^{(u)}$ which fulfills

$$\overline{BER}_{H/L}^{(u)}(f(\gamma_{1,opt}^{(u)})) \leq BER_T, \qquad (25)$$

which again can be done numerically using for example the *fzero* function in MATLAB™.

6 Numerical Results

In the following, we consider an OFDMA scheme applying WPFS with $U = 25$ users. For simplicity, we assume that the average SNR in the system is $\bar{\gamma} = 10$ dB for all users. The target BER is set to $BER_T = 10^{-3}$. First, we assume perfect CQI and $U_H = 3$ high priority users and consequently $U_L = 22$ low priority users, i.e., $1 \leq g \leq 8.33$. In Fig. 1a, the average number of transmitted bits per allocated subcarriers is depicted as a function of the priority gain g, which is related to the priority factor p according to (8). As one can see from the figure, the total system data rate

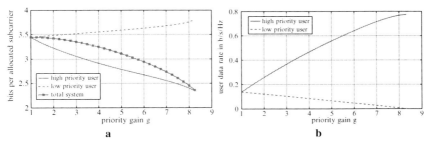

Fig. 1 (a) Number of transmitted bits per allocated subcarrier vs. priority gain, (b) user data rate vs. priority gain with $U_H = 3$ high and $U_L = 22$ low priority users

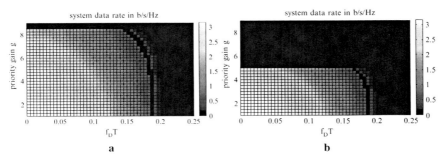

Fig. 2 System data rate vs. time delay $f_D T$ and priority gain g with (a) $U_H = 3$, (b) $U_H = 5$ high priority users

decreases when increasing the priority gain, since favouring the high priority users even if they are in bad channel conditions results in a performance degradation. From the solid line, representing the number of transmitted bits per subcarrier when allocated to a high priority user, one can see that the number of bits decreases with increasing gain g. In Fig. 1b, the user data rate of a high priority user is depicted as a function of the priority gain. One can see that the user data rate increases due to the increased access to the channel. For the low priority users it is vice versa, i.e., the number of bits per subcarrier, when allocated to a low priority user, increases with increasing g, see Fig. 1a, since only strong channels of low priority users can compete successfully with the favoured channels of high priority users. However, the user data rate of a low priority user decreases due to the reduced channel access, see Fig. 1b.

Next, the CQI is assumed to be outdated expressed by the normalised time delay $f_D T$, where the Doppler frequency f_D is assumed to be the same for each user. Furthermore, the SNR values are noisy estimates with $\sigma_E^2 = (1+\bar{\gamma})^{-1}$ for each user. In the following, we investigate the joint impact of imperfect CQI and user priority on the performance of the system. In Fig. 2a, the number U_H of high priority users remains $U_H = 3$. The average system data rate indicated by different colours is depicted as a function of the time delay $f_D T$ and the priority gain g. As one can see, the achievable data rate is highest for small time delays and low priority gains. When increasing g for a given $f_D T$, the system data rate decreases as seen in Fig. 1a. At a certain value of g_{max}, the priority gain cannot be achieved any more, i.e., the data rate is zero, where g_{max} is upper bounded by $g_{max} \leq \frac{U}{U_H}$ as shown in Section 4. When increasing $f_D T$ for a given priority gain g, the data rate also decreases, since a more robust modulation scheme is required to cope with the outdated CQI in order to fulfill the BER requirements. In Fig. 2b, the number U_H of high priority users is changed to $U_H = 5$. One can see that the region in which a transmission is possible diminishes, since the upper bound of g_{max} decreases with an increasing number U_H of high priority users.

7 Conclusions

In this paper, we analyse the performance of an adaptive OFDMA system applying WPFS with high priority users in the presence of imperfect CQI. Closed form expressions for the average data rate and BER are analytically derived taking into account the joint impact of imperfect CQI and user priority. From the numerical results one can conclude that serving users with different priorities comes at the expense of reduced system data rate and less robustness against outdated CQI.

References

1. R.D.J. van Nee and R. Prasad, *OFDM for Wireless Communications*, Artech House, Boston, 2000.
2. P. D. Morris and C. R. N. Athaudage, "Fairness based resource allocation for multi-user MIMO-OFDM systems," in *Proc. of Vehicular Technology Conference*, Melbourne, Australia, May 2006.
3. S. Ryu, B. Ryu, H. Seo, and M. Shin, "Urgency and efficiency based packet scheduling algorithm for OFDMA wireless system," in *Proc. of IEEE ICC*, vol. 4, Seoul, Korea, May 2005.
4. J. Kim, E. Kim, and K. S. Kim, "A new efficient BS scheduler and scheduling algorithm in WiBro systems," in *Proc. of International Conference Advanced Communication Technology*, Phoenix Parc, Korea, Feb. 2006.
5. H. Kim, K. Kim, Y. Han, and J. Lee, "An efficient scheduling algorithm for QoS in wireless packet data transmission," in *Proc. of IEEE PIMRC*, Lisboa, Portugal, Sept. 2002.
6. A. Fernekess, A. Klein, B. Wegmann, and K. Dietrich, "Influence of High Priority Users on the System Capacity of Mobile Networks," in *Proc. of IEEE Wireless Communications & Networking Conference*, Hong Kong, China, Mar. 2007.
7. B. Hassibi and B. Hochwald, "How much training is needed in multiple antenna wireless links ?," *IEEE Trans. Inf. Theory*, vol. 49, pp. 951–963, April 2003.
8. S. T. Chunk and A. Goldsmith, "Degrees of freedom in adaptive modulation: A unified view," *IEEE Trans. Communications*, vol. 49, pp. 1561–1571, Sept. 2001.
9. Q. Ma and C. Tepedelenlioglu, "Practical Multi-user diversity with outdated channel feedback," *IEEE Trans. on Vehicular Technology*, vol. 54, no. 4, pp. 1334–1345, July 2005.

Adaptive BICM-OFDM Systems

Carsten Bockelmann, Dirk Wübben, and Karl-Dirk Kammeyer

Abstract Link Adaptation for Orthogonal Frequency Division Multiplex (OFDM) systems is an important topic to achieve higher spectral efficiencies or better error rate performance. However, channel coding has been neglected previously, only gaining some attention recently. In this paper, an overview over existing techniques and algorithms to adapt Bit Interleaved Coded Modulation (BICM) systems are given and new results regarding the extension of known approaches to multiple antenna OFDM systems are presented.

1 Introduction

The adaptation of power and modulation of a set of parallel channels, e.g., using Orthogonal Frequency Division Multiplex (OFDM), is an important topic in research. Early works focused on uncoded BER optimisation [1–3], but nowadays the consideration of channel coding into bit and power loading schemes has become a major aspect. Several contributions to this field have been made, especially considering the influence of Bit Interleaved Coded Modulation (BICM) on such systems. With regard to the channel capacity of BICM systems, respectively the average mutual information (AMI) of a code word, strategies have been devised to either maximise the transmittable data rate [4–6] or minimise the bit or frame error rate at a fixed data rate [7–10]. In order to compare and analyse these approaches, the respective optimisation problems for coded systems will be pointed out and different solutions will be discussed with regard to their complexity, performance and specific differences of the given approaches.

One very interesting extension of the coded bit and power loading field is the application to multiple-antenna systems. Besides the approach of diagonalisation, which requires perfect channel state information (CSI), few contributions have been made. Especially, regarding suboptimal receivers (e.g., linear receivers), new results will be discussed.

C. Bockelmann (✉), D. Wübben, and K.-D. Kammeyer
Department of Communications Engineering, University of Bremen, Germany
e-mail: {bockelmann, wuebben, kammeyer}@ant.uni-bremen.de

The remainder of this paper is organised as follows. In Section 2 the system model and important parameters are shortly stated, which are then used in Section 3 to formulate the examined optimisation problems and discuss the respective solutions. In Section 4 the single antenna results are then extended to multiple antenna systems. Finally, Section 5 concludes this paper.

2 System Model

We consider an OFDM system with N_T transmit and N_R receive antennas, which is assumed to be perfectly synchronised and free of intersymbol and inter-channel interference (ISI, ICI). Furthermore, CSI at both transmitter and receiver side shall be perfectly known. Thus, the equivalent baseband system in frequency domain is given by

$$\mathbf{y}_k = \mathbf{H}_k \mathbf{P}_k^{1/2} \mathbf{d}_k + \mathbf{n}_k \,, \tag{1}$$

where $\mathbf{H}_k \in \mathbb{C}^{N_R \times N_T}$ and $\mathbf{P}_k^{1/2} = \text{diag}\{[\sqrt{p_{1,k}}, \cdots, \sqrt{p_{N_T,k}}]\}$ denote the channel and power allocation matrix on subcarrier $k = 1, \ldots, N_C$ and $\mathbf{y}_k \in \mathbb{C}^{N_R}$, $\mathbf{d}_k \in \mathbb{C}^{N_T}$, $\mathbf{n}_k \in \mathbb{C}^{N_R}$ are the receive vector, transmit vector and noise vector, respectively. The total transmit power is given by $\mathcal{P} = \sum_{i=1}^{N_T} \sum_{k=1}^{N_C} p_{i,k}$ and the power of the i.i.d. noise $\mathbf{n}_k \sim \mathcal{N}_C(0, \sigma_n^2 \mathbf{I}_{N_R})$ is fixed to $\sigma_n^2 = 1$. To each subcarrier k and antenna i an individual alphabet (M-QAM) of cardinality $M_{i,k}$ may be assigned. In terms of forward error correction (FEC) non-systematic non-recursive convolutional encoders of rates $R_C \in \{1/4, 1/3, 1/2, 2/3, 3/4\}$ and constraint length $L_C = 3$ are applied throughout this paper leading to a variety of possible *transmission modes* (code/modulation combinations).

As a special case, the system model for a single antenna at receiver and transmitter $N_T = N_R = 1$ results

$$y_k = h_k \cdot \sqrt{p_k} \cdot d_k + n_k \,, \tag{2}$$

which will be used throughout Section 3.

3 Link Adaptation for Single Antenna Systems

In general two classes of link adaptation problems can be defined: rate maximisation and error rate enhancement. The latter can be achieved directly by optimisation of some measure of the bit or frame error rate (BER/FER) or by optimising the required power for a certain error constraint. The specific approach to each of these optimisation problems, however, varies according to the assumptions on performance measures and constraints. Two widely used quality indicators are the capacities and signal-to-noise-ratios (SNRs) of a set of parallel channels. SNR requirements can be formulated for the allowed *modes* and capacity results are generally a good indicator for the performance of strong channel codes. Considering single-antenna systems both measures are exchangeable to some extend. However, regarding finite input alphabets in contrast to Gaussian input alphabets, a saturation at the maximum rate of the input alphabet occurs. Thus, higher SNRs may provide no gain in terms of capacity, whereas lower BER/FER may still be achieved if the actual performance of a coded system is considered. Furthermore, the resulting capacity expression can no longer be solved in closed form, even though an efficient numerical solution via Gauss-Hermite quadrature [11] is possible. Nevertheless, considering the bitwise capacity the BER/FER performance of a system can be described independently from the applied alphabet by the AMI of a code word [4, 6]. This observation has motivated a class of algorithms, which will be described in Section 3.2.

On the other hand, the issue of saturation still remains, especially for non-capacity approaching codes, justifying another approach, which utilises the AWGN performance of an equivalent coded system to describe the SNR requirements of certain *modes*. The following section will discuss this idea in more detail.

3.1 Simulation Supported Methods

In the following the description will be focused on the rate maximisation problem as the problem solutions of both rate optimisation and error enhancement are almost identical except for some realisation aspects. Aiming at the maximisation of the data rate, we have to solve

$$\text{maximize } R_{\text{Total}} = \sum_{k=1}^{N_C} r_k$$
$$\text{subject to } \sum_{k=1}^{N_C} p_k = \mathcal{P} \quad \text{and} \quad \text{P}_{b/f} < \text{P}_{\text{Target}}, \qquad (3)$$

where r_k and p_k denote the rate and power on subcarrier k, respectively. $\mathcal{P}$ is the overall available power and P_{Target} denote a service of quality target in form of a minimum BER P_b or FER P_f, which should be obtained.

One important question arises in this context: how are the rates and powers of the parallel subchannels connected? In order to carry out the optimisation, the rates have to be formulated as explicit functions of the powers, such that $r_k = f(p_k)$, which in case of coded systems with finite symbol alphabets is a non-trivial problem. If channel coding is neglected, analytical BER/FER expressions, e.g., for M-QAM modulations, can be used [1–3]. Due to the difficulties in finding analytical expressions for the FER of coded systems, however, the simulated performance of different code and modulation combinations is often utilised to quantify the performance [8, 10]. The error rate constraint P_{Target} provides the necessary connection of power and rate to calculate a look-up table of the rate-power function, where only discrete points can be achieved. This provides an operating point allowing for some greedy or heuristic procedure to solve the optimisation problem approximately.

Equation (3) has been solved by the authors via expansion to coded systems of the well known bisection approach used by Krongold et al. [1]. The set of all rate-power pairs at a given BER/FER P_{Target} of all allowed transmission modes, employing SNR thresholds obtained from AWGN simulation as well as analytical expressions for uncoded M-QAMs, has been used to find a convex set allowing for a convex optimisation solution of (3) via Lagrangian multipliers [10]. The important difference to previous approaches are a fine grained power allocation and the optimality of the bisection solution under the given constraints.

Stiglmayr et al. [8] developed a very similar approach, however, solving a capacity maximisation problem via linearisation, which resulted in a simple heuristic. Here, also a look-up table of SNR thresholds from AWGN simulations of equivalent frame length has been applied. One important difference to the previously mentioned solution is, that out of all available *modes*, only these with the same rate but higher SNR threshold than the minimum one at that rate are excluded, whereas a convex set over all possible points may exclude modes viable in terms of the SNR threshold.

3.2 Capacity Based Approaches

The derivation of bit and power loading algorithms employing mutual information (MI) with Gaussian signalling as a quality indicator is widely used in literature and capacity losses due to finite symbol alphabets have been incorporated by application of the well known gap approximation, e.g. [12]. However, this approach still neglects the BICM system structure widely adapted in communication systems [13]. Recent research, hence, considers the BICM or parallel decoding capacity of a system promising a better performance estimate due to consideration of the interleaving and negligence of dependencies between the bits of a symbol. An equivalent optimisation problem to (3) in capacity terms is

$$\text{maximize} \quad R_{\text{Total}} = \sum_{k=1}^{N_C} r_k$$

$$\text{subject to} \quad \sum_{k=1}^{N_C} p_k = \mathcal{P} \quad \text{and} \quad \text{MI}_{\text{avg}} \leq \text{MI}_{\text{req}}, \tag{4}$$

where MI_{avg} and MI_{req} denote the AMI of one code word and the minimum required AMI, respectively. Furthermore, here the r_k are limited to integer number of bits as available through M-QAM constellations.

For nearly capacity achieving codes (Turbo codes, LDPC) it was found out, that the AMI of a code word is a good indicator for the error rate performance and approximately independent of the applied modulation, e.g. [4]. Even if the assumptions of perfect codes and infinitely long code words used to derive the corresponding capacity expression are never exactly achievable with practical codes, the mean BER/FER can be described quite accurately also for weaker codes and shorter code word lengths. Unfortunately, the BICM capacity is a nonlinear function of the SNR and has to be calculated and stored as look-up tables beforehand. Then, interpolation can be used to calculate the capacities of the subchannels. To ease memory consumption, the number of different entries in such a look-up table can be limited by thresholding at the cost of accuracy. Also, similar to the previous section and indicated by (4) a look-up table of SNR thresholds can be derived from AMI requirements using a certain target BER/FER.

Based on these observations, e.g., Li et al. [4] proposed three algorithms, each determining power and bit allocations for a set of parallel channels in order to maximise the rate given a BER/FER requirement. The BER/FER requirement is translated to a minimum AMI requirement, which is employed to allocate suitable modulations and powers to the subchannels in a greedy fashion. Exemplary, considering Algorithm 1 of [4] a subchannel's allocated bits are incremented if it has the smallest mutual information loss of all subcarriers. The overall AMI requirement has to be checked after each allocation to ensure the constraints. This approach employs the construction of SNR look-up tables as pointed out before. A related solution has been proposed for MIMO systems by the authors [6].

Another idea has been developed by Sankar et al. [5], who approximated the BICM capacities by simple exponential functions allowing the closed form derivation of a waterfilling scheme, which can be used to adapt the rate and power of a set of parallel channels. However, their approach is hard to apply to real systems as optimised code rates may vary strongly, which leads to severe performance losses if rounding to available code rates is exercised. Similarly, Lozano et al. [14] developed a closed form waterfilling solution dependent on the MMSE functions of the finite alphabets, which have to be numerically determined. This so called mercury waterfilling provides the optimal power allocation for a fixed bit allocation complementing schemes that neglect power loading.

The equivalent problem of error rate minimisation has been tackled by the authors with an algorithm that maximises the AMI of a code word for a single-input-single-output (SISO) system given an overall data rate [9]. Through this approach, no error rate requirement is needed as no power allocation beyond the reallocation of power from switched off subchannels is applied. Independently, Stiersdorfer et al. [7] presented a fast bit loading algorithm using a coarse approximation of the bit level capacities to distribute bits, which is very fast, but also results in slight to sever performance losses depending on the application scenario.

3.3 Results

Figure 1 shows the Monte Carlo simulations of several algorithms for rate optimisation, including their respective constraints. As can be observed, the FER requirement $P_{\text{Target}} = 10^{-2}$ is roughly fulfilled by all approaches, with the tightest fulfilment up to medium SNR by the coded bisection solution. Only the "Coded Bisection" and

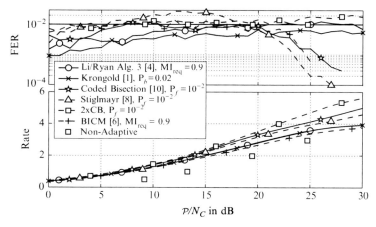

Fig. 1 FER and Rate vs average power per subcarrier $\mathcal{P}/N_C$ for convolutional codes of constraint length $L_C = 3$; $N_C = 1024$ subcarriers, $L_F = 10$ channel taps, target FER of $P_{Target} = 10^{-2}$

"Stiglmayr" algorithms apply code rate adaptation, thus, achieving an overall better performance than all fixed code rate solutions in terms of the supported rate. "Li/Ryan", "BICM" and "Krongold" show very similar results. At some point all methods reach saturation – near the maximum achievable rate no further improvement is possible – thereby overfulfilling the FER constraint with growing power. In case of the "Coded Bisection" the performance can be further enhanced by a two step process, running the algorithm once, fixing the code rate of the outer code, then running it a second time considering the fixed code rate, which is shown in "2xCB". The additional gain due to the new degree of freedom, the choice of the code rate, which is offered by the simulation supported methods, obviously allows for even higher transmission rates.

Figure 2 shows results regarding the error rate enhancement at different spectral efficiencies of 1, 2 and 3 bit/s/Hz. Similar to the previous results, the additional adaptation of code rates allows for far greater performance gains. For this example the approximation used for "Stiersdorfer" is obviously not valid. Applying stronger codes, however, will lessen the overall loss considerably as will more diversity, i.e. more channel taps.

4 Extension to Multiple Antenna Systems

MIMO-OFDM is a promising technique considered for future communication systems. The extension of the aforementioned link adaptation schemes to multiple antennas, however, is not straight forward and depends on the MIMO transmission scheme. If perfect transmit CSI can be assumed, channel diagonalisation via singular value decomposition (SVD) is a well known approach shown to be almost

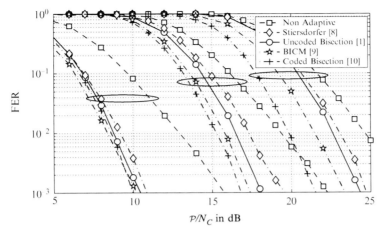

Fig. 2 FER vs average power per subcarrier $\mathcal{P}/N_C$ for convolutional codes of constraint length $L_C = 3$ at spectral efficiencies of 1, 2 and 3 bit/s/Hz (*left* to *right*); $N_C = 1024$ subcarriers, $L_F = 10$ channel taps

optimal utilising waterfilling with the gap approximation [15] under the assumption of an uncoded transmission. Independent parallel spatial channels would again allow for the application of all previously discussed approaches.

However, achieving full transmit CSI in a MIMO-OFDM system requires vast amounts of feedback. Thus, the authors proposed a solution for spatial multiplexing without spatial power and bit loading, that shows good performance [6]. Considering MIMO channels poses the problem, that the BICM capacity is not only depended on the transmit power, but also the channel matrix $\mathbf{H}_k$, ruling out efficient look-up table solutions. Instead the capacity has to be numerically calculated for every subcarrier by Monte-Carlo integration, which is computationally prohibitive. Using approximations, however, allows for a near optimal efficient solution [6]. Recent results on the capacity of MIMO systems with finite alphabets, e.g., [16] and [17], may either be used to achieve lower complexity solutions adapting [6] to suboptimal receivers or to provide the necessary BICM capacities to apply SISO link adaptation schemes.

Another quite straightforward approach is the consideration of linear zero-forcing (ZF) receivers leading to N_T equivalent AWGN channels with noise power $\sigma_{k,\ell}^2 = \sigma_n^2 \{(\mathbf{H}_k^H \mathbf{H}_k)^{-1}\}_{\ell,\ell}$; $\ell = 1, \cdots, N_T$ on each subcarrier k. Simply applying one of the SISO loading strategies, though, neglects the colouring of the noise and assumes independence of the equivalent AWGN channels on a subcarrier. Due to this, the solution will inherently be suboptimal, as antennas will be switched off by the algorithm, which is unconsidered by the receive filter. A joint receive filter design and loading should be developed in such a case.

Figure 3 shows results for ZF receive filtering, where the author's BICM loading approach has been used to adapt the resulting set of "parallel" channels. Even though channel dependencies are neglected, huge performance gains can be achieved.

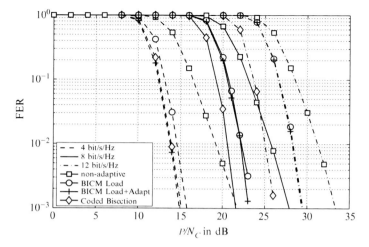

Fig. 3 FER vs average power per subcarrier $\mathcal{P}/N_C$ for ZF receive filtering and convolutional codes of constraint length $L_C = 3$; $N_C = 1024$ subcarriers, $N_T = N_R = 4$ antennas, $L_F = 10$ channel taps

"BICM Load+Adapt" additionally shows results for adapted ZF receive filters with respect to switched off antennas. As can be seen, only a relatively small enhancement is possible, which may be expanded by a few iterations of link and filter adaptation. A similar approach regarding rate optimisation is presented in [18].

5 Conclusion

A variety of solutions to the resource allocation problem of coded adaptive systems has been discussed in this contribution, where two principle approaches have been identified. The BICM capacity and simulated FER rates both offer multiple algorithmic solutions, which show performance differences due to their heuristic nature. The best approaches to the knowledge of the authors, however, adapt modulation, power and channel code jointly. Application of the discussed approaches is restricted to parallel SISO channels, necessitating some form of channel diagonalisation in the case of multiple antennas. Depending on the available transmitter side CSI, this may be carried out as a singular value decomposition or some suboptimal linear filter leading to self-interference. Regarding multi-antenna OFDM systems, a vast amount of feedback would be needed, so that alternative solutions are required. To this end, an approach for MIMO systems utilising ZF receive filters has been presented. Even though, dependencies of the spatial channels are neglected, good performance can be achieved.

References

1. B.S. Krongold, K. Ramchandran, and D.L. Jones. Computationally Efficient Optimal Power Allocation Algorithms for Multicarrier Communication Systems. *IEEE Trans. on Communications*, 48(1):23–27, January 2000.
2. R.F.H. Fischer and J.B. Huber. A New Loading Algorithm for Discrete Multitone Transmission. In *IEEE Global Telecommunications Conference*, pages 724–728, London, England, November 1996.
3. H. Hartogs. Ensemble modem structure for imperfect transmission mead. US Patent No. 4,731,816, March 1988.
4. Y. Li and W.E. Ryan. Mutual-Information-Based Adaptive Bit-Loading Algorithms for LDPC-Coded OFDM. *IEEE Trans. on Wireless Communication*, 6(5):1670–1680, May 2007.
5. H. Sankar and K.R. Narayanan. Design of Near-Optimal Coding Schemes for Adaptive Modulation with Practical Constraints. In *Int. Conference on Communications*, Istanbul, Turkey, June 2006.
6. C. Bockelmann, D. Wübben, and K.-D. Kammeyer. Mutual Information based Rate Adaptation for MIMO-OFDM System with Finite Signal Alphabets. In *Int. ITG Workshop on Smart Antennas*, Darmstadt, Germany, February 2008.
7. C. Stierstorfer and R.F.H. Fischer. Rate Loading in OFDM Based on Bit Level Capacities . In *Int. Symposium on Information Theory*, Toronto, Canada, July 2008.
8. S. Stiglmayr, M. Bossert, and E. Costa. Adaptive Coding and Modulation in OFDM Systems Using BICM and Rate-Compatible Punctured Codes. In *European Wireless*, Paris, France, April 2007.
9. C. Bockelmann and K.-D. Kammeyer. Adaptive Interleaving and Modulation for BICM-OFDM. In *Int. OFDM-Workshop (InOWO 2008)*, pages 40–44, Hamburg, Germany, August 2008.
10. C. Bockelmann, D. Wübben, and K.-D. Kammeyer. Efficient Coded Bit and Power Loading for BICM-OFDM. In *IEEE Vehicular Technology Conference (VTC2009-Spring)*, Barcelona, Spain, April 2009.
11. W.H. Press, S.A. Teukolsky, W.T. Vetterling, and B.P. Flannery. *Numerical Recipes 3rd Edition: The Art of Scientific Computing*. Cambridge University Press, Cambridge, 2007.
12. P.S. Chow, J.M. Cioffi, and J.A.C. Bingham. A practical discrete multitone transceiver loading algorithm for data transmission over spectrally shaped channels. *IEEE Trans. on Communications*, 43(234):773–775, Feb/Mar/Apr 1995.
13. G. Caire, G. Taricco, and E. Biglieri. Bit-Interleaved Coded Modulation. *IEEE Trans. on Information Theory*, 44(3):927–946, 1998.
14. A. Lozano, A.M. Tulino, and S. Verdu. Optimum Power Allocation for Parallel Gaussian Channels With Aribitrary Input Distributions. *IEEE Trans. on Information Theory*, 52(7):3033–3051, July 2006.
15. D.P. Palomar and S. Barbarossa. Designing MIMO Communication Systems: Constellation Choice and Linear Transceiver Design. *IEEE Trans. on Signal Processing*, 53(10):3804–3818, October 2005.
16. S. Stiglmayr, J. Klotz, and M. Bossert. Mutual Information of V-BLAST Transmission. In *IEEE Int. Symposium on Wireless Communication Systems*, Reykjavik, Iceland, October 2008.
17. P. Fertl, J. Jaldén, and G. Matz. Capacity-based performance comparison of MIMO-BICM demodulators. In *IEEE Workshop on Signal Processing Advances in Wireless Communications*, pages 166–170, Recife, Brasilien, July 2008.
18. S. Stiglmayr, J. Klotz, and M. Bossert. Adaptive Coding and Modulation in MIMO OFDMA Systems. In *IEEE Int. Symposium on Personal, Indoor and Mobile Radio Communications*, Cannes, France, September 2008.

Power Allocation with Interference Constraint in Multicarrier Based Cognitive Radio Systems

Musbah Shaat and Faouzi Bader

Abstract The cognitive radio technology aims to increase the spectrum utilization by accessing opportunistically the unused spectrum bands. Multicarrier communications are considered to be used in cognitive radio systems. In cognitive radio systems, the power allocation should be carried out within the cognitive radio network so that no excessive interference is caused to the primary user. In this paper, an iterative power allocation algorithm is proposed to maximize the downlink capacity of the cognitive radio system under both the total power budget and interference introduced to the primary user constraints. Moreover, two complexity reduced algorithms are presented. The problem of channel allocation in multiuser cognitive radio networks is discussed. Finally, the performance of the proposed algorithms is investigated using computer simulations.[1]

1 Introduction

Rapid development of wireless communication makes the spectrum scarcity as a one of the serious problems. The Federal Communication Commission (FCC) has reported that the most of the licensed spectrum is currently underutilized [1].

Cognitive radio (CR) [2, 3], which is an intelligent wireless communication system capable of learning from its radio environment and dynamically adjusting its transmission characteristics accordingly, is considered to be one of the possible solutions to solve the spectrum efficiency problem. By CR, a group of unlicensed users [referred to as secondary users (SU's)] can detect and use idle frequency channels (spectrum holes) without causing a harmful interference to the licensed users [referred to as primary users (PU's)] and thus implement efficient reuse of the licensed channels.

M. Shaat (✉) and F. Bader
Centre Tecnològic de Telecomunicacions de Catalunya-CTTC, Parc Mediterrani de la Tecnología,
Avd. Canal Olimpia s/n. 08860 Castelldefels-Barcelona, Spain
e-mail: {musbah.shaat, faouzi.bader}@cttc.es

[1] This work was partially supported by the European ICT-2008-211887 project PHYDYAS

Multicarrier (MC) communication system has been suggested as a candidate for the future CR systems due to its flexibility to fill the spectrum holes left by PU's as well as allocate resources between the different SU's [3, 4]. In [4], the authors proposed that when the band of the SU covers multiple PU's, then the SU modulates zero on the subcarriers belong to the detected PU's bands while utilizing other subcarriers for the transmission.

As the SU and PU bands may be exist side by side and their access technologies may be different, the mutual interference between the two systems is considered as a limiting factor affects the performance of both networks. In [5], the mutual interference between PU and SU was studied. This mutual interference depends on the transmitted power as well as the spectral distance between PU and SU.

In traditional multicarrier systems, optimal power allocation that maximizes the channel capacity is waterfilling on the subcarriers with the total power constraint. In cognitive radio multicarrier systems, an additional constraint should be introduced due to the interference caused by the sidelobes in different subcarriers. The transmit power of each subcarrier should be adjusted according to the channel status and the location of the subcarrier with respect to the PU spectrum [6]. In [7], an iterative partitioned waterfilling was presented. This algorithm aims to maximize the capacity of the CR system under a total power constraint with the consideration of the per subcarrier power constraint caused by the PU's interference limit. The interference caused by the side lobes of SU's subcarriers was not considered. In [8], the authors proposed an optimal power loading scheme using the Lagrange formulation. This loading scheme maximizes the downlink transmission capacity of the CR system while keeping the interference induced to the PU below a pre specified interference threshold. The total power constraint was not considered in this scheme.

Motivated by the work in [8], this paper proposes a power loading scheme to maximize the downlink capacity of the CR system under the interference and power constraints. In addition, reduced complexity algorithms are presented and compared with the proposed one. Furthermore, suggested metric guiding the subcarrier allocation in multi user CR systems is discussed. The rest of this paper is organized as follow: Section 2 gives the system model and formulates the problem. The proposed algorithm is described in Section 3. Section 4 discusses the multi user subcarrier allocation problem. Selected numerical results are given in Section 5. Section 6 concludes the paper.

2 System Model and Problem Formulation

The side by side distribution of the PU and SU will be assumed. As in Fig. 1, the frequency band B has been occupied by the PU. The CR band, the spectrum hole that can be used by the SU, is divided into N subcarriers each having a Δf bandwidth and lies next to the PU bands.

Assuming that $\Phi_i(f)$ is the power spectrum density (PSD) of the ith subcarrier in the CR band. The expression of the PSD depends on the used multicarrier tech-

Fig. 1 Distribution of primary and cognitive users

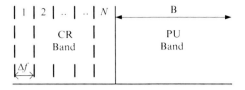

nique. If an orthogonal frequency division multiplexing (OFDM) based CR system is assumed, the PSD of the ith subcarrier in CR band can be written as [5]

$$\Phi_i(f) = P_i T_s \left(\frac{\sin \pi f T_s}{\pi f T_s} \right)^2 \quad (1)$$

where P_i is the total transmit power emitted by the ith subcarrier and T_s is the symbol duration. The interference introduced by the ith subcarrier to PU, $I_i(d_i, P_i)$, is the integration of the PSD of the ith subcarrier across the PU band, B, and can be expressed as [5]

$$I_i(d_i, P_i) = \int_{d_i - B/2}^{d_i + B/2} |g_i|^2 \Phi_i(f) df = P_i |g_i|^2 K_i \quad (2)$$

where g_i is the ith subcarrier channel gain from the cognitive base station (BS) to the PU. d_i is the spectral distance between the ith subcarrier of the CR band and the PU band. K_i denotes the interference factor of the ith subcarrier.

It will be assumed that each subcarrier goes under frequency flat fading gains and the instantaneous fading gains are perfectly known at the transmitter. Our objective is to maximize the total capacity of the SU subject to that the instantaneous interference introduced to the PU constraint and total transmit power constraint. Therefore, the problem can be formulated as follow

$$C = \max_{P_i} \sum_{i=1}^{N} \log_2 \left(1 + \frac{P_i |h_i|^2}{\sigma^2} \right)$$

$$\text{s.t.} \quad \sum_{i=1}^{N} I_i(d_i, P_i) \leq I_{th} \quad (3)$$

$$\sum_{i=1}^{N} P_i \leq P_T, \quad P_i \geq 0$$

where C denotes the transmission capacity of SU, N is the total number of subcarriers, h_i is the channel fading gain from the BS to the SU, I_{th} denotes the interference threshold prescribed by the PU, σ^2 is the mean variance of the additive white Gaussian noise (AWGN) and P_T is the total SU power budget.

3 Proposed Algorithm

The aim of the proposed algorithm is to maximize the channel capacity under the total power constraint as well as the total interference constraint. The problem is formulated mathematically in (3). If we ignore the interference constraint, the solution of the problem will be the well known waterfilling algorithm. On the other side, if the total power constraint is ignored, the optimal solution will be as derived in [8] and can be written as

$$P_i^* = \frac{1}{\lambda K_i |g_i|^2} - \frac{\sigma^2}{|h_i|^2}$$

$$\lambda = \frac{N}{I_{th} + \sum_{i=1}^{N} \frac{\sigma^2 K_i |g_i|^2}{|h_i|^2}} \quad (4)$$

Its obvious that if the summation of the allocated power under the interference condition only is lower than or equal the available total power budget, i.e. $\sum_{i=1}^{N} P_i^* \leq P_T$, (4) will be the optimal solution for capacity maximization problem under the total power and interference constraint. In most cases, the total power budget is quite lower than this summation and hence, we propose the following iterative **P**ower **I**nterference constraint algorithm named as *PI-Algorithm* to iteratively allocate the subcarrier power under both the total power and the interference constraint.

3.1 PI-Algorithm

Let $A = \{1, 2, \ldots, N\}$ to be the set of all the CR available subcarriers and P_i to be the solution of the optimization problem.

Step 1.

(a) Find the power allocation vector P_i^* according to interference constraint only using (4) and make it to be the maximum power that can be allocated to each subcarrier $P_{i\,max} = P_i^*$.
(b) If $\sum_{i=1}^{N} P_{i\,max} \leq P_T$, then the solution of the problem is found and hence $P_i = P_{i\,max}$, else continue.

Step 2. Find the power allocation vector P_{WFi} using a traditional iterative waterfilling under the constraints of total power, P_T, and the maximum power that can be allocated to each subcarrier, $P_{i\,max}$.

Step 3. Find the set of subcarriers $B \subset A$ (see Fig. 2) in which $P_{WFi} = P_{i\,max}$ and evaluate the left available interference $I_{left} = I_{th} - \sum_{i \in A} I_i(d_i, P_{WFi})$.

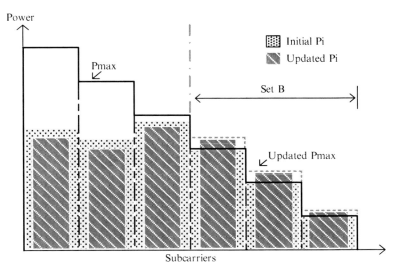

Fig. 2 An Example of SU's allocated power after two iterations

Step 4. If $I_{left} = 0$ or $B = \phi$, the solution of the problem is found where $P_i = P_{WFi}$, else, update $P_{i\,max}$ by applying (4) on the subcarriers in the set B only under the interference constraint $I' = I_{left} + \sum_{i \in B} I_i(d_i, P_{WFi})$ and go to step 2.

This algorithm converges to the required solution after several iterations. A graphical description of this algorithm can be found in Fig. 2.

In what follows, reduced complexity algorithms named as *(PI-A)-Algorithm* and *(PI-B)-Algorithm* are proposed.

3.2 (PI-A)-Algorithm

In this algorithm the power allocated will be done using only the steps 1 and 2 from the PI-Algorithm. The maximum power that can be allocated in each subcarrier is determined using (4) and then the iterative waterfilling algorithm is applied with the total power budget and maximum power per subcarrier constraints.

3.3 (PI-B)-Algorithm

This algorithm is different from (PI-A) Algorithm in the way of determining the maximum power that can be allocated in each subcarrier. If the total interference is divided on the number of the available subcarriers, equal interference threshold per subcarrier will be determined and hence from (2), the maximum power that can be allocated to each subcarrier will be

$$P_{i\,\max} = \frac{I_{th}}{N\,|g_i|^2\,K_i} \quad (5)$$

After the maximum power that can be allocated in each subcarrier is determined, the iterative waterfilling algorithm is applied with the total power budget and maximum power per subcarrier constraints.

4 Multi User Resource Allocations in Downlink

The problem of power and subcarrier allocation for multiuser multicarrier systems has been widely studied. The optimal allocation of these resources among users is combinatorial optimization problem and its complexity grows exponentially with respect to the input size. In order to reduce its computational complexity, the problem is solved in two steps by many of the proposed suboptimal algorithms: firstly the subcarriers are assigned for the users and then the power is allocated for these subcarriers. In traditional multicarrier systems, the maximum aggregated data rate in downlink can be obtained by simply allocating each subcarrier to the user with the high signal to noise ratio (SNR), i.e. $|h_i|^2/\sigma^2$. Once the subcarriers are allocated to the users, the multiuser system can be viewed virtually as a single user multicarrier system and therefore the power allocated to the subcarriers using any of the single user multicarrier algorithms [9].

Using of the subcarrier allocation based on maximum SNR criteria is not always efficient in CR systems especially under low interference and total power constraints. Referring to (2), one can find that the interference to the PU at a given subcarrier depends on the power allocated to the subcarrier which will be determined in the second step, and on the interference factor of the subcarrier which is constant for the users competing for the subcarrier and also on the subcarrier channel gain from the CR BS to PU. The higher the channel gain, i.e. $|g_i|^2$, the more interference can be introduced to the PU. Hence, the following subcarrier to user assignment metric, $\Delta_{i,u}$, is proposed

$$\Delta_{i,u} = \frac{|h_{i,u}|^2}{|g_{i,u}|^2} \quad (6)$$

The ith subcarrier will be allocated to the user u who's the highest value of the proposed metric $\Delta_{i,u}$ among all the competing users. Using this metric, the CR BS makes a balance between the SU channel quality and the interference that can be introduced to the PU. Our proposed PI-Algorithm can be used as a second step after finishing the subcarrier to user assignment step. Simulation results will discuss in detail the efficiency of using this metric.

5 Simulation Results

For the results presented in this section, all the simulations are performed under the scenario given in Fig. 1. An OFDM system is assumed and the values of T_s, Δf and B is assumed to be 4μ s, 0.3125 MHz and 5 MHz respectively. Additive white Gaussian noise (AWGN) variance of 1×10^{-6} is assumed. The channel gains h_i and g_i are assumed to have a Rayleigh distribution and assumed to be perfectly known at the CR BS. The CR is assumed to have $N = 128$ available subcarriers. All the results have been averaged over 1,000 trials.

5.1 Single User

In this section, the capacity achieved using PI-algorithm is simulated and compared with the reduced complexity PI-A and PI-B algorithms. In Fig. 3, the achievable capacity of the SU is plotted versus the interference threshold prescribed by the PU. It can be noted that under a given power total constraint; the PI algorithm achieves the highest capacity. The capacity achieved using PI-A algorithm is slightly higher that achieved by PI-B algorithm. The main reason behind that is while the PI algorithm can use the entire available interference threshold, the reduced algorithms guarantee only that the total interference is under the prescribed interference threshold as can be seen from Fig. 4. The number of iteration required for the PI algorithm to reach the solution is plotted in Fig. 5.

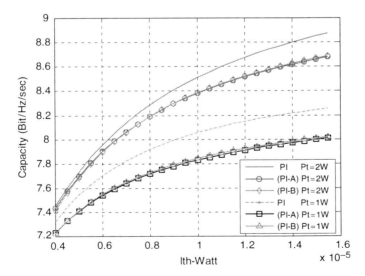

Fig. 3 Achieved SU's capacity vs allowed interference threshold

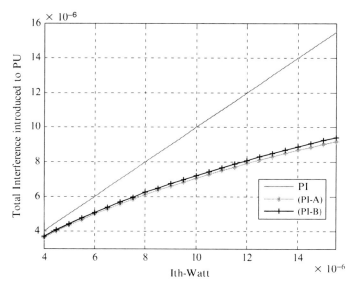

Fig. 4 Total interference introduced to PU vs interference threshold

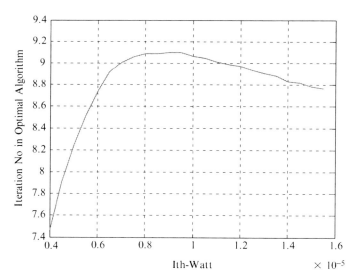

Fig. 5 Number of iterations needed to reach the solution

5.2 Multi User

In this section, the efficiency of using our proposed metric in subcarrier to user assignment is discussed. The proposed user selection criteria are compared with maximum SNR (MAX-SNR) criteria used in conventional multicarrier systems that

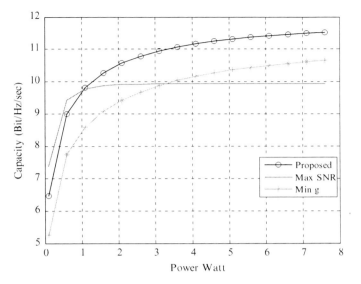

Fig. 6 Achieved capacity vs power constraint for different subcarrier allocation criteria

assign the subcarrier to user with highest SNR value. Moreover, the proposed criteria is compared with the another criteria called (MIN-g) that assign the subcarrier to the user with the minimum gain g_i, i.e. the user who cause minimum interference to the PU. Figure 6 plots the total capacity achieved for ten users under interference threshold $I_{th} = 5\mu$ W versus different total power constraints for the different selection criteria. All the results are obtained using PI-Algorithm.

It can be noted that under the given interference constraint, the achieved capacity by using the proposed subcarrier assignment criteria is greater than that achieved by using MAX-SNR or MIN-g criteria except in low total power constraint where the MAX-SNR is the best. This result can be justified because with such a small total power constraint, very low interference will be introduced to PU and hence the CR system is very close to the conventional multicarrier system where MAX-SNR criteria is optimal.

6 Conclusion

An iterative power allocation algorithm is proposed to maximize the downlink total capacity of multicarrier based CR systems. The proposed algorithm considers the total power and the maximum allowable interference induced to primary user constraints. Moreover, a new metric for guiding the channel allocation in multiuser cognitive radio systems is proposed and compared with the channel selection criteria in conventional multicarrier systems. It is found that by using the new metric, the CR system mostly achieve more capacity and has more flexibility in the total power constraint. Enhanced channel selection algorithms could be the guideline of our future research work towards better resource allocation algorithms.

References

1. Federal Communication Commission, "Spectrum Policy Task Force," Report of ET Docket 02–135, Nov. 2002.
2. J. Mitola, "Cognitive radio for flexible mobile multimedia communications", Mobile Multimedia Communications, MoMuC'99, pp. 3–10, 15–17 Nov. 1999.
3. S. Haykin, "Cognitive Radio: Brain-Empowered Wireless Communications," IEEE Journal on Selected Areas in Communications, vol. 23, no. 2, pp. 201–220, Feb. 2005.
4. T. Weiss and F. K. Jondral, "Spectrum Pooling: An Innovative Strategy for the Enhancement of Spectrum Efficiency," IEEE Communications Magazine, Radio Communications Supplement, vol. 42, pp. S8–S14, Mar. 2004.
5. T. Weiss, J. Hillenbrand, A. Krohn and F. K. Jondral, "Mutual interference in OFDM-based spectrum pooling systems," Vehicular Technology Conference, VTC 2004-Spring.
6. B. Farhang-Boroujeny and R. Kempter, "Multicarrier Communication Techniques for Spectrum Sensing and Communication in Cognitive Radios," IEEE Communications Magazine. (Special Issue on Cognitive Radios for Dynamic Spectrum Access), vol. 48, no. 4, pp. 80–85, Apr. 2008.
7. W. Peng, Z. Ming, X. Limin, Z. Shidong and W. Jing, "Power allocation in OFDM-Based cognitive radio systems," IEEE Global Telecommunications Conference, GLOBECOM '07, pp. 4061–4065, Nov. 2007.
8. G. Bansal, M. J. Hossain and V. K. Bhargava, "Adaptive power loading for OFDM-based cognitive radio systems," IEEE International Conference on Communication, ICC'07, pp. 5137–5142, Aug. 2007.
9. J. Jang and K. B. Lee, "Transmit Power Adaptation for Multiuser OFDM Systems", IEEE Journal on Selected Areas in Communications, vol. 21, no. 2, pp. 171–178, Feb. 2003.

Limited-Feedback Multiuser MIMO-OFDM Downlink with Spatial Multiplexing and Per-Chunk/Per-Antenna User Scheduling

Mohsen Eslami and Witold A. Krzymień

Abstract In this paper, we compare two spatial multiplexing MIMO-OFDM downlink transmission schemes based on limited feedback: I. Per-chunk user scheduling; II. Zero-forcing receiver processing with per-antenna user scheduling. We use average net throughput and average sum rate as our comparison benchmarks. We show while Scheme II achieves a higher average sum rate over Scheme I, Scheme I has a greater average net throughput. For Scheme I, analytical derivations and a semi-analytical approach are used to obtain the optimum (net throughput maximizing) time-frequency chunk size.

1 Introduction

Multiple-input multiple-output (MIMO) downlink transmission schemes based on limited feedback have recently attracted much attention as practical alternatives to much more complex schemes based on full channel state information at the transmitter (CSIT). Spatial multiplexing with opportunistic feedback and linear receivers [1], and opportunistic beamforming with signal-to-interference-plus-noise ratio (SINR) feedback [2] are some examples.

On the downlink and over frequency selective fading channels multiuser orthogonal frequency division multiplexing (OFDM) is very promising due to its low complexity and high flexibility, as the data on individual subcarriers can be modulated independently [3]. Furthermore, MIMO-OFDM has been selected as the underlying transmission method for many emerging high data-rate wireless standards, e.g., WiMAX (IEEE 802.16e), and 3GPP Long Term Evolution (LTE).

In MIMO-OFDM systems, the control overhead and signal processing complexity is quite large. Therefore, to reduce the overhead and complexity, the available time-frequency resources are divided into tiles or chunks [4]. The chunks are considered two-dimensional, and each chunk consists of a number of adjacent subcarriers

M. Eslami (✉) and W.A. Krzymień
University of Alberta/TRLabs, Edmonton, Alberta, Canada
e-mail: {meslami, wak}@ece.ualberta.ca

in frequency domain and a number of consecutive OFDM symbols in time domain. For all subcarriers and all OFDM symbols within the chunk, the same spatial signal processing is applied, reducing the signal processing complexity and the feedback overhead considerably.

In this paper, we show that per-antenna scheduling on each subcarrier based on limited feedback from users with zero forcing (ZF) linear processors (Scheme II) leads to a higher average sum rate than limited feedback per-chunk user scheduling (Scheme I) when the number of users in the system is sufficient. However, Scheme I has a much higher net throughput, as it requires far less feedback, especially when optimum chunk size is used.

The paper is organized as follows: In Section 2, the system model for multiuser MIMO downlink, and the net throughput definition are described. The two transmission schemes considered in this paper are outlined in Section 3. Numerical results are provided in Section 4, and Section 5 concludes the paper.

2 Preliminaries

2.1 Channel and System Model

We consider MIMO-OFDM downlink in a single cell, in which the base station is equipped with M antennas and there are K homogeneous users each equipped with N ($N = M$) antennas.

The channel is assumed to be frequency selective and is modeled as a length P finite impulse response (FIR) filter [5]. The space frequency channel at the lth tone for the kth user is then obtained as

$$\mathbf{H}_l^{(k)} = \sum_{p=0}^{P-1} \sigma_p \mathbf{H}[k,p] \exp\left(-j2\pi \frac{l}{L} p\right), \quad 0 \leq l \leq L-1 \qquad (1)$$

where $\exp(\cdot)$ denotes the exponential function, L is the total number of subcarriers in the system, and the size $M \times N$ $\mathbf{H}[k,p]$ matrices for $p = 0, \ldots, P-1$, represent the MIMO channel impulse responses of the kth user. The $\mathbf{H}[k,p]$ matrices are assumed to be mutually uncorrelated. It is also assumed that the channel elements exhibit spatially uncorrelated Rayleigh fading, in which case each $\mathbf{H}[k,p]$ contains independent $\mathcal{CN}(0,1)$ elements. Also, σ_p for $p = 0, \ldots, P-1$, represents the channel power delay profile and is normalized according to $\sum_{p=0}^{P-1} \sigma_p^2 = 1$. The channel is assumed to be quasi-static, remaining constant for duration of one frame, but changing independently between frames. The correlation coefficients between the channel elements on two arbitrary subcarriers l and u are obtained according to [6],

$$\beta_{l,u} = \mathrm{E}[(\mathbf{H}_l^{(k)})_{i,j}(\mathbf{H}_u^{(k)})_{i',j'}^*]$$
$$= \sum_{p=0}^{P-1} \sigma_p \exp(-j2\pi(l-u)p/L)\delta[i-i']\delta[j-j'] \quad (2)$$

where $(\mathbf{A})_{i,j}$ denotes the element of matrix $\mathbf{A}$ at the ith row and jth column, $(\cdot)^*$ denotes complex conjugate and $\delta(\cdot)$ denotes the Dirac's delta function. The closer the subcarriers are, the larger the magnitude of the correlation coefficient, $|\beta_{l,u}|$, is.

The kth user receives the following signal vector on the lth subcarrier

$$\mathbf{y}_l^{(k)} = \mathbf{H}_l^{(k)} \mathbf{x}_l + \mathbf{n}_l^{(k)}, \quad (3)$$

where $\mathbf{H}_l^{(k)}$ is given by (1), and $\mathbf{n}_l^{(k)}$ is the noise vector with $\mathcal{CN}(0,1)$ elements. The vector $\mathbf{x}_l$ is the transmitted signal vector on the lth subcarrier. Hence, the average signal to noise ratio (SNR) denoted by ρ is $\rho = \frac{P_T}{M}$ where $\frac{1}{L}\sum_{l=0}^{L-1} \mathrm{Tr}(\mathrm{E}[\mathbf{x}_l \mathbf{x}_l^H]) = P_T$ defines the power constraint of the base station. Let the total delivered rate by the base station to the users during one time slot be R_T. Then the expected throughput of the system is obtained by taking ensemble average of R_T over $\mathbf{H}_l^{(k)}$'s, i.e., $R = E_{\mathbf{H}_0^{(k)},\ldots,\mathbf{H}_{L-1}^{(k)},k=1,\ldots,K}[R_T]$. Throughout the paper, the terms system throughput and sum rate are used interchangeably.

2.2 Net Throughput

It is assumed that the feedback overhead reduces the transmission rate, R, to the net (effective) throughput, R_{net}, given by [4]

$$R_{net} = R\left(1 - N_f \frac{\zeta}{R_d T}\right) \quad (4)$$

where N_f is the average number of feedback terms per subcarrier sent back by all users during time T; T is the channel's coherence time (in seconds). ζ is the number of bits used to quantize each feedback term, and R_d is the feedback channel's bit rate (in bits/second).

3 Multi-User MIMO-OFDM Downlink Transmission Schemes Based on Limited Feedback

In the following subsection, we outline two low complexity limited feedback schemes for multi-user MIMO-OFDM downlink and compare their average sum rate and net throughput.

3.1 Scheme I: Multiuser MIMO-OFDM Downlink with Per-Chunk User Scheduling

Assuming the number of subcarriers in each chunk is L_c (for simplicity we assume all chunks have the same number of subcarriers and L is an integer multiple of L_c), the highest instantaneous reliable rate on the qth chunk in bits per second per Hertz (b/s/Hz) for user k is given by

$$r_{chk.q}^{(k)} = \frac{1}{L_c} \sum_{l_c=(q-1)L_c}^{qL_c-1} r_l^{(k)} \quad 1 \leq q \leq Q \tag{5}$$

where $Q = \frac{L}{L_c}$ denotes the number of chunks in the system and $r_l^{(k)}$ is defined as [7] $r_l^{(k)} = \log_2 \det(\mathbf{I} + \rho \mathbf{H}_l^{(k)} \mathbf{H}_l^{(k)^H})$ for $0 \leq l \leq L-1$. This rate is achieved by spatial multiplexing transmission and successive interference cancellation at each receiver [7]. In Scheme I, each user sends back the achievable rate of each chunk, Q in total, to the base station and the base station assigns each chunk to the user, which has reported highest rate for that chunk. It is expected that by grouping subcarriers into chunks, especially chunks with small number of adjacent subcarriers, which have higher correlation compared to subcarriers further apart, the sum rate would not significantly decrease, while the amount of feedback is reduced by a factor of L_c.

The average sum rate of Scheme I is given by

$$R_I = \mathrm{E}\left[\frac{1}{Q} \sum_{q=1}^{Q} \max_{1 \leq k \leq K} r_{chk.q}^{(k)} \right]. \tag{6}$$

As $r_l^{(k)}$ is well approximated to be Gaussian [8], $r_{chk.q}$ can also be approximated by Gaussian distribution. In [9], it has been shown that a close approximation for the expected value of the maximum of a set of K independent and identically distributed (i.i.d.) Gaussian random variables each having mean μ and variance σ^2, is given by

$$\mathrm{E}\left[\max_{1 \leq k \leq K} \mathcal{N}^{(k)}(\mu, \sigma^2) \right] \approx \sigma\, \mathrm{G}(K) + \mu. \tag{7}$$

where $G(K) = \frac{1}{0.1975}\left[0.5264^{\frac{0.135}{K}} - (1 - 0.5264^{\frac{1}{K}})^{0.135}\right]$. Therefore, in order to be able to use (7) as an approximation for R_I, the mean μ and variance σ^2 of $r_{chk.q}^{(k)}$ are required. In [6] it has been shown that the mean of the rate of each chunk, $\mathrm{E}[r_{chk.q}]$ (the user index, k, has been discarded), is the same for all chunks, i.e., $\mathrm{E}[r_{chk.q}] = \mathrm{E}[r_{chk.u}]$, $q \neq u$, even if the chunk sizes are not equal and is given by [10]

$$\mathrm{E}[r_{chk.q}] = \frac{1}{\ln(2) \prod_{m=1}^{M}[\Gamma(M-m+1)]^2} \sum_{k=1}^{M} \det(\mathbf{\Psi}(k)) \tag{8}$$

where $\Psi(k)$, $k = 1, \ldots, M$, are $M \times M$ matrices whose entries are defined by

$$\{\Psi(k)\}_{i,j} = \begin{cases} u_{i,j}! \exp(1/\rho) \sum_{s=1}^{u_{i,j}} \frac{\Gamma(s-u_{i,j},\frac{1}{\rho})}{\rho^{(u_{i,j}-s)}}, & j = k \\ u_{i,j}!, & j \neq k. \end{cases} \quad (9)$$

where $u_{i,j} = 2M - i - j + 1$, $n!$ denotes factorial of n, and $\Gamma(\cdot,\cdot)$ denotes the incomplete Gamma function [11].

For a MIMO-OFDM link, the variance of the instantaneous reliable rate has been derived in [6]. We use the derivations of [6] to obtain the variance of reliable rate for individual chunks of a MIMO-OFDM link. For the variance of the rate of each chunk we provide the following theorem.

Theorem 1. *For a MIMO-OFDM link, the variance of the instantaneous reliable rate on each chunk, variance of $r_{chk,q}$, for all chunks with equal number of subcarriers, L_c, is the same and is given by*

$$\text{Var}(r_{chk,q}) = \frac{2(\log_2(e))^2 \exp(2M/\rho)}{(L_c)^2 [\prod_{m=1}^{M}(M-m+1)]^2}$$

$$\times \sum_{d=1}^{L_c-1} \left[(L_c - d) \sum_{r=1}^{M} \sum_{s=1}^{M} \det\left(\mathbf{C}_{r,s}(\beta_{0,d})\right) \right]$$

$$+ \frac{1}{L_c \ln^2(2) [\prod_{m=1}^{M} \Gamma(M-m+1)]^2} \sum_{k=1}^{M} \sum_{l=1}^{M} \det(\mathbf{\Psi}_{k,l}) \quad (10)$$

where the matrices $\mathbf{C}_{r,s}(\beta_{0,d})$ and $\mathbf{\Psi}_{k,l}$ are given in [6, eq. (14)] and [10, eq. (32)], respectively. $\beta_{0,d}$ is defined by (2).

Due to space limitation, the proof of Theorem 1 will be presented in the journal version of this paper. Using (7), (8) and (10), the average sum rate of Scheme I is approximated by

$$R_I \approx \sqrt{\text{Var}(r_{chk,q})} \, G(K) + E[r_{chk,q}] \quad (11)$$

The unique feature of (11) is that the effect of the chunk size and the effect of the number of available users in the system on the sum rate are clearly separate through $\sqrt{\text{Var}(r_{chk,q})}$ and $G(K)$. The maximum net throughput for this scheme is then approximated by

$$R_{net,I}^{max} \approx R_I \left(1 - \frac{N_f^{(max,I)} \zeta}{R_d T}\right). \quad (12)$$

where $N_f^{(max,I)} = K \frac{L_c^{max}}{L}$ is the average number of feedback terms per subcarrier. L_c^{max} is the chunk size which maximizes $R_{net,I}$, the net throughput given by (12) for an arbitrary chunk size. In our work, $R_{net,I}$ is obtained numerically.

3.2 Scheme II: Multiuser MIMO-OFDM Downlink with Per-Antenna User Scheduling on Each Subcarrier

For this scheme we consider spatial multiplexing transmission and post-processing SNR feedback from users with linear ZF processors on each subcarrier. Each user adopts a linear ZF processor given by $\mathbf{G}_l^{(k)} = (\mathbf{H}_l^{(k)})^{-1}$ on the lth subcarrier and evaluates M SNR values each obtained as [12] $\gamma_{l,m}^{(k)} = \rho / \left[(\mathbf{H}_l^{(k)H} \mathbf{H}_l^{(k)})^{-1} \right]_{mm}$, $1 \leq m \leq M$, where $[\cdot]_{mm}$ denotes the mth diagonal term of the matrix argument and $\gamma_{l,m}^{(k)}$ is the post-processing SNR of the mth data stream on the lth subcarrier of user k's channel. After receiving $M \times L$ post-processing SNRs from each user, for each spatial data stream and subcarrier, the base station selects the user which has reported the largest post-processing SNR for that data stream and subcarrier.

As mentioned in Section 2, we assume $M = N$. The average sum rate of Scheme II in b/s/Hz is [12]:

$$R_{II} = \mathrm{E}\left[\frac{1}{L} \sum_{l=1}^{L} \sum_{m=1}^{M} \log_2(1 + \rho \gamma^{max}) \right]$$

$$= \frac{KM}{\ln(2)} \sum_{q=0}^{K-1} \frac{1}{q+1} \binom{K-1}{q} (-1)^q \exp\left(\frac{q+1}{\rho}\right) E_1\left(\frac{q+1}{\rho}\right) \quad (13)$$

where $E_1(x) = \int_x^{\infty} \exp(-u)/u\, du$ is the exponential integral [11] and $\gamma^{max} = \max_{1 \leq k \leq K} [\gamma_{l,m}^{(k)}/\rho]$ (we have dropped the subscripts l and m, as the maximum post-processing SNR has the same distribution for any $0 \leq l \leq L-1$ and $1 \leq m \leq M$). The net throughput for this scheme is $R_{net.II} = R_{II} \left(1 - \frac{N_f^{(II)} \zeta}{R_d T}\right)$ where $N_f^{(II)} = KM$ is the number of feedback terms from all users, per subcarrier. By comparing $N_f^{(max.I)}$ and N_f^{II}, it becomes clear that Scheme I has a far lower feedback requirement than Scheme II. Hence, we try the following feedback reduction concept.

Scheme II with Opportunistic Feedback. Similarly to [13], we consider the case where not all users send back their per-antenna rate (or SNR) values. Only users with per-antenna post processing SNRs above a pre-defined threshold are allowed to use the feedback channel. We simply use the expected value of SNR, $\mathrm{E}[\gamma^{max}]$, as the threshold as finding the optimum threshold is beyond the scope of this paper.

$$\mathrm{E}[\gamma^{max}] = \int_0^{\infty} x f_{\gamma^{max}}(x) dx$$

$$= K \int_0^{\infty} x(1 - \exp(-x))^{K-1} \exp(-x) dx = \sum_{k=1}^{K} 1/k. \quad (14)$$

Using the cumulative distribution function (CDF) of γ^{max} given by

$$F_{\gamma^{max}}(x) = Pr(\gamma^{max} \leq x) = (1 - \exp(-x))^K, \qquad (15)$$

the average number of feedback terms for Scheme II is reduced to $N_f^{(rdc,II)} \approx KM(1 - F_{\gamma^{max}}(T[K]))$ where $T[K] = \sum_{k=1}^{K} 1/k$. The net throughput is then $R_{net,II}^{opp} \approx R_{II}^{opp}(1 - \frac{N_f^{(rdc,II)}\zeta}{R_d T})$ where R_{II}^{opp} is average sum rate of Scheme II with the feedback scheme described above.

4 Numerical Results

For the numerical results presented in this section, we model the channel using the exponential power delay profile [6]:

$$\sigma_p^2 = \frac{1 - \exp(-1/2)}{1 - \exp(-7/2)} \exp(-p/2), \quad 0 \leq p \leq 7. \qquad (16)$$

Figure 1 shows the average sum rates of Schemes I and II obtained using a semi-analytical approach for $M = N = 2$, or 3 antennas at both the BS and each user terminal plotted versus the number of users at $P_T = 10$ dB. The system considered has $L = 64$ subcarriers with chunk size of $L_c = 16$. In the semi-analytical approach, for one system realization, a set of random channel matrices is generated and used to evaluate the sum rate of each scheme according to the scheme's scheduling and signal processing approach. The average sum rate is then obtained by averaging the sum rates over a large number of realizations. Analytical results have also been plotted and are in good agreement with the semi-analytical results.

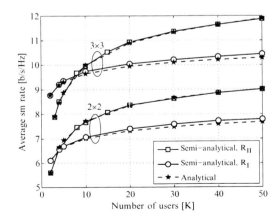

Fig. 1 Average sum rate of Schemes I and II for MIMO-OFDM downlink with $L = 64$ subcarriers, chunk sizes of $L_c = 16$, $P_T = 10$ dB, and $M = N = 2$, and 3

As expected and due to multiuser diversity, the average sum rate increases with the number of users in both schemes. Scheme II (per-antenna user scheduling), thanks to having more degrees of freedom, takes better advantage of multiuser diversity and achieves a higher sum rate.

In Fig. 2, the net throughput of Scheme I is plotted versus the chunk size, L_c. The system considered has $L = 2048$ subcarriers with $M = N = 2$ as the number of antennas at the base station and each user terminal. Similarly to [4] we set $\frac{\zeta}{R_d T} = 0.002$ for all the results presented in this section. As seen in this figure, $L_c^{max} = 128$ is the optimum chunk size that maximizes the net throughput.

In Fig. 3, the average net throughput of Schemes I and II is plotted versus the number of users for a system with $L = 2048$ subcarriers. Chunk size of $L_c^{opt} = 128$ is used for Scheme I and $R_{net.I}$ is obtained using the analytical approximations of (11) and (12). For Scheme II with opportunistic feedback, $R_{net.II}^{opp}$ is obtained through semi-analytical approach to find R_{II}^{opp} and using $N_f^{(rdc.II)}$ to evaluate the number of feedback terms. Figure 3 shows that as the number of users increases, the net throughput of Scheme II approaches zero, even with opportunistic feedback. It is known that the sum rate of Scheme II has an asymptotically optimum increase with the number of users, when the number of users approaches infinity [14], while this is not true for Scheme I. Therefore, as we can see with a more realistic net throughput definition, the throughput superiority of Scheme II does not exist anymore.

In Fig. 3 and for small number of users we see $R_{net.II}^{opp} < R_{net.II}$. This is due to the fact that when the user pool is small, there is a high possibility that for some channel realization no user's post processing SNR is above the predefined threshold (in our case set to $E[\gamma^{max}]$), which results in no user being scheduled for data transmission and zero sum rate for those specific channel realization. This decrease in sum rate can be avoided by adopting enhanced feedback reduction policies.

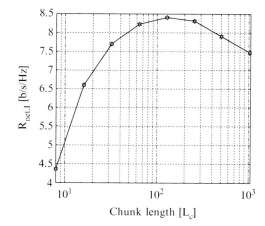

Fig. 2 Average net throughput for a system with $L = 2048$ subcarriers, $P_T = 10$ dB, $K = 100$ users, and $M = N = 2$ antennas, versus chunk size

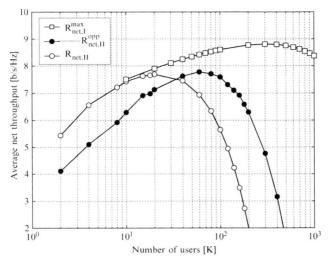

Fig. 3 Average net throughput of Schemes I ($L_c^{max} = 128$) and II for a system with $L = 2048$ subcarriers, $P_T = 10$ dB, and $M = N = 2$ antennas, versus the number of users

5 Conclusions

By comparing two low complexity spatial multiplexing MIMO-OFDM downlink transmission schemes which require partial CSIT, we show how the amount of feedback affects the MIMO-OFDM downlink transmission schemes by reducing their net throughput. Our results show that while per-antenna user scheduling with ZF receiver processing achieves a higher average sum rate compared to per-chunk user scheduling, the latter scheme has a much greater average net throughput. In fact, as the number of users in the system increases, net throughput for per-antenna user scheduling approaches zero much faster than per-chunk user scheduling.

Acknowledgement Funding for this work has been provided by TRLabs, Huawei Technologies, the Rohit Sharma Professorship, and the Natural Sciences and Engineering Research Council (NSERC) of Canada.

References

1. T. Tang, R. W. Heath, S. Cho, and S. Yun, "Opportunistic feedback for multiuser MIMO systems with linear receivers," *IEEE Trans. Commun.*, vol. 55, no. 5, May 2007.
2. P. Viswanath, D. Tse, and R. Laroia, "Opportunistic beamforming using dumb antennas," *IEEE Trans. Infor. Theory*, vol. 48, no. 6, June 2002.
3. C. Y. Wong, R. S. Cheng, K. B. Lataief and R. D. Murch, "Multiuser OFDM with adaptive subcarrier, bit and power allocation," *IEEE Journ. Select. Area Commun.*, vol. 17, no. 10, pp. 1747–1758, Oct. 1999.

4. E. Jorswieck, A. Sezgin, B. Ottersten, and A. Paulraj, "Feedback reduction in uplink MIMO OFDM systems by chunk optimization," *EURASIP Journal on Advanced Signal Proc.*, Article ID 597072, 14 pages, 2008.
5. H. Bölcskei, D. Gesbert, and A. Paulraj, "On the capacity of OFDM based spatial multiplexing systems," *IEEE Trans. Commun.*, vol. 50, no. 2, pp. 225–234, Feb. 2002.
6. M. R. McKay, P. J. Smith, H. A. Suraweera, and I. B. Collings, "On the mutual information distribution of OFDM-based spatial multiplexing: exact variance and outage approximation," *IEEE Trans. Infor. Theory*, vol. 54, no. 7, pp. 3260–3278, July 2008.
7. D. Tse and P. Viswanath, *Fundamentals of Wireless Communication*, Cambridge: Cambridge University Press, 2005.
8. P. J. Smith, and M. Shafi, "On a Gaussian approximation to the capacity of wireless MIMO systems," in *Proc. IEEE Int. Conf. Commun. (ICC)*, vol. 1, pp. 406–410, April 2002.
9. M. Eslami, and W. Krzymień, "Efficient transmission schemes for multiuser MIMO downlink with linear receivers and partial side information," submitted to *IEEE Trans. Veh. Tech.*, Feb. 2009.
10. M. Kang, and M. S. Alouini, "Capacity of MIMO Ricean channels," *IEEE Trans. Wireless Commun.*, vol. 5, no. 1, pp. 112–122, Jan. 2006.
11. M. Abramowitz, and I. A. Stegun, *Handbook of Mathematical Functions with Formulas, Graphs, and Mathematical Tables*, New York: Dover, 1972
12. C. J. Chen, and L. C. Wang, "Performance analysis of scheduling in multiuser MIMO systems with zero-forcing receivers," *IEEE Journal on Selected Areas in Commun.*, vol. 25, no. 7, pp. 1453–1445, Sept. 2007.
13. T. Tang, R. W. Heath, S. Cho and S. Yun, "Opportunistic feedback for multiuser MIMO systems with linear receivers," *IEEE Trans. Commun.*, vol. 55, no. 5, pp. 1020–1032, May 2007.
14. M. Airy, R. W. Heath, and S. Shakkottai, "Multi-user diversity for the multiple antenna broadcast channel with linear receivers: asymptotic analysis," in *Proc. Asilomar Conf. Signals, Systems, and Computers*, pp. 886–890, Nov. 2004.

Part VII
Performance Evaluation

Simple Series Form Formula of BER Performance of M-ary QAM/OFDM Signals Over Nonlinear Fading Channels

Yuichiro Goto, Akihiro Yamakita, and Fumiaki Maehara

Abstract This paper proposes a simple series form formula of the bit error rate (BER) for M-ary QAM/OFDM signals under nonlinear fading channels. In the proposed derivation, the effect of the nonlinear distortion due to the soft envelope limiter is approximated by the additive Gaussian noise with the variance equal to the power of the clipped portion of the composite waveform. Since the BER is represented by the simple series form formula, the proposed approach frees us from the time-consuming computer simulations. Moreover, the proposed formula is applicable to arbitrary M-ary quadrature amplitude modulations (QAM). The validity of the proposed approach is confirmed by the agreement with the computer simulation results with parameters of the input back-off (IBO) and average CNR.

1 Introduction

Orthogonal frequency division multiplexing (OFDM) has become one of the most promising techniques in broadband wireless communications systems because of its high bandwidth efficiency as well as its robustness against the severe frequency selective fading [1,2]. However, OFDM has an inherent drawback that the time domain waveform suffers from a very high peak-to-average power ratio (PAPR). This large envelope variation causes the nonlinear distortion in the output of the high-power amplifier (HPA) at the transmitter. Therefore, it is important to investigate the transmission performance considering the effect of the nonlinear distortion introduced by the HPA from the practical point of view.

In most cases, the performance evaluation has been done by computer simulations. This is because computer simulations can provide the performance irrespective of the complexity of a target system. However, the drawback of the computer simulations is to cause a very time-consuming task. Therefore, it is preferable to

---

Y. Goto (✉), A. Yamakita, and F. Maehara
Graduate School of Fundamental Science and Engineering, Waseda University,
4-1 Ohkubo 3-chome Shinjuku-ku, Tokyo 169-8555, Japan
e-mail: fumiaki@toki.waseda.jp

analyze the performance theoretically as much as possible. Once the transmission performance is represented by mathematical formulae, it is possible to easily clarify the relationship between the transmission performance and the various kinds of systems parameters.

Considering the background described above, we derive the theoretical bit error rate (BER) of OFDM signals under nonlinear fading channels. So far, the impact of the nonlinear distortion on the OFDM transmission performance has been theoretically investigated [3–11] by using Bussgang's theorem [12]. In [3–7], the analytical BER in the presence of the nonlinear distortion and AWGN was investigated. In more realistic wireless scenarios where the nonlinear distortion and fading are jointly introduced, the theoretical results have been investigated in [8–11]. As for the BER, the analytical framework has been carried out in [9], but its expression unfortunately needs the integral calculation. In our previous work [10, 11], we have derived the simple series form formula of the BER, which does not require the analytical calculation any more. However, in [11], the modulation scheme is unfortunately limited to DQPSK and therefore, the applicability of the proposed approach to an arbitrary M-ary QAM is expected from the point of view of the recent trend for higher spectral efficiency.

This paper presents the simple series form formula of the BER for M-ary QAM/OFDM signals in the presence of nonlinear fading channels. Moreover, the theoretical results obtained by the proposed formula are verified by the agreement with the computer simulation results.

This paper is organized as follows. Section 2 describes the transmission system model of M-ary QAM/OFDM. In Section 3, we derive the series form formula of the BER in the presence of nonlinear fading channels. In Section 4, some numerical results are shown to verify the proposed theoretical approach. Finally, conclusions are drawn in Section 5.

2 System Description

Figure 1 shows the system configuration of a transmitter and a receiver for M-ary QAM/OFDM. At the transmitter, incoming information bits are mapped into the transmit samples corresponding to M-ary QAM modulation. The N-parallel modu-

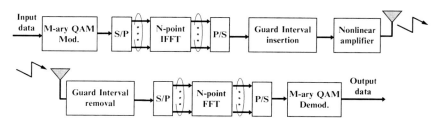

Fig. 1 System configuration of M-ary QAM transmitter and receiver

lated signals are fed into the N-point IFFT circuit so as to be converted into the time domain signals and the guard interval with the length of T_G is added to the time domain symbols through the parallel to serial (P/S) converter. After the OFDM time domain signals go through the HPA, the nonlinear distortion arises from the large dynamic range of the multicarrier signals.

At the receiver, the effective symbol duration T_s is extracted from the OFDM signals. Then the N-parallel signals are fed into the N-point FFT circuit through the S/P converter, which results in the N_c sub-channels signals. After the FFT processing, coherent detection is carried out so that the input data can be recovered.

3 Bit Error Rate Analysis

3.1 Effect of Nonlinear Distortion

The complex baseband OFDM signal composed of a large number of sub-carriers can be modeled as a complex Gaussian process with Rayleigh envelope distribution because of the central limit theorem [12]. This makes the analysis on nonlinear OFDM signals utilize the general results for nonlinear distortions of Gaussian signals.

The signal-to-distortion power ratio (SDR) at the output of nonlinear devices is derived by making use of Bussgang's theorem [13]. The extension of Bussgang's theorem to complex Gaussian inputs [3, 4] makes it possible to express the nonlinear output $s_d(t)$ as the sum of a complex useful input replica and an uncorrelated nonlinear distortion $n_d(t)$, as expressed by

$$s_d(t) = \alpha s(t) + n_d(t), \tag{1}$$

where α is the attenuation coefficient for stationary input processes and $s(t)$ is the input signal.

Let $f(r) = g(r) \cdot e^{j\phi(r)}$ be a complex nonlinear distortion function, $s_d(t)$ is rewritten as

$$s_d(t) = f[r(t)]e^{j\theta(t)}, \tag{2}$$

where $r(t)$ and $\theta(t)$ are the amplitude and the phase of the input signal $s(t)$, respectively.

The complex version of the Bussgang's theorem introduces the expression of the output autocorrelation function $R_{s_d s_d}(\tau)$, as given by

$$R_{s_d s_d}(\tau) = |\alpha|^2 \cdot R_{ss}(\tau) + R_{n_d n_d}(\tau), \tag{3}$$

where $R_{ss}(\tau)$ and $R_{n_d n_d}(\tau)$ are the input autocorrelation function and the nonlinear distortion autocorrelation function, respectively.

The attenuation coefficient of the useful component α is given by [3]

$$\alpha = \frac{R_{s_d s}(0)}{R_{ss}(0)} = \frac{E[s_d^*(t)s(t)]}{\sigma_s^2}$$

$$= \frac{E[f(r) \cdot r]}{\sigma_s^2}$$

$$= \frac{1}{\sigma_s^2} \int_0^\infty f(r) \cdot r \cdot p(r) dr, \qquad (4)$$

where $R_{s_d s}(\tau)$ denotes the input-output cross-correlation function, σ_s^2 is the input signal power, and $p(r)$ is the Rayleigh probability density function (PDF) of the input envelope, as given by

$$p(r) = \frac{2r}{\sigma_s^2} e^{-\frac{r^2}{\sigma_s^2}}. \qquad (5)$$

Hence, the SDR can be derived as the power ratio between two uncorrelated output components by the following expression:

$$\Lambda = \frac{|\alpha|^2 \cdot R_{ss}(0)}{R_{s_d s_d}(0) - |\alpha|^2 \cdot R_{ss}(0)}$$

$$= \frac{|\alpha|^2 \sigma_s^2}{R_{s_d s_d}(0) - |\alpha|^2 \sigma_s^2}. \qquad (6)$$

Assuming a soft envelope limiter, the nonlinear distortion function $f(r) = g(r) \cdot e^{j\phi(r)}$ is represented by [13]

$$g(r) = \begin{cases} r & r \leq A \\ A & r > A \end{cases} \qquad (7)$$

$$\phi(r) = 0,$$

where A is the saturation level.

Substituting (7) into (4), the attenuation of the useful component is obtained and is expressed as

$$\alpha = \frac{1}{\sigma_s^2} \left(\int_0^A r^2 p(r) dr + \int_A^\infty A \cdot r \, p(r) dr \right)$$

$$= 1 - e^{-\zeta} + \frac{\sqrt{\pi \zeta}}{2} \mathrm{erfc}\left(\sqrt{\zeta}\right), \qquad (8)$$

where $\zeta = A^2/\sigma_s^2$ denotes the input back-off (IBO).

Since the output autocorrelation is given by

$$R_{S_d S_d}(0) = \int_0^A r^2 p(r) dr + \int_A^\infty A^2 p(r) dr$$
$$= \sigma_s^2 \left(1 - e^{-\zeta}\right), \tag{9}$$

the SDR can be rewritten as

$$\Lambda = \frac{\alpha^2}{1 - e^{-\zeta} - \alpha^2}. \tag{10}$$

Here, it is noted that the above analysis is based on the assumption that the clipping is conducted on the Nyquist-rate samples and the HPA performs in a linear region. Thus, the impact of out-of-band radiation is not taken into account, which implies that the total nonlinear distortion remains within the OFDM signal bandwidth [5]. When the number of sub-carriers N_c is large, the distortion component can be seen as the Gaussian distribution in the sequel [8], [14]. Therefore, the average sub-channel power of the distortion component corresponds to the power spectral density of the distortion component. Since the sub-channel SDR is quite the same for most of the N_c sub-carriers [9] in the case of the equal transmitted sub-carrier power, the sub-channel SDR results in Λ. It is worth bearing in mind that the SDR Λ does not depend on the channel attenuation in fading channels.

3.2 Derivation of SNDR

Now, we will derive the PDF of the instantaneous signal-to-noise plus distortion power ratio (SNDR). Let β and σ_n^2 denote the instantaneous sub-channel attenuation and the variance of the sub-channel noise, respectively, the SNDR at each sub-channel can be defined as

$$\gamma = \frac{\beta^2 \alpha^2 \sigma_{sc}^2}{\frac{\beta^2 \alpha^2 \sigma_{sc}^2}{\Lambda} + \sigma_n^2}$$
$$= \Lambda - \frac{\Lambda^2}{\gamma_s + \Lambda}. \tag{11}$$

where $\sigma_{sc}^2 = \sigma_s^2/N_c$ and $\gamma_s = \beta^2 \alpha^2 \sigma_{sc}^2/\sigma_n^2$ is the input sub-channel power and the instantaneous sub-channel SNR, respectively.

Assuming that the radio channel is subject to Rayleigh fading at each sub-channel, the PDF of γ_s is expressed as

$$p(\gamma_s) = \frac{1}{\Gamma_s} e^{-\frac{\gamma_s}{\Gamma_s}}, \tag{12}$$

where Γ_s is the average sub-channel SNR.

Thus, let Γ be the average sub-channel SNR in the absence of HPA nonlinearities, the relationship between Γ_s and Γ can be seen as

$$\Gamma_s = |\alpha|^2 \Gamma. \tag{13}$$

The PDF of the instantaneous SNDR γ can be derived by changing the variable γ_s into γ and is given by

$$\begin{aligned} p(\gamma) &= p(\gamma_s)\frac{d\gamma_s}{d\gamma} \\ &= \frac{\Lambda^2}{\Gamma_s(\gamma-\Lambda)^2} e^{\frac{\Lambda\gamma}{\Gamma_s(\gamma-\Lambda)}}. \end{aligned} \tag{14}$$

3.3 Derivation of Bit Error Rate

In general, the average BER in fading channels is obtained by averaging the BER in AWGN as a function of the received SNR over $p(\gamma)$, as given by

$$\bar{P}_e = \int_0^\infty P_e(\gamma) p(\gamma) d\gamma. \tag{15}$$

Strictly speaking, in order to obtain the mean BER for OFDM systems, averaging $\bar{P}_e$ over all sub-channels is required. However, the averaged BER can be approximated as (15) because the average sub-channel SNR is quite the same for most of the N_c sub-channels [3].

The BER of M-ary QAM modulation in AWGN is represented by a complementary error function. However, this expression is not easily applicable to the computation of the integral in (15). Once the BER is expressed by an exponential function such as $ae^{-b\gamma}$, the result of (15) can be modified into a simple series expression by using a Taylor series, which enable us to analyze easily the BER of M-ary QAM signals under nonlinear fading channels. The BER for an arbitrary M-ary QAM modulation can be fortunately agrees well with the form of $ae^{-b\gamma}$ where coefficients (a, b) for QPSK, 16QAM, and 64QAM are (0.25, 0.57), (0.27, 0.12), and (0.26, 0.029), respectively. Figure 2 shows the tightness of this approximation to the standard formula using the complementary error function. From Fig. 2, it can be observed that the approximation provides excellent fit to the standard formula irrespective of the modulation scheme. Here, it is noted that the CNR can be referred to as γ because the SNR γ is so far defined in the sense of the CNR.

Substituting both this approximation and (14) into (15), the BER for M-ary QAM under nonlinear fading channels can be derived in a simple series formula, as given by

Fig. 2 BER approximation for M-ary QAM

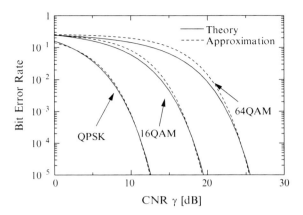

Table 1 Simulation parameters

Modulation / Detection	M-ary QAM / Coherent detection
FFT size (N)	512
Number of sub-channels (N_c)	512
Guard interval length (T_G)	$128\Delta T$
Channel model	6-ray exponentially-decaying
r.m.s. delay spread (τ_{rms})	$8\Delta T$

$$\bar{P}_e = ae^{-b\Lambda}\left\{1 + e^{\frac{\Lambda}{\Gamma_s}}E_1\left(\frac{\Lambda}{\Gamma_s}\right)\sum_{i=1}^{\infty}\frac{(-1)^{i-1}b^i\Lambda^{2i}}{i!(i-1)!\Gamma_s^i}\right.$$
$$\left. + \sum_{i=1}^{\infty}\frac{(-1)^{i-1}b^{i+1}\Lambda^{2i+1}}{(i+1)!i!\Gamma_s^i}\sum_{j=0}^{i-1}j!\left(-\frac{\Gamma_s}{\Lambda}\right)^j\right\}, \qquad (16)$$

where $E_1(\ )$ is the exponential integral.

4 Numerical Results

We compare the theoretical results with the computer simulation results to confirm the validity of the proposed theoretical approach. Therefore, the simulated results have been carried out by Nyquist-rate sampling ($f_s = 1/\Delta T$), not by oversampling, which is the same condition as the theoretical approach. Table 1 shows the simulation parameters. In this paper, the radio channel model is assumed as a 6-ray exponentially decaying multipath channel, where the amplitude and phase of each path are characterized by Rayleigh and uniform distributions, respectively. Moreover, since the multipath delay is assumed not to exceed the guard interval length T_G, there is no effect of the inter-symbol interference (ISI). Although the proposed formula in (16) is represented by the infinite series, the theoretical BER can actually be obtained by the practical truncation with the reasonable number of terms.

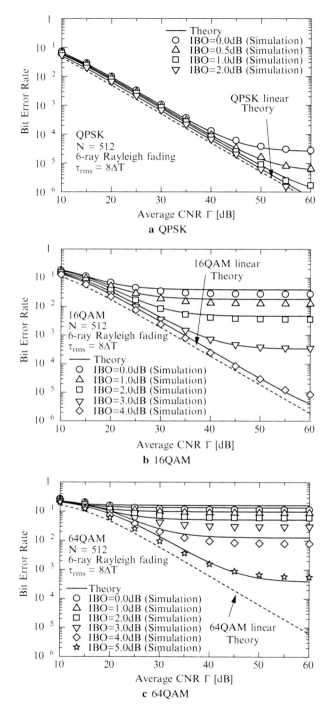

Fig. 3 BER versus average CNR Γ with a parameter of input back-off (IBO)

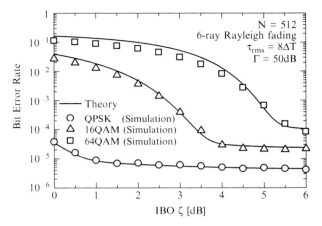

Fig. 4 BER versus IBO for QPSK, 16QAM and 64QAM

Figure 3 shows the BER performance versus average CNR Γ with a parameter of input back-off (IBO). It is found that the proposed theoretical result for QPSK agrees well with the simulated result. This is because there is the good agreement between the standard formula and the approximation as shown in Fig. 2. On the other hand, the theoretical BER for 16QAM or 64QAM gets a little worse than the simulated BER especially in a low IBO. The reason comes from the approximation error that can be seen in a low CNR.

Figure 4 shows the BER performance versus IBO with a parameter of modulation scheme. From Fig. 4, it is observed that the theoretical results agree well with the simulated results. Moreover, since the higher level modulation is sensitive to the nonlinear distortion, it can be seen that the BERs of 16QAM and 64QAM are degraded significantly with the decrease in the IBO.

From these results, we can conclude that the proposed formula is applicable to M-ary QAM signals under nonlinear fading channels.

5 Conclusion

This paper proposed a simple series form formula of the BER performance for M-ary QAM/OFDM signals in nonlinear fading channels. Although the proposed formula consists of the infinite terms, the BER performance can actually be obtained by the reasonable number of terms. Hence, the proposed formula takes advantage of easily analyzing the impact of the nonlinear distortion and fading effects on the BER performance. The validity of the proposed formula has been demonstrated by the agreement between the theoretical results and the simulated results.

Acknowledgment This work was supported by Grant-in-Aid for Young Scientists (B) 18760287 from the Ministry of Education, Culture, Sports, Science and Technology-Japan.

References

1. V. Nee and R. Prasad, *OFDM for Wireless Multimedia Communications*, Artech House, London, 2000.
2. S. Hara and R. Prasad, *Multicarrier Techniques for 4G Mobile Communications*, Artech House, London, 2003.
3. P. Banelli and S. Cacopardi, "Theoretical analysis and performance of OFDM signals in nonlinear AWGN channels," *IEEE Trans. Commun.*, vol. 48, no. 3, pp. 430–441, Mar. 2000.
4. D. Dardari, V. Tralli, and A. Vaccari, "A theoretical approach to error probability evaluation for OFDM systems with memoryless non-linearities," *IEEE Trans. Commun.*, vol. 48, pp. 1755–1764, Oct. 2000.
5. H. Ochiai and H. Imai, "Performance of the deliberate clipping with adaptive symbol selection for strictly band-limited OFDM systems," *IEEE J. Select. Areas Commun.*, vol. 18, no. 11, pp. 2270–2277, Nov. 2000.
6. E. Costa and S. Pupolin, "M-QAM-OFDM system performance in the presence of a nonlinear amplifier and phase noise," *IEEE Trans. Commun.*, vol. 50, no. 3, pp. 462–472, Mar. 2002.
7. P. Zillmann, H. Nuszkowski, and G. Fettweis, "A novel receive algorithm for clipped OFDM signals," *WPMC2003*, pp. 380–384, Oct. 2003.
8. H. Ochiai and H. Imai, "Performance analysis of deliberately clipped OFDM signals," *IEEE Trans. Commun.*, vol. 50, no. 1, pp. 89–101, Jan. 2002.
9. P. Banelli, "Theoretical analysis and performance of OFDM signals in nonlinear fading channels," *IEEE Trans. Wireless Commun.*, vol. 2, no. 2, pp. 284–293, Mar. 2003.
10. F. Maehara, "Series form expression of BER performance for DQPSK/OFDM signals employing selection combining diversity reception over nonlinear fading channels," *Proc. IEEE VTC2005-Spring*, May 2005.
11. F. Maehara, A. Taira, and F. Takahata, "Simple series form formula of BER performance for DQPSK/OFDM signals in comprehensive nonlinear fading channels," *Proc. IEEE RWS2008*, pp. 37–40, Jan. 2008.
12. A. Papoulis, *Probability random variables, and stochastic processes*, 4th ed. NewYork: McGraw-Hill, 2002.
13. H. E. Rowe, "Memoryless nonlinearities with Gaussian inputs: Elementary results," *Bell Syst. Tech. J.*, vol. 61, no. 7, pp. 1520–1523, Sept. 1982.
14. G. Santella and F. Mazzenga, "A model for performance evaluation in M-QAM-OFDM schemes in presence of nonlinear distortions," *Proc. IEEE VTC95*, pp. 830–834, July 1995.

A Novel Exponential Link Error Prediction Method for OFDM Systems

Ivan Stupia, Filippo Giannetti, Vincenzo Lottici, and Luc Vandendorpe

Abstract Coded multicarrier techniques combined together with link resources adaptation algorithms are the key technologies toward efficient high-data-rate communications over wireless fading channels. This paper contributes with a novel accurate yet simple method to predict at the transmitter the link performance of bit-interleaved coded (BIC) orthogonal frequency division multiplexing (OFDM) links, which offers appealing features when compared with conventional techniques. Its effectiveness is confirmed through extensive simulation results obtained over typical wireless channel environments.

1 Introduction

Link resource adaptation (LRA) is an effective technique to improve the spectral efficiency of high-speed wireless links over fading channels [1]. In LRA, data-rate, transmit power and code rate or scheme, are adaptively modified according to channel fading dynamics in order to get the most of the available link resources. However, the potential of the LRA schemes is fully achieved only on condition that an accurate prediction of link level performance, e.g., the packet error rate (PER), is available at the transmitter. In a multicarrier (MC) system, such as orthogonal frequency division multiplexing (OFDM), the frequency selective fading over the transmission channel induces large SNR variations across the bandwidth, thus making PER prediction a demanding task. The need for an accurate yet low-complexity technique for predicting the link-level performance figure has therefore gained recently an ever increasing interest [2–5]. One of the most promising methods to be

---

I. Stupia (✉), F. Giannetti, and V. Lottici
University of Pisa, Department of Information Engineering, Via G. Caruso, 16, I-56122 Pisa, Italy
e-mail: {ivan.stupia,filippo.giannetti,vincenzo.lottici}@iet.unipi.it

L. Vandendorpe
Université Catholique de Louvain, Place du Levant, 2, B-1348 Louvain-la-Neuve, Belgium
e-mail: luc.vandendorpe@uclouvain.be

mentioned is the effective SNR mapping (ESM) concept, which is based on the idea of mapping the instantaneous SNRs of the received subcarries into a single scalar value from which the actual PER can be easily derived. Among the several ESM approaches proposed in literature, the exponential ESM (EESM) [2] offers a rather simple evaluation of link PER performance by using a convex mapping function based on the pairwise error probability (PEP) Chernoff bound for the case of binary signalling. Although it can be expressed in a closed form, the generalization for high order modulations does not exist, and a tuning factor is thus necessary for adjusting the PER estimate for each combination of modulation and coding schemes. Exploiting mapping functions different from the exponential one give other variants of the EESM, as the logarithmic effective SNR metric (LESM) and the capacity effective SNR metric (CESM) [3]. The ESM based upon the mutual information approach (MIESM) proposed in [4], instead, allows to separate the modulation and the coding models, so that good prediction performance are possible for the mixed-modulation case whenever different modulation orders are employed over the subcarriers. However, the mapping function is not convex, a feature particularly required in the context of adaptive modulation and coding (AMC), and the lack of a closed expression for the mutual information requires a polynomial approximation, as proposed in [5].

In this paper, a novel link performance prediction method, tagged as "cumulant generating function based" ESM, or κESM for short, is proposed, which combines together the simplicity of the EESM with the accuracy of MIESM by exploiting the *Gaussian approximation* of the PEP for a coded OFDM system [6]. The κESM (i) allows the transmitter to acquire the actual effective SNR through a simple low-rate feedback channel only, instead of the whole channel state information (CSI), (ii) guarantees better accuracy with respect to the conventional EESM in view of a more accurate formulation of the PEP metric, (iii) separates the modulation and coding models thus making the code adjusting factor independent on the particular modulation used on each subcarrier, (iv) unlike the MIESM, provides also a convex mapping function which can be expressed in a simple closed form.

2 Channel Model

Consider a bit interleaved coded (BIC) OFDM system over frequency selective fading channels. The binary coded symbols b_k are obtained by first encoding and then interleaving the information bit stream originating from the binary source. The sequence of b_ks are thus Gray coded and associated to the label of QAM symbols that are transmitted through the channel both along the time and frequency domains according to the OFDM multicarrier format. Specifically, each block of QAM symbols with size N is processed at the transmitter by a Inverse Fourier Transform (IDFT) unit, and properly extended by a cyclic prefix (CP) to avoid interference between consecutive blocks and maintain the subcarriers orthogonal with each others. Designating with $x_{l_k}^{(n_k)}$ the symbol mapped to the n_kth subchannel in the l_kth OFDM block within the transmitted frame corresponding to the coded symbol b_k, with $0 \leq n_k \leq N - 1$ and $0 \leq l_k \leq L - 1$, it is well known that the output of the receiver DFT unit (under the assumption of ideal synchronization) can be written as

$$z_{l_k}^{(n_k)} = A^{(n_k)} x_{l_k}^{(n_k)} + w_{l_k}^{(n_k)}, \qquad (1)$$

where $A^{(n_k)}$ is a complex-valued coefficient depending on the channel frequency response and $w_{l_k}^{(n_k)}$ is the noise contribution modelled as a zero-mean Gaussian random variable (RV) with variance σ_w^2. For the sake of simplicity, in the sequel we will assume that the multipath channel is invariant within each frame but randomly varies from frame to frame, and the channel state information (CSI) knowledge is without errors.

3 PER Prediction Based on Effective SNR Mapping

Generally speaking, the aim of the ESM technique is to predict the PER of a coded multicarrier transmission link employing a given modulation and coding scheme (MCS) through a single (scalar) SNR value γ_{eq} related to an "equivalent" coded system operating over a AWGN channel. The quantity γ_{eq} compresses the multicarrier channel state, or in other words, the set of the instantaneous received SNR levels $\boldsymbol{\gamma} \triangleq \left[\gamma^{(0)}, \gamma^{(1)}, \cdots, \gamma^{(N-1)}\right]^T$, with $\gamma^{(n)} \triangleq \frac{|A^{(n)}|^2}{\sigma_w^2}$, $0 \leq n \leq N - 1$, experienced by the active subcarriers, so that the equality condition

$$\text{PER}_{\text{AWGN}}(\gamma_{eq}) = \text{PER}(\boldsymbol{\gamma}) \qquad (2)$$

holds, where PER_{AWGN} and PER refer to the AWGN equivalent system and that to be modelled, respectively. Hence, according to the specific rule adopted to map the array $\boldsymbol{\gamma}$ into the scalar γ_{eq}, a number of ESM-based abstraction methods may come up with different level of performance capability.

In this section, after a brief outline of the two currently most known ESM techniques, namely the exponential ESM (EESM) and mutual information ESM (MIESM), we focus on a novel PER prediction scheme for BIC-OFDM transmission links over frequency-selective channels based on the novel concept of κESM. Differently from the conventional ESM methods, the κESM relies on an accurate evaluation of the PEP figure through the statistical description of the BIC log-likelihood metrics, thus offering an efficient accuracy versus manageability tradeoff.

3.1 EESM

The equivalent (scalar) SNR γ_{EESM} of the EESM is derived applying the Chernoff bound for coded binary transmissions over AWGN channel. In [2], it is illustrated that

$$\gamma_{\text{EESM}} \stackrel{\Delta}{=} -\beta \ln \frac{1}{|\mathcal{S}|} \sum_{n \in \mathcal{S}} e^{-\frac{\gamma_n}{\beta}}. \tag{3}$$

where $\gamma_n \stackrel{\Delta}{=} \frac{|A^{(n)}|^2}{\sigma_w^2}$ represents the generic entry of $\boldsymbol{\gamma}$, $\mathcal{S}$ is the set of indexes of the active subcarriers and $\beta \stackrel{\Delta}{=} \beta_{\text{mod}} \beta_{\text{cod}}$, β_{mod} and β_{cod} being tuning parameters optimized for the adopted combination of modulation scheme and coding rate. The EESM method has been shown to produce good results even at reasonable complexity, especially for low modulation orders. However, the dependence of the adjusting factors β_{mod} and β_{cod} on the specific MCS mode makes difficult the application of this approach in the context of hybrid ARQ (HARQ), where several codewords with different MCS levels may be combined within consecutive retransmissions.

3.2 MIESM

The main limitation of the EESM can be overcome by resorting to the MIESM method whose mapping law is based on two distinct models, one for modulation and the other for coding [4]. In the first model, the measure of the mutual information $\text{I}(\cdot)$ is evaluated as a function of the SNR on the subcarriers on a symbol-by-symbol basis for a given modulation order regardless of the coding scheme. Such a step is based on polynomial approximation since simple closed forms are not available [5], which adds to the overall complexity. Then, the so-called received bit information rate (RBIR) is derived as

$$\text{RBIR} \stackrel{\Delta}{=} \frac{1}{|\mathcal{S}|} \sum_{n \in \mathcal{S}} \text{I}(\gamma_n/\beta), \tag{4}$$

i.e., by accumulating the total mutual information within one coding block, β being a scaling factor depending on the adopted coding scheme only. As final step, the RBIR is mapped into the PER quality metric so that the equivalent SNR γ_{MIESM} can be provided as

$$\gamma_{\text{MIESM}} \stackrel{\Delta}{=} \beta \text{I}^{-1}(\text{RBIR}), \tag{5}$$

where γ_ns are the quality measures, i.e., the SNRs on the active subcarriers. It is worth remarking that the mapping related to the coding model is independent of the modulation scheme and allows the MIESM link quality prediction to be defined also for systems wherein the coded block may include symbols with mixed modulation order. However, the measure of the mutual information is based on polynomial approximation since simple closed forms are not available.

4 κESM

Differently from both the EESM and MIESM, the κESM approach is based on an in-depth evaluation of the PEP figure through the cumulant generating function (CGF) of a RV. Toward this end, consider two distinct sequences of coded binary symbols originating from the same state of the trellis code that merge after d steps, denote them as the d-dimensional arrays $\mathbf{b}$ and $\mathbf{a}$, whose kth elements are b_k and a_k, respectively, and recall that

$$\text{PEP}(d) \triangleq \Pr\left\{\hat{\mathbf{b}} = \mathbf{a}|\mathbf{b}, \mathbf{A}\right\}, \tag{6}$$

where $\hat{\mathbf{b}}$ is the decoder output and $\mathbf{A}$ collects the complex-valued coefficients defined in (1) affecting the symbols transmitted over the subcarriers. Under the assumption of ideal interleaving, ideal CSI and that the OFDM subchannels behave as a memoryless binary-input output-symmetric (BIOS) channel, it has been shown in [6] that the PEP can be computed as the tail probability

$$\text{PEP}(d) = \Pr\left\{\sum_{k=1}^{d} \mathcal{L}_k > 0\right\}, \tag{7}$$

where

$$\mathcal{L}_k \triangleq \log \frac{\sum_{\tilde{x} \in \chi_{a_k}^{(i_k, n_k)}} \exp\left(-\left|z_{l_k}^{(n_k)} - A^{(n_k)} \tilde{x}\right|^2\right)}{\sum_{\tilde{x} \in \chi_{b_k}^{(i_k, n_k)}} \exp\left(-\left|z_{l_k}^{(n_k)} - A^{(n_k)} \tilde{x}\right|^2\right)} \tag{8}$$

is the log-likelihood ratio (LLR) metric for the kth coded binary symbol and $\chi_y^{(i,n)}$ is the subset of all the M-QAM constellation symbols on the nth subcarrier whose ith label bit is equal to y.

Taking into account that, thanks to the presence of the interleaver, the variables $\mathcal{L}_k$ can be considered i.i.d., the PEP can be approximated as [6]

$$\text{PEP}(d) \simeq Q\left(\sqrt{-2d\,\kappa_\mathcal{L}(\hat{s})}\right), \tag{9}$$

where $\kappa_\mathcal{L}(s) \triangleq \log M_\mathcal{L}(s)$ is the CGF, $M_\mathcal{L}(s) \triangleq E_\mathcal{L}\{e^{s\mathcal{L}}\}$, is the moment generating function (MGF), and $\hat{s}$ is the saddlepoint defined as the value for which $\kappa'_\mathcal{L}(\hat{s}) = 0$.

Now, it can be observed that at high SNRs the MGF can be approximated as

$$M_\mathcal{L}(s) \simeq E_\varphi\left\{\exp\left[-\gamma^{(n_k)} d^2\left(x_{l_k}^{(n_k)}, \bar{x}\right)(s - s^2)\right]\right\}, \tag{10}$$

i.e., considering the dominant term relevant to the nearest neighbor $\bar{x}$ (in the sense of Euclidean distance) of the symbol $x_{l_k}^{(n_k)}$ in the complementary subset $\chi_{a_k}^{(i_k,n_k)}$. In (10), $d(z,x)$ is the Euclidean distance between the complex-valued symbols z and x, $\gamma^{(n_k)}$ is the SNR defined just before (2), and the subscript φ indicates that the expectation is taken with respect to all the nuisance parameters, i.e., the noise, the symbols $x_{l_k}^{(n_k)}$ transmitted on the n_kth subchannel in the l_kth OFDM block, and the position of the coded binary symbols b_k in the label of the QAM symbols of the complementary subset $\chi_{a_k}^{(i_k,n_k)}$. To compute the expectation in (10) the following remarks are of interest: (i) due to the Gray mapping rule, $d\left(x_{l_k}^{(n_k)}, \bar{x}\right) = \Delta^{(n_k)} d_{\min}^{(n_k)}$, where $d_{\min}^{(n_k)}$ is the minimum Euclidean distance between the symbols in the complete QAM set associated with the n_kth subchannel and $\Delta^{(n_k)}$ is a positive integer; (ii) for each subcarrier n_k there are $m^{(n_k)}$ label bits, each of which has $2^{m^{(n_k)}-1}$ symbols on its complementary subset, so that the total number of terms to be averaged results $m^{(n_k)} 2^{m^{(n_k)}-1}$; (iii) the possible distinct values for $d_{\min}^{(n_k)}$ are $2^{m^{(n_k)}-1}/2$; (iv) the number of symbols at distance $\Delta^{(n_k)} d_{\min}^{(n_k)}$ in the complementary subset is $\psi^{(n_k)}(\Delta)$, such that $\forall n_k \in \mathcal{S}$ we get $\sum_{\Delta=1}^{2^{m^{(n_k)}-1}/2} \psi^{(n)}(\Delta) = m^{(n_k)} 2^{m^{(n_k)}-1}$, $\mathcal{S}$ being the set of indexes of the active subcarrier; (v) the probability that the coded bit b_k is sent through the subchannel n_k in the case of ideal random interleaving is $\zeta^{(n_k)} = m^{(n_k)} / \sum_{n_k \in \mathcal{S}} m^{(n_k)}$. Hence, collecting the above facts together, the average in (10) over φ turns out to be

$$M_{\mathcal{L}}(s) \simeq \sum_{n \in \mathcal{S}} \frac{\zeta^{(n)}}{m^{(n)} 2^{m^{(n)}-1}} \sum_{\Delta=1}^{2^{m^{(n)}-1}/2} \psi^{(n)}(\Delta) e^{-\gamma^{(n)} \left[\Delta d_{\min}^{(n)}\right]^2 (s-s^2)}, \quad (11)$$

and after recalling that the saddlepoint in the case of BIOS channels is placed at $\hat{s} = 1/2$, the PEP in (9) results

$$\text{PEP}(d) \simeq Q\left(\sqrt{-2d \, \log \sum_{n \in \mathcal{S}} \sum_{\Delta=1}^{2^{m^{(n)}-1}/2} \frac{\zeta^{(n)} \psi^{(n)}(\Delta)}{m^{(n)} 2^{m^{(n)}-1}} e^{-\gamma^{(n)} \frac{\left[\Delta d_{\min}^{(n)}\right]^2}{4}}}\right). \quad (12)$$

Since (9) can be interpreted as the PEP of an equivalent binary system that experiences a AWGN channel with SNR equal to $-\kappa_{\mathcal{L}}(\hat{s})$, the expression (12) suggests a simple way to derive a novel model for the prediction of link performance, that will be referred in the sequel as κESM. First, for each subcarrier experiencing the SNR $\gamma^{(n)}$ and employing a given modulation scheme, the weighted sum of exponential functions is computed as

$$\Omega^{(n)} \triangleq \sum_{\Delta=1}^{2^{m^{(n)}-1}/2} \frac{\psi^{(n)}(\Delta)}{2^{m^{(n)}-1}} e^{-\gamma^{(n)} \frac{\left[\Delta d_{\min}^{(n)}\right]^2}{4\beta}}, \quad (13)$$

A Novel Exponential Link Error Prediction Method for OFDM Systems

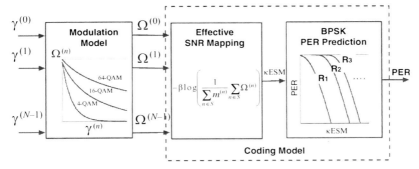

Fig. 1 κESM link prediction model

β being a parameter to be optimized independent of the modulation order. Then, the above quantities are the input of the coding model that provides the resulting equivalent SNR value

$$\kappa\text{ESM} \triangleq -\beta \log\left(\frac{1}{\sum_{n\in\mathcal{S}} m^{(n)}} \sum_{n\in\mathcal{S}} \Omega^{(n)}\right). \quad (14)$$

that is eventually mapped into the required PER level. Let us remark that the tuning parameter β depends only on the adopted coding scheme, and accordingly, the κESM method allows an accurate PER prediction also in the case of mixed modulation over the subcarriers. Figure 1 depicts the κESM illustrated above, giving emphasis on the two separate models, one for modulation and the other for coding.

5 Numerical Results

The performance of the proposed κESM is quantified and compared to that of the EESM and MIESM conventional methods. We focus on a coded OFDM scheme with $N = 64$ subcarriers, M-QAM as modulation format with $M = 16, 64$, and a 64-state punctured convolutional encoder with rate $R = 1/2, 3/4$. In the numerical simulations, a total number of $P = 60$ different multipath channel realizations are taken into account, each with 6-taps modeled as independent Rayleigh RVs. The tuning factor β is optimized through the least-squares fit

$$\beta_{\text{opt}} = \arg\min_{\beta} \sum_{i=1}^{P} |\gamma_{\text{eff},i}(\beta, \boldsymbol{\gamma}_i) - \gamma_{\text{AWGN}}(\overline{\text{PER}_i})|^2 \quad (15)$$

where P denotes the number of different channel realization, while $\overline{\text{PER}_i}$ and $\boldsymbol{\gamma}_i$ are the PER level and the received subcarrier SNRs obtained from a link-level simulation for the multipath channel realization i, respectively. In (15), the

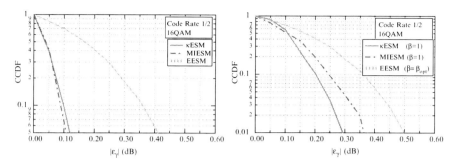

Fig. 2 CCDF of the ESNR error for MCS1

PER interval of interest is $\overline{\text{PER}}_i \in [0.01, 0.95]$. The accuracy of the link performance prediction models EESM, MIESM and κESM equipped with the optimum value of β given by (15) is quantified evaluating (i) the complementary cumulative density function (CCDF) of the effective SNR (ESNR) error defined as $\varepsilon_\gamma \triangleq 10\log_{10}\gamma_{\text{eff},i}(\beta,\boldsymbol{\gamma}_i) - 10\log_{10}\gamma_{\text{AWGN}}(\overline{\text{PER}}_i)$, for the $R = 3/4$ with 64-QAM (MCS1) and $R = 1/2$ with 16-QAM (MCS2), (ii) the PER versus the equivalent SNR for the MCS1.

CCDF performance. The left side plot (LSP) and right side plot (RSP) in Fig. 2 show the CCDF for the MCS1. The tuning factor for both the MIESM and κESM is set as $\beta = \beta_{\text{opt}}$ (LSP) and $\beta = 1$ (RSP), while for the EESM β is optimized in both plots. The better accuracy offered by the κESM over the EESM is evident. Indeed, in the LSP (RSP) for a CCDF level of 0.05 (0.01), the κESM yields an absolute error around 0.12 dB (0.3 dB), whereas for the EESM we get an error greater than 0.4 dB (0.5 dB). In addition, while the κESM and the MIESM are almost equivalent in the LSP, the former has an edge of about 0.06 dB over the latter in the RSP, thus indicating a greater robustness against the errors in the optimization of the tuning factor. Figure 3 refers, instead, to the MCS2. Focusing on the case when all the models are optimized (LSP), we observe that the κESM improves both on the MIESM and EESM, while for the RSP the gap between the the κESM and EESM at the CCDF of 0.05 is around 0.15 dB. This means that κESM is particularly adapt to be employed over the EESM and MIESM whenever modulation order gets higher.

PER performance. Figure 3 shows the accuracy on the PER prediction when the MCS1 scheme is considered. The superiority of the κESM over the EESM for the PER levels of interest is apparent despite 64-QAM is employed as modulation format. The κESM method guarantees the same slope of the actual PER curve, while this condition does not hold for the EESM where a "cross point" occurs. On the other side, similar performance is obtained if compared to the MIESM, but the price to be paid in the latter is that it is not based on closed expressions.

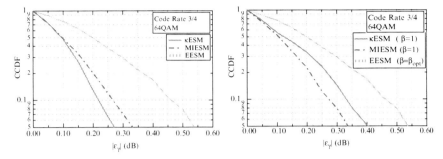

Fig. 3 CCDF of the ESNR error for MCS2

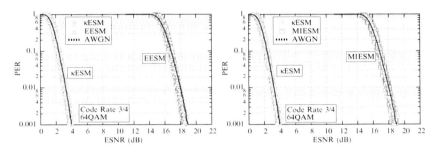

Fig. 4 PER of EESM, MIESM and κESM as a function of the ESNR

6 Conclusions

A novel method for the link performance prediction of BIC-OFDM links has been proposed and testified through numerical simulations on typical multipath fading channels. Its inherent low-complexity structure offers better accuracy with respect to the conventional EESM, especially for higher modulation orders, and allows the separation between the modulation and the coding models as the MIESM but with a convex mapping function expressed in a simple closed form. These remarkable advantages make the proposed technique the basis for efficient adaptive modulation and coding algorithms.

Acknowledgement This work was supported by the European Commission in the framework of the FP7 Network of Excellence in Wireless COMmunications NEWCOM++ (contract n.216715)

References

1. S.T. Chung and A.J. Goldsmith, "Degrees of Freedom in Adaptive Modulation: A Unified View," *IEEE Trans. on Commun.*, vol. 49, pp. 1561–1571, Sept 2001.
2. Y. Blankenship, P.J. Sartori, B. K. Classon, V. Desai and K. L. Baum, "Link Error Prediction Methods for Multicarrier Systems," *VTC 2004–Fall*, vol. 6, pp. 4175–4179, Sept. 2004.

3. K. Brueninghaus et al., "Link Performance Models for System Level Simulations of Broadband Radio Access Systems," *PIMRC 2007*, Sept 2005.
4. L. Wan, S. Tsai, and M. Almergn, "A Fading-Insensitive Performance Metric for a Unified Link Quality Model," *IEEE WCNC 2006*, vol. 4, pp. 2110–2114, Apr 2006.
5. IEEE, "Project 802.16m Evaluation Methodology Document (EMD)," Apr. 2008.
6. A. Martinez, A. Guillén i Fàbregas and1 G. Caire, "Error Probability Analysis of Bit-Interleaved Coded Modulation," *IEEE Trans. Inf. Theory*, vol. 52, pp. 262–271, Jan. 2006.

WiMAX Performance in the Airport Environment

Paola Pulini and Snjezana Gligorevic

Abstract In this paper, the multicarrier physical layers of WiMAX are evaluated in the context of airport data links. The orthogonal frequency-division multiplexing (OFDM) and orthogonal frequency-division multiple-access (OFDMA) cases are applied to the forward link (FL) and reverse link (RL), respectively. The performance of the so called parking and taxi scenarios is presented for airport communications in C-band. Numerical results show that the proposed scheme brings good performance for both the FL and the RL. For the OFDMA case a structure changing called double-tile is also proposed to improve the system performance.

1 Introduction

In the framework of the European project EMMA2 [1], a preliminary investigation on the performance of multicarrier modulation techniques for data links in the airport environment has been carried out. The scope of this investigation deals with the development of technologies able to support the increasing data traffic coming from vehicles operating on the airport surface, i.e. mainly aircraft, but also vehicles providing luggage handling, fueling, etc. Currently, only VDL Mode 2 [2], which operates in the highly congested very high frequency (VHF)-band, can be used for air-traffic data link communications. Thus, this data link technology cannot satisfy the demand of robustness, security and efficiency requirement for the future aeronautical communications. New bands and new data communication systems have to be investigated. For these reasons, at the ITU World Radio Conference in 2007 [3], new allocations to the Aeronautical Mobile Route Service (AMRS) were defined, especially in C band (between 5.091 and 5.150 GHz) for airport surface operations. Furthermore, a proposal for the adoption of a new communication scheme was

---

P. Pulini (✉) and S. Gligorevic
German Aerospace Center (DLR), Institute of Communications and Navigation, Oberpfaffenhofen, 82234 Wessling, Germany
e-mail: paola.pulini@dlr.de and Snjezana.Gligorevic@dlr.de

introduced in [4]. The proposal relies on the adoption of an IEEE 802.16e [5]. The IEEE 802.16e standard, which will be referred in the following as mobile WiMAX or simply WiMAX, provides a large variety of profiles for the physical layer coding and modulation, and hence enables the choice of the one which may be more suitable for the airport environment.

In this paper, we recall the main results on this investigation, with particular reference to the IEEE 802.16e standard. We focus on the OFDM and OFDMA modes. The analysis is carried out considering realistic channel models, mainly derived from a measurement campaign at the airport of Munich (Germany) [6]. Numerical results related to the system performance in the most relevant airport scenarios are presented, considering both the forward and the return links in C-band.

The paper is organized as follows. In Section 2, the system characteristics, i.e. modulation and coding, are recalled, with emphasis on the OFDM and OFDMA modes. A novel framing called double-tile is also introduced in this section. Section 3 provides a description of the used channel model. In Section 4 we presents the simulation results and, finally, conclusions follow in Section 7.

2 System Characteristics

The WiMAX standard includes three different physical layers, i.e. one single carrier and two multicarrier modes based on OFDM. This work focuses on the OFDM and the OFDMA cases and evaluates their performance in the aeronautical airport environment.

2.1 FL–OFDM

For the broadcast transmissions in the FL, the WiMAX OFDM mode is investigated. The WiMAX OFDM uses 256 subcarriers with only the 200 central effectively being used for data and pilots. The other subcarriers, i.e. DC and two lateral guard-bands of 27 and 28 subcarriers are left unused (Fig. 1). In this mode, it is possible to change the subcarrier spacing to increase the bandwidth up to a maximum of 5 MHz.

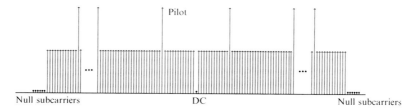

Fig. 1 Symbol structure for the OFDM case

The WiMAX OFDM case includes a concatenated Reed-Solomon convolutional code (RS-CC) with different coding rates. Optional codes are the convolutional turbo code (CTC) and the block turbo code (BTC). The modulation set for the subcarriers includes binary phase shift keying (BPSK), quadrature phase shift keying (QPSK), quadrature amplitude modulation (QAM) with 16 constellation points (16-QAM) and optionally with 64 (64-QAM). The cyclic prefix may be chosen within 1/4, 1/8, 1/16 or 1/32 of the symbol duration. The simulation parameters adopted in this work include a bandwidth of 5 MHz with subcarrier spacing of roughly 20 kHz, a basic coding scheme with rate 1/2 convolutional code, and a quadrature phase shift keying (QPSK) subcarrier modulation. The cyclic prefix has been set 1/8 of the symbol length. Frames of 24 symbols have been considered, with all available subcarriers allocated to one user.

2.2 RL–OFDMA

For the RL, the OFDMA physical mode is investigated. The OFDMA [7] represents one of the more interesting multiple access schemes for broadband wireless systems and was included in [8] as an alternative access scheme with respect to frequency/time-division multiple access (FDMA/ TDMA). In the OFDMA case the subchannelization concept is introduced. The subcarriers are divided into subsets of subcarriers called subchannels which constitute the minimum allocation unit. The OFDMA mode supports two different ways to permutate the subcarriers: the *contiguous permutation* that considers adjacent groups of subcarriers and the *diversity permutation* that distributes the subcarriers pseudo randomly [5]. In the partial usage of subcarriers (PUSC) case, the RL subchannel allocation is performed by means of entities called tiles. The tiles are composed of 12 subcarriers over three symbols as depicted in Fig. 2. The pilot and data subcarriers are allocated within these units. The pilot symbols are positioned in the corners of the tile and are present only in the even OFDMA symbols. The subchannels are composed of six tiles scattered in the frame by an interleaver as illustrated in Fig. 3.

The OFDMA, differently from the OFDM case, supports a variety of subcarrier numbers. The included FFT (fast Fourier transform) sizes are 128, 512, 1024 and 2048. Focusing on the 1,024 subcarriers case, only 840 subcarriers are effectively used for data and pilots. We used for our simulation a 1,024 subcarriers system with

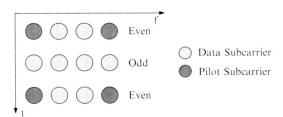

Fig. 2 OFDMA-PUSC tile structure

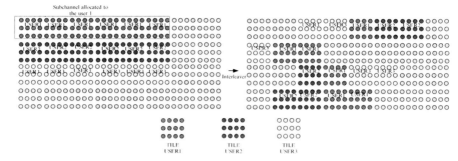

Fig. 3 Tiles permutation scheme (OFDMA RL-PUSC)

Table 1 Simulation parameters

Parameters	RL–OFDMA	FL–OFDM
Bandwidth	10 MHz	5 MHz
FFT	1024	256
Guard Subcarriers	183	56
Pilot Subcarriers	280	8
Subcarriers Spacing	10 kHz	20 kHz
Frame Size	24	24
Coding	Convolutional, 1/2	Convolutional, 1/2
Modulation	QPSK	QPSK
Pilot Boosting	2.5 dB	2.5 dB
Permutation Mode	PUSC	-
Data Allocation	Subchannels (15, 30, 80)	All frame
Channel	Aeronautical [9]	Aeronautical [9]
Channel Estimation	Linear / Ideal	Linear / Ideal

a bandwidth of 10 MHz and a frame size of 24 symbols. The subcarriers allocation to the users is performed within the tile and subchannel concepts. Different numbers of subchannels have been allocated to the users to evaluate the performance with variable numbers of subcarriers.

The OFDMA case includes for the forward error correction (FEC) several coding options. The convolutional code (CC) constitutes the basic coding scheme, optional codes are low density parity check code (LDPC), convolutional turbo code (CTC) and block turbo code (BTC). The modulation set for the subcarriers includes BPSK, QPSK, 16-QAM and optionally 64-QAM. The cyclic prefix may be chosen within 1/4, 1/8, 1/16 and 1/32 of the symbol length.

In this work we used a convolutional channel coding with rate 1/2 and a QPSK subcarrier modulation. The cyclic prefix is 1/8 of the symbol duration. Table 1 provides all the simulation parameters.

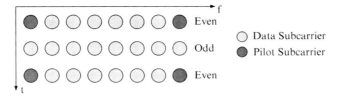

Fig. 4 OFDMA-PUSC double-tile structure

2.3 Double-tile Framing

The OFDMA-RL provides a very small distance between two pilot subcarriers. However, the pilot spacing in the frequency direction can be increased without considerable performance degradation when the coherence bandwidth of the channel is large enough. The distance between two pilot subcarriers in the frequency domain (N_f) should be less than the coherence bandwidth of the channel (B_{co}). Similar rule holds for the time spacing (N_t) and the coherence time of the channel (T_s). Approximating $B_{co} = 1/\tau_{max}$, where τ_{max} is the maximum delay of the channel, the spacing between two pilot subcarriers in both frequency and time domains should be selected as $N_f < 1/(\tau_{max} \cdot \Delta f)$ and $N_t < 1/(2 f_D \cdot T_s)$. Considering the channel parameters and the tile structure, it is possible to double the spacing between subcarriers in the frequency domain without loosing performance in the channel estimation (at least for the taxi scenario). We therefore introduced a novel framing which introduces the double tiles. A double tile is composed of two adjacent tiles in which the central pilots are replaced with data (see Fig. 4). In this way the single subchannel, composed of three double tiles instead of six single tiles, would contain 60 data subcarriers. The creation of bigger tiles with less pilot subcarriers increases the net data rate. In particular for 1,024 FFT size this solution would lead to a reduction of the number of pilot sub-carriers from 420 to 210 (per even OFDMA symbol). The simulation parameters adopted for the double-tile case include a 1,024 FFT size with a bandwidth of 10 MHz, a basic CC scheme and a QPSK modulation. The frame is composed of 24 symbols.

3 Channel Model

The channel is assumed to be a time-variant, frequency-selective fading channel according to the model presented in [9]. Two different scenarios are evaluated: the so-called taxi and the parking scenarios. The first represents the phase during which the aircraft is traveling toward or from the gate and is characterized by slow fading. The parking scenario applies when the aircraft is moving at a very slow speed close to the gate. In this case usually there is no line of sight (LOS) component. For the taxi scenario, considering that in this phase the aircraft is moving with an

Table 2 Channel parameters

Scenario	Taxi	Parking
Number of Echoes	5	12
Doppler Spread	10 Hz	10 Hz
Delay max	2 μs	1.3 μs

approximative velocity of 60 km/h, we assumed a Doppler spread of 10 Hz. We then considered five taps and a maximum delay of 2 μs. For this delay and our subcarrier spacings of 20 and 10 kHz, the cyclic prefix set to 1/8 of the symbol length should avoid inter-symbol interference (ISI). For the parking scenario we considered a Doppler spread of 10 Hz. The presence of buildings increases the multipath, thus we used 12 taps. The maximum delay of 1.3 μs shall be compensated by the cyclic prefix length. The channel parameters used for the simulation (Table 2) are derived from a measurement campaign at the Munich airport [6].

For both the FL and RL cases we implemented a linear channel interpolation using the pilot tones. For the OFDM-FL we performed frequency domain pilot interpolation while for the OFDMA-RL case we used a frequency and time interpolation tailored to the tile structure.

4 Results

This section provides the results on the simulated performance of the OFDM and OFDMA WiMAX physical layers in the airport environment.

4.1 FL–OFDM

Figure 5 shows the performance of the OFDM-FL case and provides the bit error rate (BER) vs E_b/N_0 for the taxi and the parking scenarios. The results in the taxi case are better than those of the parking scenario, since in the taxi scenario the channel is more flat in time and frequency. The channel estimation with linear interpolation compared to the ideal case brings a loss of performance of roughly 2 dB in the parking scenario and of 1 dB in the taxi scenario.

4.2 RL–OFDMA

The following figures provide the performance of the OFDMA-RL case. Figure 6 compares the results obtained for the taxi and parking scenarios by one user with 15 allocated sub-channels, i.e. 720 data subcarriers. This number of subchannels corresponds to the transmissions of the smallest RL message according to the

WIMAX Performance in the Airport Environment 307

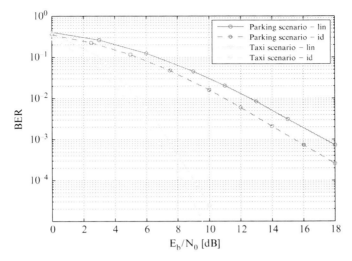

Fig. 5 OFDM-FL (5 MHz). Performance evaluation for the taxi and the parking scenarios

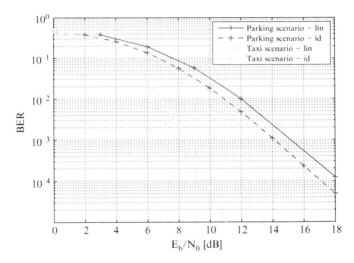

Fig. 6 OFDMA-RL (10 MHz). Performance evaluation for the taxi and the parking scenarios

Communications Operating Concept and Requirements (COCR) document [10]. As expected, also in this case the taxi scenario outperforms the parking scenario.

Figure 7 provides a performance evaluation of the taxi scenario within different numbers of subchannels assigned to one user. The number of subchannels allocated to one user is 15, 30 and 80, all resulting in the same performance. The explanation of this results is in the nature of the taxi scenario. The taxi case is characterized by a slow fading and the diversity gain provided by the different subchannel allocation

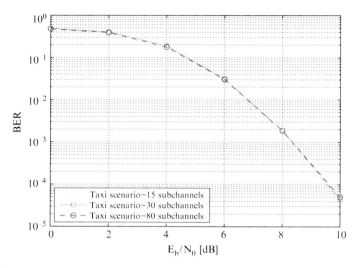

Fig. 7 OFDMA-RL (10 MHz). Performance evaluation for different number of allocated subchannels

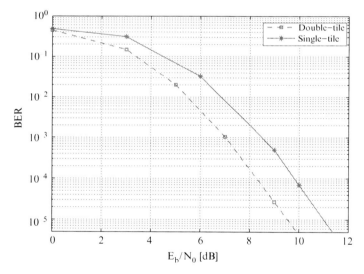

Fig. 8 OFDMA-RL (10 MHz). Performance evaluation of the double-tile structure for the taxi scenario

considered here does not provide sufficient improvements. The channel estimation with linear interpolation introduces a performance loss of roughly 1 dB respect to the ideal channel estimation.

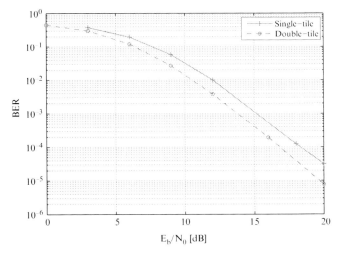

Fig. 9 OFDMA-RL (10 MHz). Performance evaluation of the double-tile structure for the parking scenario

4.3 Double-tile Framing

Figures 8 and 9 show the performance of the double-tile case respectively for the taxi and parking scenario. The proposed scheme, with respect to the single tile one, brings to a gain of more than 1 dB in both cases.

5 Conclusions

In this paper we investigated the performance of the IEEE 802.16e standard in the context of airport data links. The investigation has been focused on the physical layer multicarrier modes, taking into account realistic channel models for the airport environment. We proposed an adaptation of the subchannel structure which allows to reduce the transmission overhead due to the pilot sub-carriers. The numerical results confirm the suitability of OFDM/OFDMA for airport data links, especially for the taxi scenario, while for the parking scenario further investigations shall be carried out to identify solutions for improving system performance.

References

1. European Airport Movement Management by A-SMGCS, Part 2 project (EMMA2) http://www.dlr.de/emma2
2. RTCA, "Signal-In-Space Minimum Aviation System Performance Standards (MASPS) for Advanced VHF Digital Data Communications including compatibility with Digital Voice Techniques," September 2000.

3. "Report on the Results of the ITU World Radiocommunication Conference (2007) (WRC-07)", Air Navigation Commission. http://www.icao.int/anb/panels/acp/repository/ AN.2007.WP.8284.en[1].pdf
4. "Report of the second Meeting", Aeronautical Communications Panel, Montreal, 21–25 April 2008. http://www.icao.int/anb/panels/acp/wg/w/wgw2/ACP-WGW02-Report
5. IEEE 802.16-2005, "Part 16: Air Interface for Fixed and Mobile Broadband Wireless Access Systems, Amendment 2: Physical and Medium Access Control Layers for Combined Fixed and Mobile Operation in Licensed Bands and Corrigendum 1," Feb. 2006.
6. EMMA2 project, "A-SMGCS Data Link Communication Study," 2008.
7. Sari H. "Orthogonal frequency-division multiple access with frequency hopping and diversity," in Proc. International Workshop on Multi-Carrier Spread-Spectrum (MC-SS'97), Oberpfaffenhofen, Germany pp. 57–68, April 1997.
8. IEEE 802.16-2004, "Local and Metropolitan Area Networks Part 16: Air Interface for Fixed Broadband Wireless Access Systems," Oct. 2004.
9. E. Haas, "Aeronautical channel modeling," IEEE Transactions on Vehicular Technology, Volume 51, Issue 2, pp. 254–264, Mar 2002.
10. EUROCONTROL/FAA, "Communications Operating Concept and Requirements for the Future Radio System", Version 2.0, 2007.

Throughput Enhancement Through Femto-Cell Deployment

Zubin Bharucha, Harald Haas, Ivan Ćosović, and Gunther Auer

Abstract This paper studies the impact of femto-cell underlay deployment that share radio frequency resources with urban macro-cells. Due to the random and uncoordinated deployment, femto-cells potentially cause destructive interference to macro-cells and vice versa. On the other hand, femto-cells promise to substantially enhance the spectral efficiency due to an increased reuse of radio resources. The performance of systems with femto base station (FBS) deployment is compared to a system where all users, including indoor users, are served by the macro base station (MBS). In addition, the impact of closed-access and open-access femto-cell operation is examined. It is demonstrated that significant throughput gains can be achieved through such FBS deployment, regardless of whether closed-access or open-access is considered. Results clearly indicate that the benefits of FBS deployment by far outweigh their impact on the macro-cell capacity.

1 Introduction

There is a growing demand for increased user and system throughput in wireless systems. Naturally, such rapidly increasing demand is served by higher bandwidth allocation, but since bandwidth is scarce and expensive, a key to substantial throughput enhancement is to improve the reuse of radio frequency resources. One of the most powerful methods of boosting wireless capacity is by shrinking the cell size. The reason for this is that smaller cell sizes enable the efficient spatial reuse of spectrum [1]. Furthermore, the shorter transmission distances allow for the use of

---

Z. Bharucha (✉) and H. Haas
Institute for Digital Communications, School of Engineering and Electronics,
The University of Edinburgh, Edinburgh EH9 3JL, UK
e-mail: {z.bharucha, (✉) h.haas}@ed.ac.uk

I. Ćosović and G. Auer
DOCOMO Euro-Labs, Landsberger Strasse 312, 80687 Munich, Germany
e-mail: auer@docomolab-euro.com

higher order modulation schemes. While decreasing the cell size boosts system capacity, the cost involved is becoming increasingly prohibitive, due to the required installation of new network infrastructure.

Studies indicate that a significant proportion of data traffic originates indoors [2]. Poor signal reception caused by penetration losses through walls severely hampers the operation of indoor data services. Therefore, the concept of 3rd generation (3G) femto-cells has recently attracted considerable interest. FBSs are low-cost, low-power, short-range, plug-and-play base stations, which aim to extend and improve macro-cell coverage in indoor areas. FBSs are directly connected to the backbone network and user equipment (UE) located indoors communicates directly with FBSs. FBSs therefore offload indoor users from the macro-cell, thus potentially enhancing the capacity both, indoors as well as outdoors.

In [3, 4], the authors propose the *TDD underlay* concept. Owing to the asymmetric nature of traffic, one of the frequency division duplex (FDD) bands (the underloaded one) can be split in time such that the FBS transmits and receives information from its associated femto user equipment (FUE) in a time division duplex (TDD) fashion. In this work, both macro and femto cells operate in the same radio frequency spectrum in frequency division duplex (FDD) mode, compliant with the specifications for beyond 3G mobile communication systems [5]. Like in the original TDD underlay concept [3], the FBS backhauls data through a dedicated broadband gateway (DSL/cable/Ethernet/etc.) to the cellular operator network. There are several obvious advantages from femto-cell deployment, the most important of which is that the operator is able to concentrate on providing better service to the outdoor macro-cell UEs (MUEs). Another selling point for operators is that high throughput coverage is extended to the indoor environment.

However, it must be kept in mind that FBSs are deployed without network planning. Since FBSs operate on the same bands as MBSs, their deployment introduces additional interference. Furthermore, questions regarding security are raised with the deployment of FBSs, such as "should the FBS allow any UE in its vicinity to connect to it?" or "should the FBS maintain a list of UEs that are exclusively allowed to use its resources?" and "who should control access to the FBS?". The discussion is therefore taken in the direction of open-access versus closed-access systems. Open-access allows any UE to connect to the FBS whereas closed-access only allows a specified set of UEs to connect to the FBS. Closed-access systems are therefore susceptible to higher interference from indoor UEs that lie geographically in a femto-cell but are not connected to its FBS.

Recent studies on femto-cells shows the geographic distribution of throughput for a deployment of open-access femto-cells in a cellular network [6] and the impact of FBS deployment on macro-cell performance [7]. In this work, system-level simulations are carried out for three different scenarios. In all systems, indoor and outdoor users exist and are randomly distributed. A system without FBS deployment, where all users (indoors or outdoors) are exclusively served by macro-cells serves as benchmark system. Systems with FBS deployment are distinguished between open and closed access (see Fig. 1). We demonstrate through system level

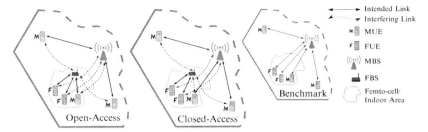

Fig. 1 The different access methods and the benchmark system. Interference scenarios are highlighted with dashed lines

simulations that FBS deployment, no matter what the access method is, has a clear advantage over the benchmark system in terms of user and overall system throughput.

The remainder of this paper is organised as follows. Section 2 describes the system model and simulator setup. Simulation results are presented in Section 3 and Section 4 highlights the key findings.

2 System Model and Simulation Setup

2.1 User Distribution

The simulation area comprises a two-tier, tessellated hexagonal cell distribution. In order to eliminate edge effects, an additional two tiers are simulated, however statistics are taken only from the first two tiers. Femto-cells have a circular area, each with a fixed radius and they are uniformly distributed over the four-tier structure with a given density. A random number of users and one FBS are uniformly distributed in each femto-cell. In the macro-cell, the MUEs are uniformly distributed over the entire region. This ensures that there is a certain probability (dependent on femto-cell and MUE density) that MUEs lie within femto-cells. In the case of open-access systems, such MUEs are *assimilated* into the femto-cell and are served by the associated FBS. In the case of closed-access systems, these MUEs are still served by the MBS. For the benchmark system, all UEs are MUEs (and are located indoors or outdoors).

2.2 Path Loss Models

Three path loss models (along with their respective delay profiles) are used – the urban micro (UMi) model for the macro-cell channel, the indoor hotspot (InH) model [8] for the femto-cell and the indoor-to-outdoor model to simulate the channel between entities lying indoors and outdoors. For each link in the system, the probability of line of sight (LoS) is calculated as specified in the appropriate path loss model and based on the LoS condition, the associated path loss model is applied to that link. Furthermore, a wall penetration loss of 20 dB is used to model outdoor-to-indoor links (and vice-versa). The delay profiles associated with these path loss models used to generate frequency-selective fading are provided in [8].

2.3 Interference and SINR Calculation

Once the users are distributed, log-normal shadowing maps containing correlated shadowing values in space are generated. Using these and the path loss models, MUEs are associated with the MBS to which they have the least path loss. Depending on whether open or closed access is considered, MUEs that lie within femto-cells are either assimilated or not.

A resource block (RB) represents one basic time-frequency unit. The RBs are distributed equally among the MUEs of a macro-cell and the same resources are reused within each femto-cell. In order to allow for a fair comparison, the transmit power per RB is computed *a priori* and is kept constant for all systems.

The signal-to-interference-plus-noise ratio (SINR) for user i on RB j (γ_i^j) is calculated as

$$\gamma_i^j = P_i^j / (I_i^j + N), \qquad (1)$$

where P_i^j is the useful received power for user i on RB j, I_i^j is the interference seen by that user on that RB and N is the thermal noise in the system. There are eight interference scenarios: MBS↔MUE, FBS↔FUE, MBS↔FUE, FBS↔MUE, where "↔" represents bi-directional interference. The first four represent "inter-site" interference scenarios, i.e., the interference originates from neighboring cells (macro or femto). The last four interference scenarios can be caused within the same cell. Since all the resources are reused within the femto-cell, FUEs are affected by the MBS of the macro-cell within which they lie (and vice-versa) and FBSs are interfered with by the MUEs which lie in the same macro-cell. The last two interference scenarios are potentially the most detrimental in closed-access systems. This is because for closed-access, an MUE may lie within a femto-cell while still being connected to the MBS, thereby causing severe MUE→FBS interference in the uplink (UL). Only the first two interference scenarios exist for the benchmark system, since all users are served by macro-cells.

3 Results

Simulations are run for a full-buffer traffic model, i.e., all users in the system are active simultaneously. Furthermore, a quasi-static channel model is assumed, where channel variations due to mobile velocities of mobile users are neglected. Perfect synchronisation in time and frequency is assumed, so that interference between neighboring RBs is avoided. Based on the achieved SINR per RB, the aggregate throughput for user j is calculated as

$$T^j = \sum_{i \in \mathcal{R}^j} W_{RB} \log_2 \left(1 + \gamma_i^j \right), \qquad (2)$$

where $\mathcal{R}^j$ is the set of RBs allocated to user j, W_{RB} is the bandwidth of one RB and γ_i^j is the achieved SINR on the ith RB of user j. The sum system throughput is generated as the sum of the individual user throughputs. The number of FUEs per femto cell is uniformly distributed with the minimum being one and the maximum being four. The rest of the parameters used in the simulation are taken from [5,9–11] and are shown in Table 1.

The overall sum system throughput for both uplink (UL) and downlink (DL) is depicted in Fig. 2. For systems with FBS deployment the aggregate femto and macro user throughputs is compared to the benchmark system where all users are served

Table 1 Simulation parameters

Parameter	Value
Femto-cells per macro-cell	20
Macro-cell major radius	200 m
Femto-cell radius	10 m
Tot. Number of available RBs	100
RB Bandwidth	100 kHz
Thermal Noise per RB	−144 dBm
MBS Tx power per RB	24 dBm
FBS Tx power per RB	−50, −30 & 0 dBm
MUE Tx power per RB	18.8 dBm
FUE Tx power per RB	−46, −26 & 4 dBm

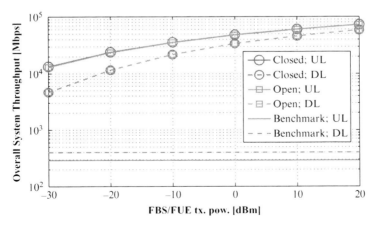

Fig. 2 Sum system throughput for all three systems

by the macro-cell. These results highlight the achieved gains through FBS deployment. It is seen that the UL system throughput for systems with FBS deployment are approximately two orders of magnitude higher than for the benchmark system. While the corresponding DL gains through FBS deployment are still impressive, the increase in DL system throughput is significantly lower compared to the UL. The reason for this is twofold: firstly, in the DL the MBS transmit power per RB is higher than in the UL, which results in an increased DL macro-cell throughput, which in turn cause more DL interference to femto-cells. Hence the DL throughput of femto-cells is lower compared to the UL. Secondly, in the UL, even if there is an MUE very close to a femto-cell, it typically does not transmit on all RBs. Thus, fewer RBs are affected.

In Fig. 2 we also observe that the system throughput for closed and open-access systems are nearly identical in DL and UL despite the fact that MUEs lying indoors suffer in terms of throughput. This is because in these simulations, the number of MUEs lying indoors is small in comparison to the total number of UEs in the cell and therefore this does not significantly affect the system throughput of closed-access. For the parameters used in this simulation, the probability of an MUE lying indoors is less than 5%, which is reflected in the results.

Figure 3a shows the throughput for only femto users in the UL for varying FUE transmit powers. It can clearly be seen that as the FUE transmit power is increased, the user throughput is boosted. The close-up also shows that closed-access outperforms open-access by a slim margin. This is because more FUEs are served by the FBS for open-access than for closed-access, as a result of the fact that some users lying inside femto-cells for closed-access are served by the MBS. Due to this, for open-access, each user, on average, is allocated fewer RBS, thus marginally bringing down the user-throughput for open-access, despite the increased interference caused by MUEs lying inside femto-cells for closed-access. The two different slopes in the

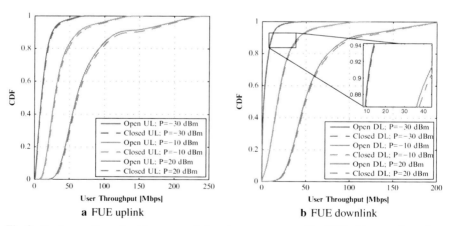

Fig. 3 Femto user throughput for open and closed-access systems. The CDF of the user throughput for FUEs is plotted for UL (**a**) and for DL (**b**)

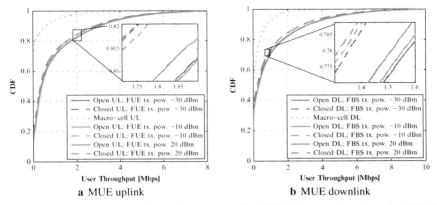

Fig. 4 Macro user throughput for the three systems. The CDF of the user throughput for MUEs is plotted in parts (**a**) for UL and (**b**) for DL

CDF arise from the fact that there can be FUEs having LoS and non LoS conditions with the associated FBS.

The throughput for only femto users in the DL for varying FBS transmit powers is shown in Fig. 3b. Here, the trends are the same as in the previous case. However, it must be noted that the user throughput is reduced in comparison to those seen in Fig. 3a. The reason is that the DL interference is, on average, higher than in the UL, due to the higher available transmit power at the MBS.

Figure 4a shows the cumulative distribution function (CDF) of macro-user throughput in the UL for varying FUE transmit powers. For the benchmark system, all users are included in the statistics because they are all served by the MBS. Interestingly, FBS deployment achieves an approximately fivefold gain at the 90th percentile compared to the benchmark system. In case of FBS deployment, the FBS offloads the MBS from serving femto-cell users, thus freeing resources that can be given to macro users. Looking closer (inset), it is seen that open-access outperforms closed-access. This is expected because for open-access, there are no MUEs lying inside femto-cells, avoiding destructive interference. On the other hand, the signal reaching the MBS from indoor MUEs for closed-access is highly attenuated (due to wall penetration losses, etc.), thus significantly bringing down the UL throughput for such users. Finally, it is seen that varying the FUE transmit powers does not significantly affect the macro-cell performance for systems with femto-cell deployment.

The same trends continue in Fig. 4b which depicts the CDF of user throughput served by the macro-cell in the DL for varying FUE and FBS transmit powers. As in the previous case, for the benchmark system the statistics of all users in the system are gathered, while only macro users are measured for systems with femto-cell deployment. Here, all systems exhibit slightly higher user throughput owing to the higher transmit power per RB in the DL. Hence, in Fig. 4b the difference between open and closed-access is slightly higher than in Fig. 4a (see inset).

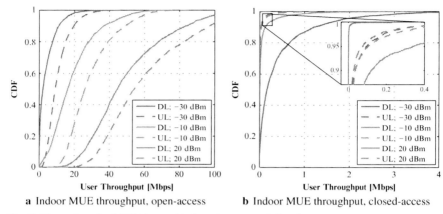

Fig. 5 Throughput for MUEs located indoors in open and closed-access systems

In order to assess the difference in performance between open and closed-access systems, those MUEs lying indoors (for closed-access) are identified and their throughput performance is compared against open-access. Figures 5a and b show the throughputs of such MUEs in UL and DL (notice the difference in scale of the x-axes). As expected for all FBS/FUE transmit powers, open-access significantly outperform closed-access, as for open-access, UEs are assimilated by the FBS. Therefore, for closed-access systems, wall penetration losses and longer transmit distances drastically reduce throughput. Furthermore, for open-access the UL performance is superior to the DL. This is attributed to the fact that interference is lower in UL than in DL. The reverse holds true for closed-access systems. Only for low FBS transmit powers of −30 dBm, a median throughput of ≈ 200 kbps is achieved. At the other powers, the interference from the femto-cell is too high to support reasonable throughputs.

4 Conclusion

Femto-cell deployment poses a viable complement to cellular networks. Operators need to bear low cost in their deployment, since they are installed directly by the users themselves. Furthermore, since they share both the radio access scheme and the frequency band with MBSs, they are compatible with legacy UEs. Aside from these benefits, a cellular network stands to significantly gain in overall system throughput through the widespread deployment of FBSs. Not only do FBSs improve indoor coverage, bringing broadband-like experience directly to the handset, but they also offload resources from the MBS which can be utilised to improve coverage to outdoor users.

In terms of overall throughput, the difference between *closed-access* and *open-access* systems is almost negligible, because of the relatively low probability that a MUE is inside a femto-cell. However, even if a MUE is within a femto-cell, it might be able to receive from its MBS if the FBS transmit power is sufficiently low – which indeed is possible given the short distances in femto-cells. This means power control and intelligent resource scheduling in femto-cells are envisaged to further reduce the interference that femto-cell entities cause to macro-cell entities.

Acknowledgement Initial parts of this work were supported by **DFG grant** HA 3570/2-1 as part of program SPP-1163 (adaptability in heterogeneous communication networks with wireless access – AKOM) while some latter parts of this work have been performed within the framework of the CELTIC project CP5-026 WINNER+.

References

1. M.-S. Alouini and A. Goldsmith, "Area Spectral Efficiency of Cellular Mobile Radio Systems," *IEEE Transactions on Vehicular Technology*, vol. 48, no. 4, pp. 1047–1066, 1999.
2. V. Chandrasekhar, J. Andrews, and A. Gatherer, "Femtocell Networks: A Survey," *IEEE Communcations Magazine*, vol. 46, no. 9, pp. 59–67, 2008.
3. H. Haas and G. J. R. Povey, "Capacity Analysis of a TDD Underlay Applicable for UMTS," in *Proc. of the 10th IEEE International Symposium on Personal, Indoor and Mobile Radio Communications (PIMRC)*, Osaka, Japan, 12–15 Sep. 1999, pp. A6–4.
4. Z. Bharucha and H. Haas, "Application of the TDD Underlay Concept to Home NodeB Scenario," in *Proc. of the Vehicular Technology Conference (VTC)*. Marina Bay, Singapore: IEEE, May 11–14, 2008, pp. 56–60.
5. 3rd Generation Partnership Project (3GPP), Technical Specification Group Radio Access Network, *Physical Channels and Modulation (Release 8)*, 3GPP TS 36.211 V 8.2.0 (2008–03), 3GPP Std., Mar. 2008.
6. H. Claussen, "Performance of Macro- and Co-Channel Femtocells in a Hierarchical Cell Structure," in *Proc. of the 18th IEEE International Symposium on Personal, Indoor and Mobile Radio Communications (PIMRC)*, Athens, Greece, Sep. 3–7 2007, pp. 1–5.
7. L. Ho and H. Claussen, "Effects of User-Deployed, Co-Channel Femtocells on the Call Drop Probability in a Residential Scenario," in *Proc. of the 18th IEEE International Symposium on Personal, Indoor and Mobile Radio Communications (PIMRC)*, Athens, Greece, Sep. 3–7 2007, pp. 1–5.
8. ITU-R Working Party 5D (WP5D) - IMT Systems, "Report 124, Report of correspondence group for IMT.EVAL," May 2008, United Arab Emirates.
9. 3rd Generation Partnership Project (3GPP), Technical Specification Group Radio Access Network, *Base Station (BS) Radio Transmission and Reception (FDD)*, 3GPP TS 25.104 V 8.3.0 (2008–05), 3GPP Std., May 2008.
10. 3rd Generation Partnership Project (3GPP), Technical Specification Group Radio Access Network, *User Equipment (UE) Radio Transmission and Reception (FDD)*, 3GPP TS 25.101 V 8.3.0 (2008–05), 3GPP Std., May 2008.
11. 3rd Generation Partnership Project (3GPP), Technical Specification Group Radio Access Network, *3G Home NodeB Study Item Technical Report*, 3GPP TR 25.820 V 8.1.0 (2008–05), 3GPP Std., May 2008.

Part VIII
Modulation & Demodulation

Phase Rotation/MC-CDMA for Uplink Transmission

Koichi Adachi and Masao Nakagawa

Abstract In the uplink transmission of a cellular system, multi-access interference (MAI) is a big problem. To eliminate the MAI, in this paper, a new multi-carrier-code division multiple access using phase rotation (PR/MC-CDMA) is proposed. In the proposed method, each user's data symbol is spread using a user-specific phase-rotated spreading code. The user-specific phase rotation is set so that the transmitted symbols are not overlapped in the delay time domain and the MAI can be completely removed through the despreading process at a receiver. At the receiver, the received signal in the frequency domain is multiplied by the inverse phase rotation based on the path-specific time delay. The transmitted signals received via different propagation paths are obtained by despreading using the use-specific phase rotated spreading code. They are coherently combined using the maximum ratio combining (MRC) method. The bit error rate (BER) performance of the proposed PR/MC-CDMA is evaluated by computer simulation and is compared to that of the conventional MC-CDMA.

1 Introduction

Multi-carrier-code division multiple access (MC-CDMA) has been considered as a promising candidate for the future wireless communication system [1]. MC-CDMA can obtain the frequency diversity gain through frequency domain spreading/despreading and frequency domain equalization (FDE). The transmission performance of MC-CDMA can be improved by FDE in the case of downlink transmission [2]. However, in the case of uplink transmission, the orthogonality between the spreading codes assigned to different users is lost because the signals transmitted from difference users experience different fading channels. The performance of MC-CDMA severely degrades owing to multi-access interference (MAI)

K. Adachi (✉) and M. Nakagawa
School of Science for Open and Environmental Systems, Graduate School of Science and Technology, Keio University, 3-14-1 Hiyoshi, Kohoku, Yokohama, Kanagawa, 223-8522, Japan
e-mail: {kouichi, nakagawa}@nkgw.ics.keio.ac.jp

resulting from the orthogonality distortion. To mitigate the MAI, there have been many works such as multi-user detection (MUD) [3] and block-spreading [4], iterative interference cancellation [5, 6]. Recently a multi-access technique called delay-time/code division multiple access (DT/CDMA), which is a kind of direct-sequence (DS)-CDMA, which multiplexes the users' data in the delay time domain was proposed [7].

In this paper, we propose a new multi-access technique based on phase rotation in the frequency domain for MC-CDMA (hereafter we call this technique PR/MC-CDMA). In the proposed technique, the signal is spread by a user-specific phase rotated spreading code, and the received signal is multiplied by the inverse phase rotation according to the path-specific time delay. The frequency domain despreading using the user-specific phase-rotated spreading code can de-multiplex simultaneously accessing users in the delay time domain. Since different users' signals are not overlapped in the delay time domain, MAI can be completely eliminated. The desired user's signal is obtained by path combining through maximum ratio combining (MRC). The performance of the proposed method is evaluated by computer simulation and is compared to those of DT/CDMA and conventional MC-CDMA with FDE.

The rest of the paper is as follows. The system model which adopts the proposed PR/MC-CDMA is described and the process of the signal detection is explained in Section 2. After describing the computer simulation parameters, the average bit error rate (BER) performance of the proposed PR/MC-CDMA is compared to those of the conventional methods in Section 3. Section 4 concludes the paper.

2 Proposed System Model

The transmission system model is illustrated in Fig. 1. In this paper, we assume T_c-sample spaced discrete time representation, where T_c is the fast Fourier transform (FFT) sample length. At the transmitter, the data symbol to be transmitted from the u-th user, $u = 0 \sim (U-1)$, is multiplied by a user-specific phase rotated spreading code. In this paper, we assume spreading factor SF and the number of sub-carriers, i.e., FFT samples N_c is the same. The n-th data symbol

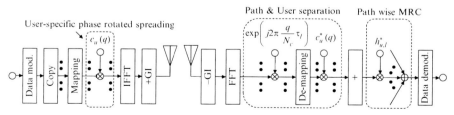

Fig. 1 Transmission system model

Fig. 2 Sub-carrier mapping

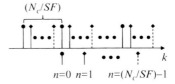

$\{d_u(n); n = 0 \sim (\lfloor N_c/SF \rfloor - 1)\}$ is frequency domain spread and then mapped onto the every (N_c/SF) sub-carriers as shown in Fig. 2, where $\lfloor x \rfloor$ is the largest integer smaller than or equal to x.

The k-th sub-carrier to which the q-th spreading chip, $q = 0 \sim (SF-1)$, of the n-th data symbol is mapped can be expressed as

$$S_u(k = (q \cdot N_c/SF + n)) = d_u(n) c_u(q) \quad (1)$$

where $\{c_u(q); q = 0 \sim (SF-1)\}$ is the user-specific phase rotated spreading code, given by

$$c_u(q) = f(q) \exp(j 2\pi u (\tau_{\max} + 1) q/SF) \quad (2)$$

with $f(q)$ being the original spreading code and $\tau_{\max}$ the maximum channel delay.

The time-domain MC-CDMA after N_g-sample guard interval (GI) insertion, $\{s_u(t \bmod N_c); t = -N_g \sim (N_c - 1)\}$, can be expressed as

$$s_u(t) = \sqrt{2P_u} \sum_{k=0}^{N_c-1} S_u(k) \exp(j 2\pi t k/N_c), \quad (3)$$

where P_u is the transmission power of the u-th user.

The channel between each user and the base station (BS) is assumed to be an L-path block Rayleigh fading channel. The impulse response of the channel between the u-th user and the BS is expressed as

$$h_u(\tau) = \sum_{l=0}^{L-1} h_{u,l} \delta(\tau - \tau_l), \quad (4)$$

where $h_{u,l}$ and τ_l are the complex path gain of the u-th user and the time delay of the l-th path. In this paper, the time delay of each path is assumed to be the same for all users.

The GI removed received signal, $\{r(t); t = 0 \sim (N_c - 1)\}$, becomes

$$r(t) = \sum_{u'=0}^{U-1} \sum_{l=0}^{L-1} h_{u',l} s_{u'}((t - \tau_l) \bmod N_c) + n(t). \quad (5)$$

where $n(t)$ is the zero-mean additive white complex Gaussian noise (AWGN) with the variance of $2\sigma^2 = 2N_0/T_c$ (N_0 is the one sided power spectrum density of AWGN).

The k-th frequency component of the received signal, $\{R(k); k = 0 \sim (N_c-1)\}$, is given as

$$R(k) = \frac{1}{N_c} \sum_{t=0}^{N_c-1} r(t) \exp\left(-j2\pi \frac{k}{N_c} t\right). \quad (6)$$

By substituting (5) into (6), we obtain

$$R(k) = \sum_{u'=0}^{U-1} \left(\begin{array}{c} \sqrt{\frac{2E_{c,u'}}{SF \cdot T_c}} d_{u'} \left(k \bmod \left(\frac{N_c}{SF}\right)\right) c_{u'} \left(\left\lfloor \frac{k}{(N_c/SF)} \right\rfloor\right) \\ \times \sum_{l'=0}^{L-1} h_{u',l'} \exp\left(-j2\pi \frac{k}{N_c} \tau_{l'}\right) \end{array} \right) + \Pi(k), \quad (7)$$

where $E_{c,u}$ is the received energy per FFT sample of the u-th user and $\Pi(k)$ is the Fourier transform of the noise, given as

$$\Pi(k) = \frac{1}{N_c} \sum_{t=0}^{N_c-1} n(t) \exp\left(-j2\pi \frac{k}{N_c} t\right). \quad (8)$$

It can be understood from (7) that the received signal is the sum of phase rotated versions of the transmitted signal according to the each path specific time delay.

The received signal is given the inverse phase rotation to extract the MC-CDMA signal received via the l-th path as

$$\hat{R}_l(k) = R(k) \exp(j2\pi k \tau_l / N_c). \quad (9)$$

By substituting (7) into (9), we have

$$\hat{R}_l(k) = \sum_{u'=0}^{U-1} \left(\begin{array}{c} \sqrt{\frac{2E_{c,u'}}{SF \cdot T_c}} d_{u'} \left(k \bmod \left(\frac{N_c}{SF}\right)\right) c_{u'} \left(\left\lfloor \frac{k}{(N_c/SF)} \right\rfloor\right) \\ \times \sum_{l'=0}^{L-1} h_{u',l'} \exp\left(-j2\pi k (\tau_{l'} - \tau_l)/N_c\right) \\ + \Pi(k) \exp(j2\pi k \tau_l / N_c) \end{array} \right). \quad (10)$$

Then, user-specific phase rotated despreading is performed as

$$\tilde{R}_{u,l}(n) = \frac{1}{SF} \sum_{q=0}^{SF-1} \hat{R}_l\left(\frac{N_c}{SF} \cdot q + n\right) f^*(q) \exp\left(-j2\pi \frac{u(\tau_{\max}+1)}{SF} q\right). \quad (11)$$

By substituting (10) into (11), we have

$$\tilde{R}_{u,l}(n) = \sqrt{\frac{2E_{c,u}}{SF \cdot T_c}} d_u(n) \sum_{l'=0}^{L-1} \left(\begin{array}{c} h_{u,l'} \exp\left(j2\pi \frac{n}{N_c}(\tau_l - \tau_{l'})\right) \times \\ (1/SF) \sum_{q=0}^{SF-1} \exp(j2\pi q(\tau_l - \tau_{l'})/SF) \\ + \mu_{MAI,u,l}(n) + \mu_{noise,u,l}(n) \end{array} \right). \quad (12)$$

The first term of (12) is the desired signal component. The second and the third terms are the MAI and the noise and they are given as

$$\begin{cases} \mu_{MAI,u,l}(n) \\ = \sum_{\substack{u'=0 \\ \neq u}}^{U-1} \sqrt{\frac{2E_{c,u'}}{SF \cdot T_c}} d_{u'}(n) \sum_{l'=0}^{L-1} \left(h_{u',l'} \exp\left(j2\pi \frac{n}{N_c}(\tau_l - \tau_{l'}) \right) \times \frac{1}{SF} \sum_{q=0}^{SF-1} \exp\left(j2\pi \frac{(u'-u)(\tau_{max}+1) + (\tau_l - \tau_{l'})}{SF} q \right) \right) \\ \mu_{noise,u,l}(n) \\ = \frac{1}{SF} \sum_{q=0}^{SF-1} \Pi(q) \exp\left(j2\pi \frac{(\frac{N_c}{SF} \cdot q + n)}{N_c} \tau_l \right) f_u^*(q) \exp\left(-j2\pi \frac{u(\tau_{max}+1)}{SF} q \right) \end{cases}$$
(13)

The inter-path interference (IPI) in the desired user's component and the MAI can be completely removed since

$$\frac{1}{SF} \sum_{q=0}^{SF-1} \exp\left(j2\pi \frac{q}{SF}(\tau_l - \tau_{l'}) \right) = \begin{cases} 1 & \tau_{l'} = \tau_l \\ 0 & \tau_{l'} \neq \tau_l \end{cases}. \tag{14}$$

and

$$\frac{1}{SF} \sum_{q=0}^{SF-1} \exp\left(j2\pi \frac{((u'-u)(\tau_{max}+1) + (\tau_l - \tau_{l'}))}{SF} q \right) = \begin{cases} 1 & u = u', \tau_l = \tau_{l'} \\ 0 & \text{otherwise} \end{cases}. \tag{15}$$

when the following condition is satisfied.

$$\max_{\substack{u, u' = 0 \sim (U-1) \\ l, l' = 0 \sim (L-1)}} \left| (u' - u)(\tau_{max} + 1) + (\tau_l - \tau_{l'}) \right| < SF. \tag{16}$$

Thus, the maximum number of users to be accommodated without MAI for the given SF becomes

$$U_{max} = \lfloor SF/(\tau_{max} + 1) \rfloor. \tag{17}$$

From (13–15), (12) becomes

$$\tilde{R}_{u,l}(n) = \sqrt{\frac{2E_{c,u}}{SF \cdot T_c}} h_{u,l} d_u(n) + \mu_{noise,u,l}(n). \tag{18}$$

As understood from the above equation, the desired user's signal received via the l-th path can be obtained without IPI and MAI. By performing the above despreading for all L paths and combining using path–wise MRC, the u-th user decision variable, $\tilde{d}_u(n)$, is obtained as

$$\tilde{d}_u(n) = \sum_{l=0}^{L-1} h_{u,l}^* \tilde{R}_{u,l}(n). \tag{19}$$

It should be noted that PR/MC-CDMA can flexibly change the data rate per user by changing spreading factor SF and mapping of the different users' spread sequences as shown in Fig. 2. The maximum number of data symbols per block per user is given by (N_c/SF). From (17), the uplink sum rate (symbols/block) which can be achieved is given by

$$\text{sum rate} = \left(\frac{N_c}{SF}\right) \times U_{\max} \leq \frac{N_c}{(\tau_{\max} + 1)}. \qquad (20)$$

3 Computer Simulation

3.1 Simulation Parameters

The spreading factor, SF, is set to $SF = 16 \sim 64$. The number of sub-carriers, N_c, is set to $N_c = 64$. QPSK (Quadrature Phase Shift Keying) and 16QAM (Quadrature Amplitude Modulation) data modulation are used. The channel is assumed to be a block Rayleigh fading channel with an $L = 16$-path uniform power delay profile. The time delay of the l-th path is assumed to be l FFT samples, thus maximum channel delay τ_{max} becomes 15. An M-sequence with a repetition period of 4,095 chips is used as a spreading factor. The same SF chips selected out of 4,095 chips are used for $u = 0 \sim (U_{\max} - 1)$ and the other SF chips are used for $u = U_{\max} \sim (U - 1)$ (Table 1).

3.2 Average BER Performance of PR/MC-CDMA

Firstly, we consider the situation that there is no received power difference among the accessing uses, i.e., ideal slow transmission power control (Slow TPC) is assumed. The spreading factor is set to $SF = 64$. The average BER performance

Table 1 Simulation parameters

Number of sub-carriers	$N_c = 64$
Spreading factor	$SF = 16, 32, 64$
Number of users	$U = 1 \sim 8$
Data modulation	QPSK/16QAM
Spreading code	M-sequence
	(repetition of 4,095 chips)
Channel model	
Fading	Block Rayleigh
Number of multipaths	$L = 16$
Time delay of the $l -$ th path	l-FFT samples
Decay factor	$\gamma = 0\,\text{dB}$
Channel estimation	Ideal

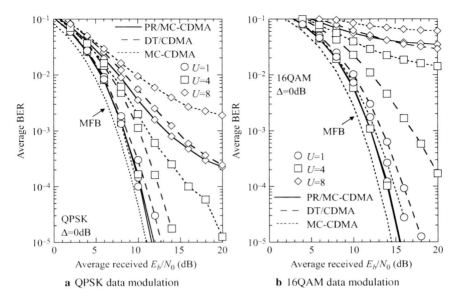

Fig. 3 Average BER performance without received power difference

of the proposed PR/MC-CDMA is plotted in Fig. 3 as a function of average received signal energy per bit-to-AWGN spectrum density ratio (E_b/N_0) with the number of simultaneously accessing users, U, as a parameter, where $E_b/N_0 = 1/C(E_s/N_0)(1 + N_g/N_c)$ with C being the constellation size, i.e., $C = 2$ (4) for QPSK (16QAM). Also plotted are the performances of DT/CDMA and conventional MC-CDMA with minimum mean squared error (MMSE)-FDE and the matched filter bound (MFB) which is the theoretical lower bound [8]. The MMSE-FDE weight for the k-th sub-carrier of the u-th user for the conventional MC-CDMA is given as

$$w_u(k) = H_u^*(k) \bigg/ \left\{ \sum_{u'=0}^{U-1} \left(\frac{1}{SF} \cdot \frac{E_{s,u'}}{N_0} |H_{u'}(k)|^2 \right) + 1 \right\}. \tag{21}$$

In PR/MC-CDMA, the maximum number of users multiplexed without causing MAI is $U_{\max} = 4$ to satisfy (16) for the case of assumed channel model. Thus PR/MC-CDMA can provide the same performance for $U = 1 \sim 4$ while the performances of DT/CDMA and conventional MC-CDMA degrade owing to the increasing residual MAI. The performance degradation of PR/MC-CDMA from MFB is due to the GI insertion loss. It can be seen from the figure that even in the case $U = 1$, PR/MC-CDMA can provide better performance compared to DT/CDMA. This is because PR/MC-CDMA can obtain the L-th order diversity gain through path-wise MRC while DT/CDMA can obtain the partial diversity gain because it uses MMSE weight in order to suppress the MAI and IPI.

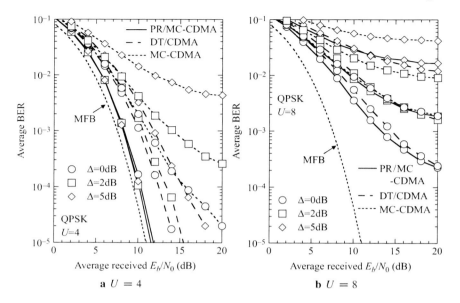

Fig. 4 Impact of received power difference

Next, we consider the situation that there is received power difference among the accessing users. The average BER performance is plotted in Fig. 4 with the power difference among the users Δ (dB) as a parameter. The spreading factor is set to $SF = 64$. The desired user is $u = 0$th user and the average received power for other users is set to Δ (dB) larger than for the $u = 0$th user. It can be seen from the figure that the average BER performance of the PR/MC-CDMA does not depend on Δ when $U \leq 4$ since all the users are orthogonally multiplexed while the performances of the DT/CDMA and conventional MC-CDMA degrades owing to the MAI. When the number of simultaneously accessing users is $U = 8$, the performance of the proposed PR/MC-CDMA degrades owing to the MAI; however, it is almost the same as that of DT/CDMA and is superior to that of the conventional MC-CDMA.

Next, we consider the impact of spreading factor SF on the average BER performance of PR/MC-CDMA. In the proposed PR/MC-CDMA, the each spread data symbol is mapped onto every (N_c/SF) sub-carriers. As it has been shown in (20), the uplink sum rate (symbols/block) that can be achieved without MAI is $(N_c/(\tau_{max} + 1)) = 4$ in the assumed channel model. The average BER performance of PR/MC-CDMA is plotted in Fig. 5 with number of simultaneously accessing users U and spreading factor SF as parameters. It can be seen from the figure that as far as (20) is satisfied, i.e., $(N_c/SF) \times U_{max} \leq 4$, there is no degradation of the BER performance irrespective of spreading factor SF. This is because as it can be understood from (18) that PR/MC-CDMA can obtain the L-th order diversity irrespective of SF.

Fig. 5 Impact of spreading factor *SF*

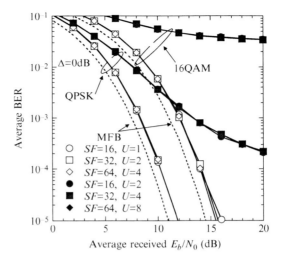

4 Conclusions

In this paper, we proposed a new multiple-access method called PR/MC-CDMA for uplink transmissions. Multiple users are spread in the frequency domain by user-specific phase rotated spreading codes. The signal received via each path is obtained by applying the inverse phase rotation according to the each path delay and by despreading using the user-specific phase rotated spreading code. The MRC-based path combining is performed through MRC to obtain the desired signal. The simulation results showed that the proposed PR/MC-CDMA can provide superior performance compared to the conventional MC-CDMA.

References

1. Y. Kim, B. J. Jeong, J. Chung, C. Hwang, J. S. Ryu, K. Kim, and Y. K. Kim. Beyond 3G: vision, requirements, and enabling technologies. *IEEE Communications Magazine*, 41(3):120–124, Mar. 2003.
2. F. Adachi, D. Garg, S. Takaoka, and K. Takeda. Broadband CDMA techniques. *Special Issue on Modulation, Coding and Signal Processing, IEEE Wireless Communication Magazine*, 12(2):8–18, Apr. 2005.
3. S. Tsumura, S. Hara, and Y. Hara. Performance comparison of MC-CDMA and cyclically prefixed DS-CDMA in an uplink channel. In *Proceedings of 60th IEEE Vehicular Technology Conference 2004 Fall (VTC'04-fall)*, 414–418, Los Angeles, USA, Sept. 2004.
4. S. Zhou, G. B. Giannakis, and C. L. Martret. Chip-interleaved block-spread code division multiple access. *IEEE Transactions on Communications*, 50(2):235–248, Feb. 2002.

5. N. Benvenuto, P. Bisaglia, and M. Finco. Soft-interference cancellation for downlink and uplink MC-CDMA systems. *In Proeedings of 15th IEEE International Symposium on Personal, Indoor and Mobile Radio Communications (PIMRC 2004)*, 1:170–174, Sept. 2004.
6. Y. Yuan-Wu and Y. Li. Iterative and diversity techniques for uplink MC-CDMA mobile systems with full load. *IEEE Transactions on Vehicular Technology*, 57(2):1040–1048, Mar. 2008.
7. F. Adachi and K. Takeda. Delay-time/code division multi-access in a frequency-selective channel. *Electronics Letter*, 43(18):984–986, Aug. 2007.
8. J. G. Proakis. *Digital Communications*, 4th edition. McGraw-Hill, New York, 2001.

Hierarchical Modulation in DVB-T/H Mobile TV Transmission

Tomáš Kratochvíl

Abstract This paper deals with an experimental laboratory assessment of utilized hierarchical modulation in the transmission of DVB-T/H mobile TV over new fading channels profiles. These profiles called PI3, PO3, VU30 and MR100 are originally from the Celtic Wing TV project. The DVB-H performance in HP (High Priority) stream of a hierarchical modulated digital terrestrial television was tested in a simulated laboratory environment. The results of the BER before and BER after Viterbi decoding, the main criteria for the DVB-H signal receptions, were evaluated using R&S test and measurements equipment.

1 Introduction

The DVB-H (Digital Video Broadcasting – Handheld) [1, 2] is a perspective standard and technology for mobile TV distribution. DVB-H is an extension of a classical standard DVB-T (Digital Video Broadcasting – Terrestrial) [3, 4] with some backwards compatibility. An important feature is that they can share the same multiplex in MPEG-2 TS (Transport Stream). It also uses a mechanism called Multi-Protocol Encapsulation (MPE), making it possible to transport data network protocols. A Forward Error Correction (FEC) scheme is used in conjunction with this to improve the robustness and thus mobility of the signal. In addition to the 2k and 8k modes available in DVB-T, a 4k mode was added to DVB-H giving increased flexibility for network design. A short in-depth interleaver was introduced for 2k and 4k modes that lead to better tolerance against impulsive noise as a help to achieve a similar level of robustness to the 8k mode. The system was built on the proven mobile performance of the DVB-T. To ensure that reliable reception is still guaranteed

---

T. Kratochvíl
Brno University of Technology (BUT), Department of Radio Electronics,
Purkyňova 118, 61200 Brno, Czech Republic
e-mail: kratot@feec.vutbr.cz

even in mobile environment, hierarchical coding [5] could be provided as an option in DVB-T/H. Without it a signal-to-noise ratio which is too low will lead to a hard "cliff-off" (or simply "blocking") and mobile TV service unavailability. In the case of the frequently used DVB-T/H transmission with 64-QAM modulation and FEC convolutional rate 3/4 or 2/3, the limit of stable reception is at signal-to-noise ratio of just under 20 dB [6].

2 Hierarchical Modulation

If the hierarchical modulation [7] is used, the DVB-T/H modulator has two transport stream inputs and two FEC blocks. One transport stream with a low data rate is fed into the so-called high priority path (HP) and provided with a large amount of error protection (selecting more robust code rate 1/2). A second transport stream with a higher data rate is supplied in parallel to the low priority path (LP) and is provided with less error protection (selecting less robust code rate 2/3). On the high priority path the QPSK modulation is used which is particularly robust type of modulation. This fits for mobile TV reception. On the low priority path a higher level of modulation is needed (16-QAM or 64-QAM) due to the higher data rate. These fit for fixed or portable TV reception. The gross data rates for LP and HP thus have a fixed ratio of 2:2 or 4:2 to one another. The net data rates are dependent on the code rate used [8]. To make the QPSK of the HP more robust and less susceptible to interference the constellation diagram can be spread at the I axis and the Q axis. A factor α equals to 1, 2 or 4 increases the distance between the individual quadrants of the M-QAM diagram (e.g. see Fig. 1). The information about the presence or absence of hierarchical modulation, the α factor and the code rates for HP and LP are transmitted to the receiver in the TPS (Transmission Parameter Signaling) COFDM carriers [7]. In principle both transport streams can contain the same program, but

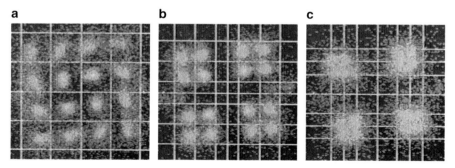

Fig. 1 Constellation analysis of the DVB-T/H hierarchical modulation with QPSK embedded in 16-QAM, used VU30 fading channel profile with introduced Doppler shift (TV channel C39) at speed 30 km/h and parameter (**a**) $\alpha = 1$, (**b**) $\alpha = 2$, (**c**) $\alpha = 4$

at different data rates which are equal to amount of data compression. On the other side, the HP can contain a DVB-H stream for mobile TV transmission with LDTV (Low Definition TV) and LP can contain a DVB-T stream with a classical SDTV (Standard Definition TV) or HDTV (High Definition TV) broadcasting [9]. All these three possibilities were experimentally tested in a simulated laboratory environment and the results were evaluated in [10].

3 Laboratory Transmission Setup

Experimental testing of the hierarchical modulation used for DVB-T/H broadcasting with higher resolution in LP (fixed or portable reception) and DVB-T/H with lower resolution in HP (mobile reception) streams was realized in the laboratory environment. The transmitter and receiver test beds (see Fig. 2) were consisted of DVB-T/H test transmitter SFU with noise generator and fading simulator up to 20 paths, MPEG-2 TS generators included in SFU and external DVRG, reference test receiver MSK-33, DVB-T receiver (set-top box) and DVB-H receiver (mobile phone). There were tested two possible scenarios of hierarchical modulation used for fixed, portable and mobile reception of digital television:

- HDTV service MPEG-4 Part 10 stream in LP (fixed or portable reception) and SDTV service MPEG-2 MP@ML stream in HP (mobile reception)
- SDTV service MPEG-2 MP@ML stream in LP (fixed or portable reception) and LDTV service MPEG-4 Part 10 stream in HP (mobile reception)

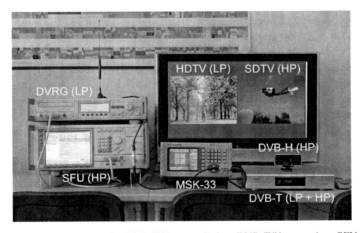

Fig. 2 Laboratory environment for DVB-T/H transmission: DVB-T/H transmitter SFU, TS players for HP (incl. in SFU) and LP (DVRG) streams, DVB-T reference test receiver MSK-33, DVB-T set-top box with LCD TV screen and DVB-H mobile phone

There were also used the new fading channel profiles models presented in Celtic Wing TV project report for the experimental transmission. These channels were PI3 (Pedestrian Indoor at speed 3 km/h), PO3 (Pedestrian Outdoor at speed 3 km/h), VU30 (Vehicular Urban at speed 30 km/h) and MR100 (Motorway Rural at speed 100 km/h) with 12 paths and Doppler spectrum characteristics [11]. These new profiles were not included in [10].

4 Experimental Results

System parameters of the analyzed DVB-T/H transmission were set to a configuration QPSK (mobile TV with LDTV resolution, compression MPEG-4 Part 10 with data rate 384 kbit/s) embedded in 16-QAM or 64-QAM (portable or fixed TV reception with SDTV or HDTV resolution, compression MPEG-2 or MPEG-4 AVC with data rate from 5.1 to 14.5 Mbits/s). The minimal carrier-to-noise ratio C/N in dB in all laboratory transmission experiments was determined at which the channel BER (Bit-Error Rate) before Viterbi for the rate of 2/3 is equal or less than 3.10^{-2}. Then the further BER after Viterbi decoding of inner error protection is equal or less 2.10^{-4}. This condition finally leads to error-free signals (QEF, Quasi Error Free) at the input of the MPEG-2 TS demultiplexer and final BER after RS decoding is equal or less than 1.10^{-11} [6]. Detailed results of referenced BER before Viterbi decoding (so called channel BER), BER after Viterbi decoding (so called BER before RS decoding) for HP path for mobile TV reception and for various hierarchical modulation setup are available in Figs. 3–6.

The approximation condition for the QEF reception was previously defined as BER after Viterbi decoding equal to 2.10^{-4} or less. This is the limit at which the subsequent FEC Reed-Solomon decoder still delivers an output BER of 1.10^{-11} or less. This presents one error per hour. There were analyzed results of a COFDM in 2k and 8k mode, with guard interval 1/4 used in large SFN (Single Frequency Network). The results for DVB-T/H transmission with both portable and both mobile reception environments of HDTV or SDTV service in LP stream, LDTV service in HP stream and required C/N ratio in dB are shown in the Table 1. The HP service (mobile TV) was not available in 8k mode and MR100 channel, where 2k mode is better to use in case of hierarchical modulation. Experimental results were compared with results presented in [11]. Theoretical C/N value in 8k mode and non-hierarchical modulated QPSK is equal to (8.9, 9.2, 9.5, 9.7) dB in the (PI3, PO3, VU30, MR100) channel profile respectively. Presented results can be used for DVB-T/H transmission distortions analysis and evaluation of the hierarchical modulation α parameters influence on portable and mobile reception of digital TV services.

Hierarchical Modulation in DVB-T/H Mobile TV Transmission

Table 1 DVB-T/H in a hierarchical modulation performance details ($\alpha = 2$)

Mode	Hierarchy	PI3	PO3	VU30	MR100
2k	QPSK in 16-QAM	7.2	7.5	10.0	10.3
8k	QPSK in 16-QAM	7.7	7.9	10.7	–
2k	QPSK in 64-QAM	9.1	9.4	11.2	12.3
8k	QPSK in 64-QAM	9.7	9.8	13.6	–

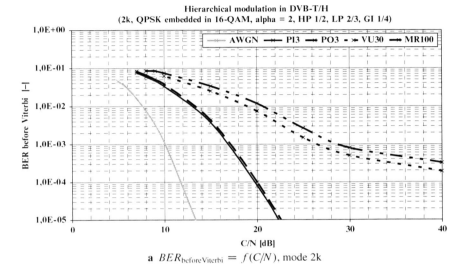

a $BER_{beforeViterbi} = f(C/N)$, mode 2k

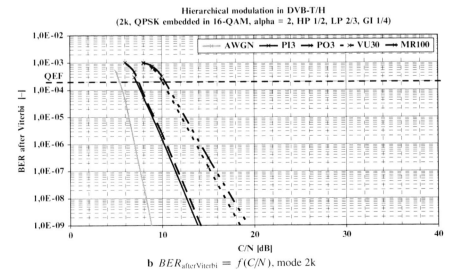

b $BER_{afterViterbi} = f(C/N)$, mode 2k

Fig. 3 DVB-T/H hierarchical modulation (QPSK in 16-QAM) performance in the transmission over fading channel models. Setup details: RX level 60 dBuV, channel C39, 8 MHz channel, OFDM mode 2k, modulation QPSK in 16-QAM, parameter $\alpha = 2$, HP code rate 1/2, LP code rate 2/3, guard interval 1/4

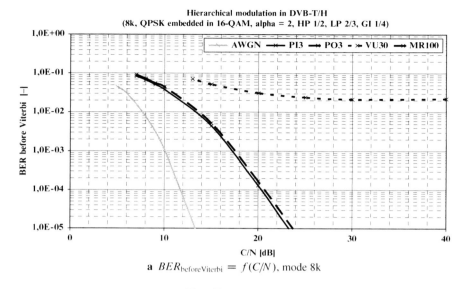

a $BER_{beforeViterbi} = f(C/N)$, mode 8k

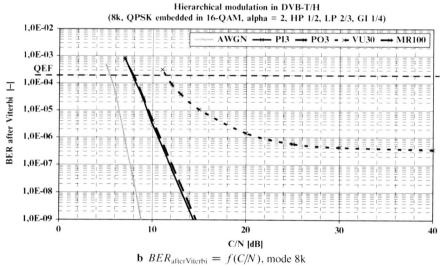

b $BER_{afterViterbi} = f(C/N)$, mode 8k

Fig. 4 DVB-T/H hierarchical modulation (QPSK in 16-QAM) performance in the transmission over fading channel models. Setup details: RX level 60 dBuV, channel C39, 8 MHz channel, OFDM mode 8k, modulation QPSK in 16-QAM, parameter $\alpha = 2$, HP code rate 1/2, LP code rate 2/3, guard interval 1/4

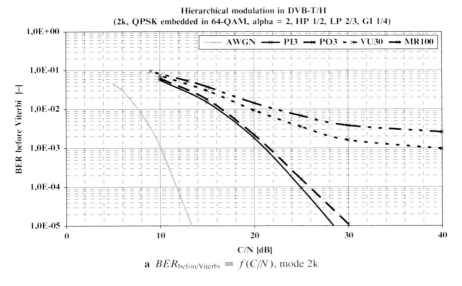

a $BER_{\text{beforeViterbi}} = f(C/N)$, mode 2k

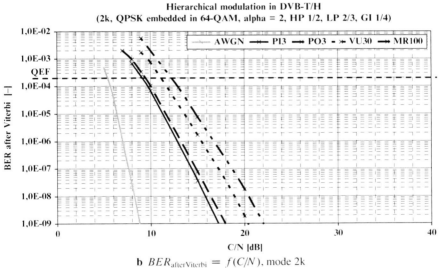

b $BER_{\text{afterViterbi}} = f(C/N)$, mode 2k

Fig. 5 DVB-T/H hierarchical modulation (QPSK in 64-QAM) performance in the transmission over fading channel models. Setup details: RX level 60 dBuV, channel C39, 8 MHz channel, OFDM mode 2k, modulation QPSK in 16-QAM, parameter $\alpha = 2$, HP code rate 1/2, LP code rate 2/3, guard interval 1/4

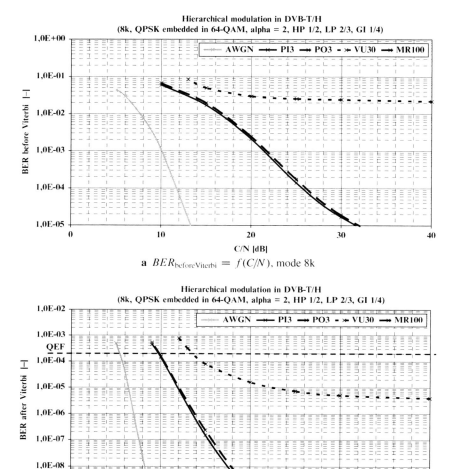

Fig. 6 DVB-T/H hierarchical modulation (QPSK in 64-QAM) performance in the transmission over fading channel models. Setup details: RX level 60 dBuV, channel C39, 8 MHz channel, OFDM mode 8k, modulation QPSK in 16-QAM, parameter $\alpha = 2$, HP code rate 1/2, LP code rate 2/3, guard interval 1/4

5 Conclusion

In case of hierarchical modulation it is necessary to supply details of the required C/N not only for the QPSK portion, but also for the M-QAM portion. When comparing corresponding data of same code rates and case of non-hierarchical and hierarchical modulation, the C/N required for QEF reception of the HP data stream transmitted in QPSK constellation points must be 0.9–4.1 dB higher than in the case of non-hierarchical modulation.

Acknowledgments This paper was supported by the Research program of Brno University of Technology no. MSM0021630513, "Electronic Communication Systems and New Generation Technology (ELKOM)" and the research project of the Czech Science Foundation no.102/08/P295, "Analysis and Simulation of the Transmission Distortions of the Digital Television DVB-T/H."

References

1. DVB Fact Sheet (2008, June). Broadcasting to Handhelds. The Global Technology Standard for Mobile Television. [Online]. Available http://www.dvb.org/technology/fact_sheets/
2. ETSI EN 302 304 V1.1.1 (2004–11). Digital Video Broadcasting (DVB); Transmission system for handheld terminals. ETSI, 2004.
3. DVB Fact Sheet (2008, June). Digital Terrestrial Television. The World's Most Flexible and Most Successful DTT Standard. [Online]. Available http://www.dvb.org/technology/fact_sheets/
4. ETSI EN 300 744 V1.5.1 (2004–11). Digital Video Broadcasting (DVB); Framing structure, channel coding and modulation for digital terrestrial television. ETSI, 2004.
5. DVB Fact Sheet (2000, March). DVB-T Hierarchical Modulation. [Online]. Available at http://www.dvb.org/technology/fact_sheets/
6. ETSI TR 101 290 V1.2.1 (2001–05). Digital Video Broadcasting (DVB); Measurement guidelines for DVB systems. ETSI, 2001.
7. W. Fisher. Digital Video and Audio Broadcasting Technology. A Practical Engineering Guide (2nd edition). Springer, Berlin, 2008.
8. U. Reimers. Digital Video Broadcasting. The Family of International Standards for Digital Television. (2nd edition). Springer, Berlin, 2004.
9. ETSI TR 102 377 V1.2.1 (2005–11). Digital Video Broadcasting (DVB); Implementation guidelines for DVB handheld services. ETSI, 2005.
10. T. Kratochvíl, R. Štukavec. Hierarchical modulation in DVB-T/H mobile TV transmission over fading channels. In Proceedings of the 2008 International Symposium ISITA2008. Auckland, New Zealand: Society of Information Theory and its Applications (SITA), 2008.
11. Celtic Wing TV project report (2006, December). Services to Wireless, Integrated, Nomadic, GPRS-UMTS & TV handheld terminals. Hierarchical Modulation Issues. D4 - Laboratory test results. Celtic Wing TV, 2006. [Online]. Available http://projects.celtic-initiative.org/WING-TV/
12. DVB Blue Book A092 rev.2 V1.3.1 (2007, May). Implementation Guidelines for DVB-H Services. [Online]. Available http://www.dvb-h.org/

Efficient Compensation of Frequency Selective TX and RX IQ Imbalances in OFDM Systems

Deepaknath Tandur and Marc Moonen

Abstract Radio frequency impairments such as in-phase/quadrature-phase (IQ) imbalances can result in a severe performance degradation in direct-conversion architectures. In this paper, a training based equalizer is developed to estimate and compensate the effects of these IQ imbalances along with channel distortions in an OFDM system. The proposed scheme provides a decoupling mechanism to estimate all the three frequency selective distortions in the communication system, namely the transmitter IQ imbalance, the receiver IQ imbalance and the channel dispersion. Once the IQ imbalance parameters are estimated, the proposed scheme then utilizes a standard one tap frequency domain equalizer to estimate and compensate the channel variations in the system. Simulation results show that the resulting calibrated equalizer requires a very small training overhead for an efficient, post-FFT equalization with performance very close to the ideal case.

1 Introduction

A direct-conversion based system is an attractive front-end radio architectural design for a communication engineer [1]. These systems are typically small in size and cheaper to implement. They also provide a very good flexibility in supporting growing number of wireless standards found in today's communication systems. However, the direct-conversion based architecture is generally very sensitive to any front-end component imperfections. These imperfections are unavoidable especially when cheaper components are used in the manufacturing process. The front-end imperfections can result in radio frequency (RF) impairments such as in-phase/quadrature-phase (IQ) imbalance, carrier frequency offset (CFO), etc. The multi-carrier based communication systems such as OFDM [2] are found to be very

---

D. Tandur (✉) and M. Moonen
Department of Electrical Engineering, Katholieke Universiteit Leuven, Kasteelpark Arenberg 10, B-3001 Leuven, Belgium
e-mail: {deepaknath.tandur, marc.moonen}@esat.kuleuven.be

sensitive to such RF impairments [3, 4]. In this paper we study the effects of only IQ imbalance in an OFDM system. The RF impairments such as IQ imbalance can result in a severe performance degradation in an OFDM system, rendering the communication system useless, unless they are adequately compensated.

Recently several articles [3–9] have been published that address these effects of IQ imbalances in an OFDM system. In [5] and [6] efficient digital compensation schemes have been developed for the case of frequency independent IQ imbalance. References [7] and [8] provide a joint compensation scheme in MIMO and SISO systems for three frequency selective distortions, namely the transmitter (Tx) IQ imbalance, the receiver (Rx) IQ imbalance and the channel dispersion. The joint compensation scheme provides a simple implementation alternative but the equalizer's training overhead is found to be large as the received signal has to be equalized at both the desired sub-carrier and its mirror component for any channel variation. Recently [9] proposed a compensation scheme based on special training symbols that decouples the frequency selective Rx IQ imbalance from the channel dispersion resulting in a reliable estimation with a small training overhead.

This paper is an extension of [8] and [9] as it targets to design a low training overhead equalizer for the general case of frequency selective Tx and Rx IQ imbalance along with channel dispersion for SISO systems. The estimation process is divided in two phases, the first phase is the calibration phase where the Tx and Rx IQ imbalance parameters are estimated. These parameters are only slowly time-varying components thus they do not have to be re-estimated frequently. In the second phase only the channel dispersion is estimated using a standard one tap equalizer. It is shown that the proposed equalizer provides an efficient and an effective compensation method for systems impaired with IQ imbalance distortions.

This paper is organized as follows: The input–output OFDM system model is presented in Section 2. Section 3 then explains the IQ imbalance compensation scheme. Computer simulations are shown in Section 4 and finally the conclusion is given in Section 5.

Notation: Vectors are indicated in bold and scalar parameters in normal font. Superscripts $\{\}^*, \{\}^T, \{\}^H$ represent complex conjugate, transpose and Hermitian respectively. $\mathbf{F}$ and $\mathbf{F}^{-1}$ represent the $N \times N$ discrete Fourier transform and its inverse. $\mathbf{I}_N$ is the $N \times N$ identity matrix and $0_{M \times N}$ is the $M \times N$ all zero matrix. Operators $\star$, . and $\div$ denote linear convolution, component-wise vector multiplication and division respectively.

2 System Model

We consider an OFDM transmission over frequency selective fading channels. We assume a single-input single-output (SISO) system, but the results can be easily extended to multiple-input multiple-output (MIMO) systems. Let $\mathbf{S}$ be an uncoded frequency domain OFDM symbol of size $(N \times 1)$. This symbol is transformed to the time domain by an inverse discrete Fourier transform (IDFT) operation. A cyclic prefix (CP) of length v is then added to the head of each symbol. The resulting time domain baseband signal $\mathbf{s}$ is then given as:

$$\mathbf{s} = \mathbf{P}_{CI}\mathbf{F}^{-1}\mathbf{S} \qquad (1)$$

where $\mathbf{P}_{CI}$ is the cyclic insertion matrix given by:

$$\mathbf{P}_{CI} = \begin{bmatrix} \mathbf{0}_{(v \times N-v)} & \mathbf{I}_v \\ \mathbf{I}_N \end{bmatrix}$$

We represent frequency selective (FS) IQ imbalance resulting from Tx front-end components by two mismatched filters with frequency responses given as $\mathbf{H}_{ti} = \mathcal{F}\{\mathbf{h}_{ti}\}$ and $\mathbf{H}_{tq} = \mathcal{F}\{\mathbf{h}_{tq}\}$. The frequency independent (FI) IQ imbalance is represented by amplitude and phase mismatch g_t and ϕ_t between the two front-end branches. Following [10], the baseband signal $\mathbf{p}$ after front-end distortions can be given as:

$$\mathbf{p} = \mathbf{g}_{ta} \star \mathbf{s} + \mathbf{g}_{tb} \star \mathbf{s}^* \qquad (2)$$

where

$$\mathbf{g}_{ta} = \mathbf{F}^{-1}\mathbf{G}_{ta} = \mathbf{F}^{-1}\left\{\frac{[\mathbf{H}_{ti} + g_t e^{-j\phi_t}\mathbf{H}_{tq}]}{2}\right\}$$

$$\mathbf{g}_{tb} = \mathbf{F}^{-1}\mathbf{G}_{tb} = \mathbf{F}^{-1}\left\{\frac{[\mathbf{H}_{ti} - g_t e^{j\phi_t}\mathbf{H}_{tq}]}{2}\right\}$$

Here $\mathbf{g}_{ta}$ and $\mathbf{g}_{tb}$ are mostly truncated to length L_t and then padded with $N - L_t$ zero elements. They represent the combined FI-FS Tx IQ imbalance. $\mathbf{G}_{ta}$ and $\mathbf{G}_{tb}$ are the frequency domain representations of $\mathbf{g}_{ta}$ and $\mathbf{g}_{tb}$ respectively.

Finally, an expression similar to (2) can be used to model IQ imbalance at the receiver. Let $\mathbf{z}$ represent the down-converted baseband complex signal after being distorted by combined FI-FS Rx IQ imbalance $\mathbf{g}_{ra}$ and $\mathbf{g}_{rb}$ of length L_r. Then $\mathbf{z}$ will be given as:

$$\mathbf{z} = \mathbf{g}_{ra} \star \mathbf{r} + \mathbf{g}_{rb} \star \mathbf{r}^* \qquad (3)$$

where

$$\mathbf{r} = \mathbf{c} \star \mathbf{p} + \mathbf{n}$$

Here $\mathbf{c}$ is the baseband representation of the multipath channel of length L and $\mathbf{n}$ is the additive white Gaussian noise (AWGN). Equation (2) can be substituted in (3) leading to

$$\begin{aligned}\mathbf{z} &= (\mathbf{g}_{ra} \star \mathbf{c} \star \mathbf{g}_{ta} + \mathbf{g}_{rb} \star \mathbf{c}^* \star \mathbf{g}_{tb}^*) \star \mathbf{s} + \mathbf{g}_{ra} \star \mathbf{n} + \\ &\quad (\mathbf{g}_{ra} \star \mathbf{c} \star \mathbf{g}_{tb} + \mathbf{g}_{rb} \star \mathbf{c}^* \star \mathbf{g}_{ta}^*) \star \mathbf{s}^* + \mathbf{g}_{rb} \star \mathbf{n}^* \\ &= \mathbf{d}_1 \star \mathbf{s} + \mathbf{d}_2 \star \mathbf{s}^* + \mathbf{g}_{ra} \star \mathbf{n} + \mathbf{g}_{rb} \star \mathbf{n}^*\end{aligned} \qquad (4)$$

where $\mathbf{d}_1$ and $\mathbf{d}_2$ are the combined Tx IQ, channel and Rx IQ impulse responses of length $L_t + L + L_r - 2$. The down-converted received signal $\mathbf{z}$ is now separated from the CP at the receiver. This CP free received symbol is then transformed to the frequency domain by an discrete Fourier transform (DFT) operation. In this paper we consider the CP length v to be sufficiently longer than both $\mathbf{d}_1$ and $\mathbf{d}_2$. The longer CP length results in a simple dot multiplication between the various frequency selective terms and the transmitted symbol in the frequency domain. This resulting signal $\mathbf{Z}$ can then be written as:

$$\begin{aligned}\mathbf{Z} &= \mathbf{FP}_{\mathbf{CR}}\{\mathbf{z}\} \\ &= (\mathbf{G}_{ra}.\mathbf{G}_{ta}.\mathbf{C} + \mathbf{G}_{rb}.\mathbf{G}^*_{tb_m}.\mathbf{C}^*_m).\mathbf{S}^{(i)} + \mathbf{G}_{ra}.\tilde{\mathbf{n}} \\ &+ (\mathbf{G}_{ra}.\mathbf{G}_{tb}.\mathbf{C} + \mathbf{G}_{rb}.\mathbf{G}^*_{ta_m}.\mathbf{C}^*_m).\mathbf{S}^{*(i)}_m + \mathbf{G}_{rb}.\tilde{\mathbf{n}}^*_m\end{aligned} \qquad (5)$$

where $\mathbf{P}_{\mathbf{CR}}$ is the cyclic removal matrix given as:

$$\mathbf{P}_{\mathbf{CR}} = \begin{bmatrix} 0_{(N \times v)} \mid \mathbf{I}_N \end{bmatrix}$$

Here $\mathbf{G}_{ra}$, $\mathbf{G}_{rb}$, $\mathbf{C}$ and $\tilde{\mathbf{n}}$ are the frequency domain representations of $\mathbf{g}_{ra}$, $\mathbf{g}_{rb}$, $\mathbf{c}$ and $\mathbf{n}$. The operator $()_m$ denotes the mirroring operation in which the vector indices are reversed, such that $\mathbf{S}_m[l] = \mathbf{S}[l_m]$ where $l_m = 2 + N - l$ for $l = 2\ldots N$ and $l_m = l$ for $l = 1$. Equation (5) shows that due to IQ imbalance, the power leaks from the signal on the mirror carrier ($\mathbf{S}^*_m$) to the carrier under consideration ($\mathbf{S}$) and thus causes inter-carrier-interference (ICI). The ICI distortion due to Tx and Rx IQ imbalance results in a severe performance degradation and thus a compensation scheme is needed in the OFDM system. In the next section, we develop a digital compensation scheme for joint Tx and Rx IQ imbalance distortions in an OFDM system.

3 IQ Imbalance Compensation Scheme

In the previous section, we obtained the equation for the received signal after being distorted by Tx IQ imbalance, channel and Rx IQ imbalance respectively. In practice both Tx and Rx IQ imbalance distortions are relatively static and change very slowly when compared to the channel characteristics. Thus for a given channel dispersion characteristic, say $\mathbf{C}^{(1)}$, the received signal (5) can be rewritten as follows:

$$\begin{aligned}\mathbf{Z}^{(1)} &= \underbrace{(\mathbf{D}^{(1)} + \mathbf{N}_r.\mathbf{N}^*_{t_m}.\mathbf{D}^{*(1)}_m)}_{\mathbf{W}_1}.\mathbf{S} + \mathbf{G}_{ra}.\tilde{\mathbf{n}} \\ &+ \underbrace{(\mathbf{N}_t.\mathbf{D}^{(1)} + \mathbf{N}_r.\mathbf{D}^{*(1)}_m)}_{\mathbf{W}_2}.\mathbf{S}^*_m + \mathbf{G}_{rb}.\tilde{\mathbf{n}}^*_m\end{aligned} \qquad (6)$$

where $\mathbf{D}^{(1)} = \mathbf{G}_{ra}.\mathbf{G}_{ta}.\mathbf{C}^{(1)}$, $\mathbf{N}_t = \frac{G_{tb}}{G_{ta}}$ and $\mathbf{N}_r = \frac{G_{rb}}{G^*_{ra_m}}$. Here $\mathbf{W}_1$ represents the scaling factor of the desired sub-carrier and $\mathbf{W}_2$ represents the amount of interference

Efficient Compensation of Frequency Selective TX and RX IQ Imbalances 347

from the mirror sub-carrier. In (6), both $\mathbf{W}_1$ and $\mathbf{W}_2$ can be estimated by designing a training based two tap frequency domain equalizer (FEQ). The first input of the FEQ is applied to the OFDM training sequence $\mathbf{S}$ and and the second input to its mirror complex conjugate $\mathbf{S}_m^*$. Similarly a two tap FEQ can also be applied to the received signal $\mathbf{Z}^{(1)}$ and its mirror conjugate $\mathbf{Z}_m^{*(1)}$ in order to estimate the transmitted symbol $\mathbf{S}$. The FEQ can be designed based on a maximum-likelihood (ML), least-square (LS) or minimum mean-square-error (MMSE) criteria [8]. Once the FEQ coefficients are trained, the equalizer can then be applied on all the forthcoming OFDM data symbols as long as the channel characteristics remain the same $\mathbf{C}^{(1)}$. This is the principle behind the joint compensation scheme in [7] and [8].

In the case of a variation in the channel characteristics, say to $\mathbf{C}^{(2)}$, the received signal $\mathbf{Z}^{(2)}$ can now be written as:

$$\mathbf{Z}^{(2)} = \underbrace{(\mathbf{D}^{(2)} + \mathbf{N}_r . \mathbf{N}_{t_m}^* . \mathbf{D}_m^{*(2)})}_{\mathbf{W}_3} . \mathbf{S} + \mathbf{G}_{ra} . \tilde{\mathbf{n}}$$
$$+ \underbrace{(\mathbf{N}_t . \mathbf{D}^{(2)} + \mathbf{N}_r . \mathbf{D}_m^{*(2)})}_{\mathbf{W}_4} . \mathbf{S}_m^* + \mathbf{G}_{rb} . \tilde{\mathbf{n}}_m^* \qquad (7)$$

Here once again, the coefficients $\mathbf{W}_3$ and $\mathbf{W}_4$ can be estimated by a training based two tap FEQ. Is is to be noted that both Tx and Rx IQ imbalance parameters $\mathbf{N}_t$ and $\mathbf{N}_r$ are small values as $\mathbf{G}_{ta} \gg \mathbf{G}_{tb}$ and $\mathbf{G}_{ra} \gg \mathbf{G}_{rb}$. This is true even for large IQ imbalance values as seen in practise. Thus for the moment we can ignore the dot multiplication of $\mathbf{N}_r$ and $\mathbf{N}_{t_m}^*$ in (6) and (7) as their contribution to the two equations are relatively small. We can now write $\tilde{\mathbf{D}}^{(1)} = \mathbf{W}_1$ and $\tilde{\mathbf{D}}^{(2)} = \mathbf{W}_3$ where $\tilde{\mathbf{D}}^{(q)}$ is the estimate of $\mathbf{D}^{(q)}$ (for q = 1,2). Now the estimates $\tilde{\mathbf{N}}_t$ and $\tilde{\mathbf{N}}_r$ of $\mathbf{N}_t$ and $\mathbf{N}_r$ can be obtained by the following equation:

$$\begin{bmatrix} \tilde{\mathbf{N}}_t \\ \tilde{\mathbf{N}}_r \end{bmatrix} \Leftarrow \begin{bmatrix} \tilde{\mathbf{D}}^{(1)} & \tilde{\mathbf{D}}_m^{*(1)} \\ \tilde{\mathbf{D}}^{(2)} & \tilde{\mathbf{D}}_m^{*(2)} \end{bmatrix}^{-1} \begin{bmatrix} \mathbf{W}_2 \\ \mathbf{W}_4 \end{bmatrix} \qquad (8)$$

Once $\tilde{\mathbf{N}}_t$ and $\tilde{\mathbf{N}}_r$ are known, we can then substitute $\tilde{\mathbf{N}}_{t_m}^*$ and $\tilde{\mathbf{N}}_r$ in $\mathbf{W}_1$ and $\mathbf{W}_3$. This results in a more precise estimate of $\tilde{\mathbf{D}}^{(q)}$, given by the following equation:

$$\begin{bmatrix} \tilde{\mathbf{D}}^{(1)} \\ \tilde{\mathbf{D}}^{(2)} \end{bmatrix} \Leftarrow \begin{bmatrix} \mathbf{W}_1 & \mathbf{W}_{1_m}^* \\ \mathbf{W}_3 & \mathbf{W}_{3_m}^* \end{bmatrix} \begin{bmatrix} 1 \\ -\tilde{\mathbf{N}}_r . \tilde{\mathbf{N}}_{t_m}^* \end{bmatrix} . \frac{1}{1 - \tilde{\mathbf{N}}_r . \tilde{\mathbf{N}}_{r_m}^* . \tilde{\mathbf{N}}_t . \tilde{\mathbf{N}}_{t_m}^*} \qquad (9)$$

The new estimates of $\tilde{\mathbf{D}}^{(q)}$ and $\tilde{\mathbf{D}}_m^{*(q)}$ are then re-substituted in (8) to obtain a more precise estimate of IQ imbalance parameters. Thus (8) and (9) can be repeated a number of times until sufficiently good estimates of $\widetilde{\mathbf{N}}_t$ and $\widetilde{\mathbf{N}}_r$ are obtained. It is observed that in the noiseless case, 2–3 iterations already lead to an accurate estimation. We will call the estimation of $\widetilde{\mathbf{N}}_t$ and $\widetilde{\mathbf{N}}_r$ as the calibration phase from now on.

Once the calibration phase is over, we can now compensate the Rx IQ imbalance by the following equation:

$$\begin{aligned} \mathbf{Z}_t &= \mathbf{Z} - \widetilde{\mathbf{N}}_r.\mathbf{Z}_m^* \\ &= \underbrace{\left(\mathbf{G}_{ra} - \frac{\mathbf{G}_{rb}.\mathbf{G}_{rb_m}^*}{\mathbf{G}_{ra_m}^*} \right)}_{\mathbf{G}_{rx}}.\mathbf{C}.(\mathbf{G}_{ta}.\mathbf{S} + \mathbf{G}_{tb}.\mathbf{S}_m^*) + \widetilde{\mathbf{n}}_1 \\ &= \mathbf{D}_c.(\mathbf{S} + \mathbf{N}_t.\mathbf{S}_m^*) + \widetilde{\mathbf{n}}_1 \end{aligned} \quad (10)$$

where $\mathbf{D}_c = \mathbf{G}_{rx}.\mathbf{C}.\mathbf{G}_{ta}$ is the composite channel and $\widetilde{\mathbf{n}}_1$ is the noise term. The superscript from the received symbol $\mathbf{Z}$ and $\mathbf{C}$ has been dropped here to make the equation concise. Equation (10) shows that the composite channel estimate $\widetilde{\mathbf{D}}_c$ can now be obtained based on a transmitted training symbol $\mathbf{S}$ and $\mathbf{S}_m^*$ as follows:

$$\widetilde{\mathbf{D}}_c = \frac{\mathbf{Z}_t}{\mathbf{S} + \widetilde{\mathbf{N}}_t.\mathbf{S}_m^*} \quad (11)$$

Once the composite channel has been estimated and then compensated from the received signal $\mathbf{Z}_t$, the final step then involves the compensation of only Tx IQ imbalance term from the received signal. The estimate of the transmitted signal $\widetilde{\mathbf{S}}$ can then be obtained by the following equation:

$$\widetilde{\mathbf{S}} = \begin{bmatrix} \mathbf{V}_1 & \mathbf{V}_2 \end{bmatrix} \begin{bmatrix} \mathbf{Z}_q \\ \mathbf{Z}_{qm}^* \end{bmatrix} \quad (12)$$

where $\mathbf{Z}_q = \frac{\mathbf{Z}_t}{\widetilde{\mathbf{D}}_c}$, $\mathbf{V}_1 = \frac{1}{1-\widetilde{\mathbf{N}}_t.\widetilde{\mathbf{N}}_{tm}^*}$ and $\mathbf{V}_2 = \frac{\widetilde{\mathbf{N}}_t}{\widetilde{\mathbf{N}}_t.\widetilde{\mathbf{N}}_{tm}^*-1}$. We call this second phase of the equalization process as channel estimation and Tx Rx IQ imbalance compensation phase. It should be noted that any channel variation requires the re-estimation of only the composite channel $\widetilde{\mathbf{D}}_c$. The remaining equalization process including the calibration phase and the IQ imbalance compensation scheme does not change. Thus once the calibration phase is over we then require only one training symbol to estimate the composite channel. The two phase scheme finally results in a lower training overhead as will be verified in the simulation section. The proposed equalization scheme when applied to the OFDM data symbol is shown in Fig. 1.

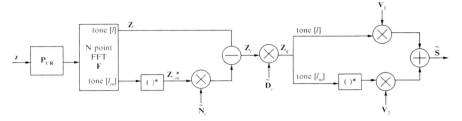

Fig. 1 Compensation scheme for Tx Rx IQ imbalance and the channel in an OFDM system

The estimates $\mathbf{W}_1 \ldots \mathbf{W}_4$ and $\widetilde{\mathbf{D}}_c$ can be further improved by utilizing a frequency smoothing operation. In this scheme we first transform $\mathbf{W}_1 \ldots \mathbf{W}_4$ and $\widetilde{\mathbf{D}}_c$ to their time domain representation by an IDFT operation. Once in time domain, the first and last L' terms at the top and bottom of their time domain sequence are left unchanged while the remaining $N - 2L'$ terms are forced to zero. This is because the impulse response length of these estimates are considered to be of finite length but they may be split at both the top and the bottom of their time domain sequence. The safe length for $2L'$ is then always the CP length. Finally the time domain sequences are again transformed to the frequency domain by a DFT operation. This simple scheme results in an overall frequency smoothing operation [9]. The smoothing operation helps in accelerating the convergence of the equalizer coefficients resulting in an overall improved performance as will be shown in the simulation section.

In the case when only Tx IQ imbalance or Rx IQ imbalance is present in the system (but not both at the same time), then the entire equalization structure can be further simplified. In this case both $\mathbf{N}_t$ and $\mathbf{N}_r$ can be directly derived from $\mathbf{W}_1$ and $\mathbf{W}_2$. The estimation of $\mathbf{W}_3$ and $\mathbf{W}_4$ is not required. Thus in the presence of only Tx IQ imbalance in the system (i.e. $\mathbf{N}_r = 0$), the estimate $\widetilde{\mathbf{N}}_t = \frac{\mathbf{W}_2}{\mathbf{W}_1}$. Also the equation (10) is now simplified to $\mathbf{Z}_t = \mathbf{Z}$. While in the case of only Rx IQ imbalance in the system (i.e. $\mathbf{N}_t = 0$), the estimate $\widetilde{\mathbf{N}}_r = \frac{\mathbf{W}_2}{\mathbf{W}_{1m}^*}$. The equation (11) is now modified as $\widetilde{\mathbf{D}}_c = \frac{\mathbf{Z}_q}{\mathbf{S}}$ and equation (12) as $\widetilde{\mathbf{S}} = \frac{\mathbf{Z}_q}{\widetilde{\mathbf{D}}_c}$. The equalization is now similar to the scheme proposed in [9].

4 Simulation Results

We have simulated a SISO based OFDM system to evaluate the performance of the proposed compensation scheme. The performance comparison is made with an ideal system with no front-end distortion and a system with joint compensation algorithm [8] included. The performance curves are also drawn for a system with no compensation algorithm in place. The parameters used in the simulation are as follows: we consider OFDM symbol of length $N = 128$, CP length $v = 16$ and the constellation size= 64QAM. The mismatched filter impulse responses at Tx and Rx are [0.01, 0.5 0.09] for the I branch and [0.09 0.5, 0.01] for the Q branch. The frequency independent amplitude and phase imbalances at both Tx and Rx are 10% and 10° respectively. It should be noted that we have considered quite large values of IQ imbalances in order to observe the robustness of the compensation scheme. The multipath channel length equals four taps. The taps of the multipath channel are chosen independently with complex Gaussian distribution. During the frequency smoothing operation, we have left the top $L' = 8$ and bottom $L' = 8$ elements in the time domain sequence of $\mathbf{W}_1 \ldots \mathbf{W}_4$ and $\widetilde{\mathbf{D}}_c$ as unchanged while the remaining $N - 2L'$ elements are forced to zero.

We have initially used 8 training symbols to estimate $\widetilde{\mathbf{N}}_t$ and $\widetilde{\mathbf{N}}_r$. The Fig. 2a shows the number of iterations required by the equalizer to estimate $\widetilde{\mathbf{N}}_t$ and $\widetilde{\mathbf{N}}_r$ accurately. The curves depict these estimates as the mean of the absolute values for all the N tones of an OFDM symbol taken together (i.e. $\Xi\{|\widetilde{\mathbf{N}}_t|\}$ and $\Xi\{|\widetilde{\mathbf{N}}_r|\}$). It is observed that even at low SNR=30dB, 3-4 iterations lead to accurate estimation. Once the equalizer is calibrated we can then obtain the performance curves with only one training symbol. In the simulation we have utilized two training symbols as this is the minimal requirement for the joint compensation scheme in [7, 8]. The Fig. 2b shows the performance curves (BER vs SNR) obtained for such a system. The BER results depicted are obtained by taking the average of the BER curves over 10^4 independent channels. It can be observed from the figure that with no compensation scheme in place, the system is completely unusable. The BER is also very high for the case of only FI Tx Rx IQ imbalance in the system. The proposed compensation scheme (with and without frequency smoothing operation) provides a very good performance results. The results obtained are very close to the ideal case even when only two training symbols are used in the system. The difference between the proposed scheme and the joint compensation scheme [8] is almost 9 dB at BER of 10^{-3}. Thus the proposed compensation scheme requires very low training overhead for an effective compensation.

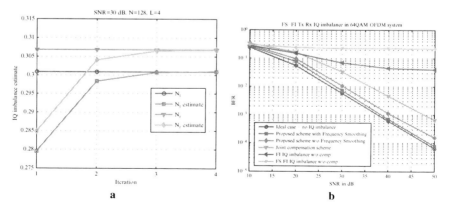

Fig. 2 (a) IQ imbalance estimation with iterative method (b) BER vs SNR of OFDM system with FS-FI Tx Rx IQ imbalance

5 Conclusion

In this paper the joint effect of Tx-Rx IQ imbalance along with multipath channel distortions has been studied. A generally applicable compensation scheme has been developed that can decouple all the three frequency selective distortions, namely the Tx IQ imbalance, Rx IQ imbalance and the channel. The proposed equalizer works in two phases. The first phase is the calibration phase where both the Tx and Rx IQ imbalance parameters are estimated. In the second phase, the equalizer estimates and compensates the channel along with IQ imbalance parameters. Once the calibration phase is over, the equalizer then has a very low training overhead requirement for effective compensation. Simulation results verify the effectiveness of the proposed scheme.

References

1. A.A. Abidi,"Direct-conversion radio transceivers for digital communications", *IEEE Journal on Solid-State Circuits*, vol. 30, pp. 1399–1410, Dec. 1995.
2. "IEEE standard 802.11a-1999: wireless LAN medium access control (MAC) & physical layer (PHY) specifications, high-speed physical layer in the 5 GHz band," 1999.
3. C.L. Liu, "Impact of I/Q imbalance on QPSK-OFDM-QAM detection", *IEEE Transactions on Consumer Electronics*, vol. 44, no. 8, pp. 984–989, Aug. 1998.
4. T. Pollet, M. Van Bladel and M. Moeneclaey, "BER sensitivity of OFDM systems to carrier frequency offset and wiener phase noise," *IEEE Transactions on Communications*, vol. 43, no. (2/3/4), pp. 191–193, 1995.
5. J. Tubbax, B. Come, L. Van der Perre, M. Engels, M. Moonen and H. De Man, "Joint compensation of IQ imbalance and carrier frequency offset in OFDM systems", in *Proceedings Radio and Wireless Conference, Boston, MA*, pp. 39–42, Aug. 2003.
6. A. Tarighat and A.H. Sayed, "OFDM systems with both transmitter & receiver IQ imbalances", in *Proceedings IEEE 6th SPAWC Workshop, New York, NY*, pp. 735–739, June 2005.
7. T. Schenck, P. Smulders and E.Fledderus, "Estimation and compensation of frequency selective Tx/Rx IQ imbalance in MIMO OFDM systems", in *Proceedings IEEE ICC, Istanbul, Turkey*, pp. 251–256, June 2006.
8. D. Tandur and M. Moonen, "Joint compensation of OFDM frequency selective transmitter and receiver IQ imbalance", *Eurasip Journal on Wireless Communications and Networking (JWCN)*, Volume 2007 (2007).
9. L. Anttila, M. Valkama and M. Renfors, "Efficient mitigation of frequency-selective I/Q imbalances in OFDM receivers", in *Proceedings IEEE 68th VTC conference, Calgary, Canada*, Sept. 2008.
10. M. Valkama, M. Renfors and V. Koivunen, "Compensation of frequency-selective IQ imbalances in wideband receivers: models and algorithms", in *Proceedings IEEE 3rd Workshop on Signap Processing Advances in Wireless Communications (SPAWC)*, Taoyuan, Taiwan, pp. 42–45, Mar. 2001.

Part IX
Spectrum & Interference

Dynamic Cross-Layer Spectrum Allocation for Multi-Band High-Rate UWB Systems

Ayman Khalil, Matthieu Crussière, and Jean-François Hélard

Abstract In this paper, we investigate a new approach for the spectrum allocation in UWB systems. This approach consists in a cross-layer scheme that takes into consideration the different users channel quality and quality of service (QoS) requirements. The new scheme is based on the WiMedia solution proposed for multiband OFDM UWB systems. The main objective is to propose a low-complexity solution for the spectrum allocation that can manage all the users constraints. Thus, we study the optimal solution of the spectrum allocation and formulate it as a convex optimization problem. Then, we show that our proposed scheme reduces significantly the complexity of the optimal solution. Moreover, we show through simulations, that the new approach and the optimal solution have close performance in term of error rate and they outperform WiMedia solution proposed for UWB systems.

1 Introduction

Ultra-wideband transmission is an emerging technology for future high-rate, short-range wireless communications. Its wide bandwidth and low transmission power density make it attractive to researchers since 2002 when the Federal Communications Commission (FCC) regulated UWB systems by allocating the 3.1–10.6 GHz spectrum for unlicensed use of UWB [1]. In order to reduce interference with other existing systems, the FCC imposed a power spectral density (PSD) limit of −41.3 dBm/MHz.

The IEEE 802.15a wireless personal area networks (WPAN) standardization group defined a very high data rate physical layer based on UWB signalling. One of the multiple-access techniques considered by the group is a multiband orthogonal

---

A. Khalil (✉), M. Crussière, and J.-F. Hélard
Institute of Electronics and Telecommunications of Rennes (IETR),
INSA, 20 avenue des Buttes de Coëmes, 35043 Rennes, France
e-mail: ayman.khalil@insa-rennes.fr

frequency division multiplexing (MB-OFDM) supported by the MultiBand OFDM Alliance (MBOA) and the WiMedia forum [2, 3] which merged in March 2005 and are today known as the WiMedia Alliance.

On December 2005, ECMA International approved two standards for UWB technology based on the WiMedia solution: ECMA-368 for high rate UWB PHY and MAC standard and ECMA-369 for MAC-PHY Interface for ECMA-368 [4].

There have been a lot of studies on the resource allocation in UWB system based on the WiMedia solution. However, to this date, most research studies on multiband UWB systems have been devoted to the physical layer issues. In [5, 6], the authors propose resource allocation solutions for OFDM-UWB systems in a single-user scheme. In [7], the author considers the multi-user context but without taking into consideration the users QoS requirements.

The aim of this paper is to propose a spectrum management scheme based on UWB signalling in a multiple medium access demand to ensure dynamic spectrum utilization that differentiates between existing users. Therefore, we propose to classify the users into two classes, the first called QoS class for real-time applications that have strict QoS requirements (video recording, A/V conferencing, interactive gaming, etc). The second class is called best effort (BE) class for non real-time applications that have tolerance to some QoS requirements (file transfer, Internet, etc.). On the other hand, we propose to represent each user channel quality by exploiting the exponential effective SINR mapping (EESM) method used in OFDM systems [8]. Consequently, we formulate the resource allocation problem as a convex optimization problem taking into account the users classification and the EESM method issues. Then, we propose our cross-layer approach that reduces the complexity of the optimal solution given by the convex optimization problem.

This paper is organized as follows. Section 2 introduces the WiMedia model by presenting the PHY and MAC layers characteristics. Section 3 derives the problem formulation as a convex optimization problem and presents the optimal solution. In Section 4, we give the proposed low-complexity cross-layer solution. Section 5 presents simulation results showing the comparison between the proposed scheme and the optimal solution, and the performance of the multiuser solution compared to the single-user WiMedia solution. Finally, Section 6 concludes this paper.

2 System Model

2.1 PHY Layer

The WiMedia solution consists in combining OFDM with a multi-banding technique that divides the available band into 14 sub-bands of 528 MHz, as illustrated in Fig. 1. An OFDM signal can be transmitted on each sub-band using a 128-point inverse fast Fourier transform (IFFT). Out of the 128 subcarriers used, only 100 are assigned to transmit data. Different data rates from 53.3 to 480 Mbits/s are obtained

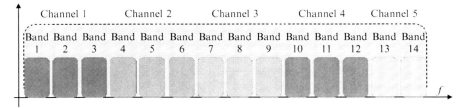

Fig. 1 Channel distribution for WiMedia solution

Table 1 WiMedia system data rates

Data rate (Mbits/s)	Modulation	Coding rate	FDS	TDS
53.3	QPSK	1/3	Yes	Yes
80	QPSK	1/2	Yes	Yes
110	QPSK	11/32	No	Yes
160	QPSK	1/2	No	Yes
200	QPSK	5/8	No	No
320	DCM	1/2	No	No
400	DCM	5/8	No	No
480	DCM	3/4	No	No

through the use of forward error correction (FEC), frequency-domain spreading (FDS) and time-domain spreading (TDS), as presented in Table 1. The constellation applied to the different subcarriers is either a quadrature phase-shift keying (QPSK) for the low data rates or a dual carrier modulation (DCM) for the high data rates. Time-frequency codes (TFC) are used to provide frequency hopping from a sub-band to another at the end of each OFDM symbol. TFC allows every user to benefit from frequency diversity over a bandwidth equal to the three sub-bands of one channel. In addition, to prevent from interference between consecutive symbols, a zero padding (ZP) guard interval is inserted instead of the traditional cyclic prefix (CP) used in the classical OFDM systems [10].

The WiMedia solution offers potential advantages for high-rate UWB applications, such as the signal robustness against channel selectivity and the efficient exploitation of the energy of every signal received within the prefix margin. However, we will see in the next section that the exploitation of the PHY layer at the MAC level is suboptimal in a multi-user context since the medium access mechanisms do not take advantage of the sub-band structure.

2.2 MAC Layer

The WiMedia MAC protocol is a distributed TDMA-based MAC protocol as defined in ECMA standard. Time is divided into superframes where each frame is composed of 256 medium access slots (MAS).

Each MAS has a length of 256 μs. Each superframe starts with a beacon period (BP) that is responsible for the exchange of reservation information, the establishment of neighbourhood information and many other functions.

WiMedia defines two access mechanisms: the prioritized contention access (PCA) and the distributed reservation protocol (DRP). PCA provides differentiated access to the medium for four access categories (ACs); it is similar to the enhanced distributed channel access (EDCA) mechanism of IEEE 802.11e standard. On the other hand, DRP is a TDMA-based mechanism which enables a device to reserve one or more MASs for the communication with neighbours.

2.3 Channel Model

The channel model used in this study is the one adopted by IEEE 802.15.3a committee for the evaluation of UWB proposals [11]. It is a modified version of the Saleh-Valenzuela model for indoor channels, fitting the properties of measured UWB channels. A lognormal distribution is used for the multipath gain magnitude. In addition, independent fading is assumed for each cluster and each ray within the cluster.

Four different channel models (CM1 to CM4) are defined for the UWB system modelling, each with arrival rates and decay factors chosen to match different usage scenarios and to fit line-of-sight (LOS) and non-line-of-sight (NLOS) cases.

3 Optimal Spectrum Allocation

Our main goal is to find the best sub-band assignment for the heterogeneous users. Therefore, we propose first to represent their channel quality in each sub-band by using the EESM method. The basic idea of this method is to find a compression function that maps a sequence of varying SINRs to a single value that is strongly correlated with the actual BER [9]. In WiMedia case, one channel is divided into three sub-bands and the allocation is made by sub-band; that means that each user is dynamically allocated one sub-band for the duration of one superframe. Therefore, the effective SINR calculated for each sub-band is given by

$$SINR_{eff} = -\lambda \ln\left(\frac{1}{N} \sum_{i=1}^{N} e^{-\frac{SINR_i}{\lambda}}\right) \quad (1)$$

where λ is a scaling factor that depends on the selected modulation and coding scheme (MCS), N is the number of subcarriers in a sub-band, and $SINR_i$ is the ratio of signal to interference and noise on the ith sub-carrier.

In our system model, we compute the effective SINR value for each user in each sub-band by using (1). For instance, in the case of one channel divided into $N_b = 3$ sub-bands, and with $N_u = 3$ users, the computation result is a matrix containing $N_b \times N_u = 9$ effective SINR values.

After representing the channel power by the effective SINR value, we formulate our allocation problem which goal is to maximize the BE users data rate while maintaining a certain data rate threshold for QoS users under a total power P_T constraint.

Let U is the total number of users, U_{QoS} is the QoS users number and $U_{BE}(U - U_{QoS})$ is the BE users number. The rate of a user u in a sub-band b is defined as

$$r_{u,b} = \log_2(1 + P_{u,b} E_{u,b}) \tag{2}$$

where $P_{u,b}$ is the allocated power of user k in the sub-band b, and $E_{u,b}$ is the effective SINR of user k in this sub-band. The optimization problem can then be formulated as follows

$$\max_{S_u} \sum_{u=U_{QoS}+1}^{U} \sum_{b \in S_u} r_{u,b}$$

$$\text{subject to } \sum_{b \in S_u} r_{u,b} \geq R_u, \ u = 1, \ldots, U_{QoS} \tag{3}$$

$$\sum_{u=1}^{U} \sum_{b=1}^{B} P_{u,b} \leq P_T$$

where B is total number of sub-bands, R_u is the QoS users required data rate, S_u is the set of sub-bands assigned to user u. In our case, $S_1, S_2, \ldots S_u$ are disjoint and each user is assigned one sub-band during one time interval. The formulated problem is a mixed integer programming problem which is hard to solve. However, we can convert this problem into a convex optimization problem by adopting a new parameter $\rho_{u,b}$ as proposed in [12]. It represents a time-sharing factor for the user u of the sub-band b. The optimization problem can be reformulated as

$$\max_{\rho_{u,b}} \sum_{u=U_{QoS}+1}^{U} \sum_{b \in S_u} \rho_{u,b} \log_2\left(1 + \frac{P_{u,b} E_{u,b}}{\rho_{u,b}}\right)$$

$$\text{subject to } \sum_{b=1}^{B} \rho_{u,b} \log_2\left(1 + \frac{P_{u,b} E_{u,b}}{\rho_{u,b}}\right) \geq R_u, \ u = 1, \ldots, U_{QoS} \tag{4}$$

$$\sum_{u=1}^{U} \rho_{u,b} = 1 \ \forall b \quad 0 \leq \rho_{u,b} \leq 1 \ \forall u, b$$

$$\sum_{u=1}^{U} \sum_{b=1}^{B} P_{u,b} \leq P_T$$

Solving this convex optimization problem using Lagrangian gives the optimal solution

$$\log_2\left(\frac{E_{u,b}}{\gamma \ln 2}\right) - \frac{1}{\ln 2}\left(1 - \frac{\gamma \ln 2}{E_{u,b}}\right) - \beta_b = 0 \quad \text{for BE users}$$
$$\alpha_u \left[\log_2\left(\frac{\alpha_u E_{u,b}}{\gamma \ln 2}\right) - \frac{1}{\ln 2}\left(1 - \frac{\gamma \ln 2}{\alpha_u E_{u,b}}\right)\right] - \beta_b = 0 \text{ for QoS users}$$
(5)

where α_u, β_b and γ are the Lagrange multipliers.

Consequently, using the KKT conditions [13], we conclude the following

$$\rho^*_{u',b} = 1, \quad \rho^*_{u,b} = 0 \quad \text{for all } u \neq u'$$

$$\text{where } u' = \begin{cases} \arg\max_u (\log_2(\frac{E_{u,b}}{\gamma \ln 2}) - \frac{1}{\ln 2}(1 - \frac{\gamma \ln 2}{E_{u,b}})) & \text{for BE users} \\ \arg\max_u (\alpha_u(\log_2(\frac{\alpha_u E_{u,b}}{\gamma \ln 2}) - \frac{1}{\ln 2}(1 - \frac{\gamma \ln 2}{\alpha_u E_{u,b}}))) & \text{for QoS users} \end{cases}$$
(6)

However, finding this optimal solution requires an intensive computation. We need to find the values of α_u that satisfy the QoS users constraints regarding the data rate. To do so, an iterative algorithm is defined; we affect small values to α_u and increase them one by one to satisfy each QoS user constraint.

4 Cross-Layer Solution

In order to reduce the complexity of the optimal solution and to exploit all useful information needed for an efficient sub-band allocation, we propose to transform the optimal solution into a simple suboptimal solution that takes into consideration all the required properties and characteristics.

Hence, since the complexity of the optimal solution resides in the research of the α_u parameter, we propose to study its characteristics in order to find an alternative solution for setting it up. Note that this parameter concerns the QoS users only. Thus, we propose to define a QoS entity that is responsible for affecting a weight or priority level for QoS users. In fact, this entity should be issued from the MAC layer which is charged for QoS support and medium access control. As a result, the proposed cross-layer approach consists in combining the information provided by the PHY layer through the exploitation of the effective SINR and the MAC layer through the weight generator entity (see Fig. 2). Consequently, the most powerful sub-band is assigned to the user u having the highest allocation level AL given by

$$AL_{u,b} = q_u + \max_b (SINR_{eff}(u,b))$$
(7)

where we have a perfect balance between the MAC and the PHY layers.

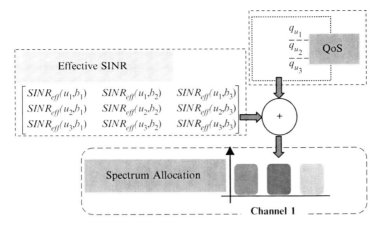

Fig. 2 Dynamic cross-layer spectrum allocation design

The advantage of this cross-layer approach is that it is simple to implement and it agrees with the MAC conditions of the system that is based on a distributed architecture where there is no central coordinator for the spectrum management, so that each device is responsible for all computation and measurements for the allocation. Consequently, this new approach provides a low-complexity self-computation mechanism and assumes a decrease in the data exchange between all the users.

5 System Performance

In this section, we present the simulation results for the proposed multiuser cross-layer allocation scheme and we compare the performance of the new scheme with that of the optimal solution as well as the single-user WiMedia solution using TFC. Therefore, we use the proposed WiMedia data rates (see Table 1). The results are performed on the first three WiMedia sub-bands (3.1–4.7 GHz) for CM1 channel model.

In Fig. 3, the case of three users transmitting simultaneously in the first channel is shown. The three users have different data rates in order to show the advantage of QoS users on BE users in terms of error rate. The QoS user is allocated a high data rate from Table 1 (400 Mbits/s) and the BE users are user outperforms the BE users although it is transmitting at a higher data allocated a lower data rate (320 Mbits/s). As illustrated in the figure, the QoS user outperforms the BE users although it is transmitting at a higher rate.

In Fig. 4, we compare the performance of a QoS user transmitting at a rate of 320 Mbps in the cross-layer solution to that in the optimal solution and to the single-user WiMedia solution with TFC. Note that for the single-user solution TFC

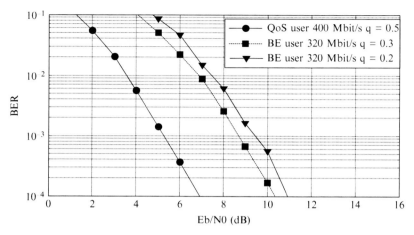

Fig. 3 Three users performance in the cross-layer solution

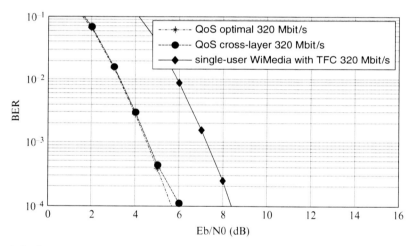

Fig. 4 Performance comparison of QoS users

is exploited because it offers better performance. As shown in the figure, for a $BER = 10^{-4}$, the cross-layer and the optimal solutions are too close and offer a 2.5 dB gain for the QoS user compared to WiMedia solution.

In Fig. 5, we consider the case of BE users transmitting at a rate of 200 Mbits/s and present their performance in the optimal and cross-layer solutions. For a $BER = 10^{-4}$, the optimal solution offers a 0.5 dB compared to the cross-layer solution. On the other hand, we note that the performance of the cross-layer in the case of BE users is close to that of the single-user WiMedia solution. This proves that the performance of the multiuser cross-layer solution performance is never degraded compared to the single-user WiMedia solution.

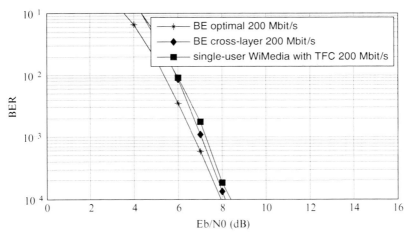

Fig. 5 Performance comparison of BE users

6 Conclusion

In this work, we proposed a dynamic cross-layer design for UWB systems. This design combines the physical layer information through the investigation of the effective SINR and the MAC layer through the QoS parameters. We studied first the optimal solution of the dynamic allocation under QoS constraints by exploiting the effective SINR and formulating an optimization convex problem. By analyzing the properties of the optimal solution we obtained, we proposed a lower complexity solution that is based on a cross-layer approach by defining a simple linear allocation function. Simulation results showed that the new multiuser cross-layer approach outperforms the WiMedia solution proposed for a single-user scheme. Moreover, we showed that the cross-layer solution performance is close to the optimal solution performance.

Acknowledgement The research leading to these results has received funding from the European Community's Seventh Framework Programme FP7/2007–2013 under grant agreement n° 213311 also referred as OMEGA.

References

1. "First report and order, revision of part 15 of the commission's rules regarding ultra-wideband transmission systems," FCC, ET Docket 98–153, Feb. 14, 2002.
2. Batra et al., "Multi-band OFDM physical layer proposal for IEEE 802.15 task group 3a," IEEE document P802.15-04/0493r1, Texas Instruments et al., Sept. 2004.
3. WiMedia Alliance, Inc., "Multiband OFDM physical layer specification," Release 1.1, July 2005.

4. Standard ECMA-368, High rate ultra wideband PHY and MAC standard, 2nd edition, Sept. 2007.
5. Z. Chen, D. Wang and G. Ding, "An OFDM-UWB Scheme with Adaptive Carrier Selection and Power Allocation," *in Proceedings of IEEE International Conference on Wireless Communications, Networking and Mobile Computing (WiCOM'06)*, pp. 1–4, China, Sept. 2006.
6. A. Stephan, J- Y. Baudais and J- F. Helard, "Efficient Allocation Algorithms for Multicarrier Spread Spectrum Schemes in UWB Applications," in *Proceedings Of IEEE International Conference on Ultra-Wideband (ICUWB '07)*, pp. 551–555, Singapore, Sept. 2007.
7. W. P. Siriwongpairat, Z. Han and K. J. Ray Liu, "Power controlled channel allocation for multi-user multiband UWB systems," *IEEE Transactions on Wireless Communications*, vol. 6, no. 2, pp. 583–592, Feb. 2007.
8. 3GPP TSG-RAN-1, "R1-030999: considerations on the system-performance evaluation of HSDP using OFDM modulations"; RAN WG1 #34.
9. 3GPP TSG-RAN-1, "R1-040090: system level performance evaluation for OFDM and WCDMA in UTRAN".
10. B. Muquet, Z. Wang, G- B Giannakis, M. de Courville and P. Duhamel, "Cylic pefix or zero padding for wireless multicarrier transmission," *IEEE Transactions Communications*, vol. 50, pp. 2136–2148, Dec. 2002.
11. J. Foester, "Channel modeling sub-committee report (final)," IEEE P802.15.-02/490rl-SG3a,2003.
12. P. W. C. Chan and R. S. Cheng, "Optimal Power Allocation in Zero-Forcing MIMO-OFDM Downlink with Multiuser Diversity," in *Proceedings of IST Mobile & Wireless Communications Summit*, Dresden, June 2005.
13. D. P. Bertsekas, *Nonlinear Programming*, 2nd Edition. Athena Scientific, Belmont, MA, 1999.

Egress Reduction for OFDM Via Transmit Windowing – Framework and Comparison

Thomas Magesacher

Abstract Every multicarrier system used in practice has to obey power spectral density limits dictated by standards or other specifications in order to maintain spectral compatibility with coexisting systems. Rectangularly-windowed FFT-based multicarrier systems exhibit significant sidelobes in their transmit spectra. An effective and low-complexity approach to amend this problem is transmit windowing. This paper investigates the most important windowing techniques (intrasymbol windowing, intersymbol windowing, and Nyquist windowing) and casts them into a framework for comparison. A case study illustrates the results.

1 Introduction

All operational and prospective communication systems are faced with two, partially contradicting, goals: On one hand, out-of-band emissions have to be kept low enough to ensure spectral compatibility with other systems. On the other hand, systems thrive for higher and higher datarates due to ever increasing datarate requirements fueled by new services and applications. Most operational OFDM systems approach the spectral-compatibility challenge with frequency-domain guard bands by simply nulling the corresponding subcarriers. This may be an acceptable solution for unlicensed and thus free spectrum. However, for licensed and thus costly parts of the spectrum, there is a need for an economic management of the available bandwidth.

Transmit windowing is an effective and computationally cheap approach to keep out-of-band egress low [1–6]. Open literature includes several distinct techniques that differ in various aspects: some preserve orthogonality among subcarriers, others require receiver-processing to do so; some use overlapping blocks, others do not. Different windowing techniques introduce different datarate penalties in form of

---

T. Magesacher
Department of Electrical and Information Technology, Lund University, Sweden
e-mail: tom@eit.lth.se

additional extensions or reduced signal-to-noise ratio (SNR) due to orthogonality-restoring receive processing. A sensible comparison has to take into account both achievable datarate and out-of-band egress. Furthermore, each technique should be applied with the best possible power loading to exploit its potential to the full.

In [7], a comparison of egress-reduction techniques including transmit windowing as well as a method referred to as spectral shaping [8] is presented. This paper focuses on three distinct transmit windowing techniques: intrasymbol windowing, intersymbol windowing, and Nyquist windowing. The paper is organized as follows. Section 2 introduces a system model suitable for the windowing techniques described in Section 3. Section 4 defines the performance measure and the optimization problem that form the basis for comparison. Section 5 illustrates the results in terms of a case study and Section 6 concludes the work.

2 System Model

This section introduces a model for a multicarrier system with N subcarriers that allows us to formulate the windowing techniques considered hereinafter. The ith transmit symbol can be written as vector

$$t_i = \underbrace{T Z_{\text{add}} W^H P}_{A} x_i, \qquad (1)$$

where the vector $x_i \in \mathbb{C}^{N \times 1}$ contains the transmit data and $P = \text{diag}\{p\}$ denotes a diagonal matrix with the power values $p \in \mathbb{R}_+^{N \times 1}$ on the main diagonal. $W \in \mathbb{C}^{N \times N}$, defined as

$$W[n,k] = \frac{1}{\sqrt{N}} e^{-j 2\pi (n-1)(k-1)/N}, \quad n,k = 1,\ldots,N,$$

is the normalized discrete Fourier transform matrix. The matrix $Z_{\text{add}} \in \mathbb{R}^{N' \times N}$ defined by

$$Z_{\text{add}} = \begin{bmatrix} \mathbf{0}_{L \times (N-L)} & I_L \\ I_N & \\ I_{L'} & \mathbf{0}_{L' \times (N-L')} \end{bmatrix} \qquad (2)$$

introduces cyclic extensions of length L and L' at the beginning and at the end of each symbol, respectively. In general, the total length of each such transmit symbol is $N' = N + L + L'$. $T = \text{diag}\{u\}$ is a diagonal matrix with the transmit window u on its main diagonal. The matrix A describes the transmit information processing.

Consecutive blocks t_i are transmitted with an overlap of L_w samples (i.e., the sum of the overlapping L_w samples of two consecutive blocks is transmitted). The effective symbol length is thus $N'' = N' - L_w$.

The time-dispersive Gaussian channel performs linear convolution of the transmit signal with the impulse response $h_n, n = 0, \ldots, M$ of length $M + 1$ ($h_0 \neq 0$, $h_M \neq 0$, $h_n = 0$ for $n < 0$ and for $n > M$). Hereinafter, we assume $L_{cp} \geq M$, which eliminates intersymbol interference and intercarrier interference after removal of the cyclic prefix (CP) so that the received time-domain symbol can be written as $r_i = Ht_i + n_i$. The noise vector $n_i \sim \mathcal{CN}(0, C_n)$ is a random vector of length N' with zero-mean proper complex Gaussian entries and covariance matrix $C_n = \mathrm{E}(n_i n_i^H)$. The channel convolution matrix $H \in \mathbb{C}^{N' \times N'}$ is defined as

$$H[n,k] = h_{n-k}, \qquad n, k = 1, \ldots, N'.$$

The frequency-domain receive signal at the output of the equalizer is given by

$$y_i = \underbrace{GRWZ}_{B} r_i, \tag{3}$$

where $Z \in \mathbb{R}^{N \times N'}$ processes the cyclic extensions introduced at the transmitter, $R \in \mathbb{C}^{N \times N}$ performs windowing-related receive processing (if required), $G \in \mathbb{C}^{N \times N}$ performs linear equalisation, and B describes the receive information processing.

For a standard receiver (no windowing), $Z = Z_{\mathrm{rem}}$ holds, where

$$Z_{\mathrm{rem}} = \begin{bmatrix} 0_{N \times L} & I_N & 0_{N \times L'} \end{bmatrix} \tag{4}$$

purges leading and trailing cyclic extensions. Typically, a standard transmitter uses a CP only ($L = L_{cp}$, $L' = 0$). No overlap occurs ($L_w = 0$) and no windowing-related receive processing is required ($T = R = I$).

3 Transmit Windowing Techniques

This section describes the windowing techniques considered hereinafter and embeds them into the model given by (1) and (3).

3.1 Intrasymbol windowing

Intrasymbol windowing [1–3] weights the CP-extended symbol ($L = L_{cp}$, $L' = 0$) with a window $u = \begin{bmatrix} u[1] & u[2] & \ldots & u[N'] \end{bmatrix}^T$ that is symmetric with respect to its center: $u[k] = u[N' - k + 1]$, $k = 1, 2, \ldots, N'$. Such weighting destroys the orthogonality among subcarriers and thus necessitates windowing-related receive processing by $R = U^{-1}$ where

$$U^{-1}[n,k] = \frac{1}{\sqrt{N}} (\begin{bmatrix} 0_{N \times L_{cp}} & W^H \end{bmatrix} u)[((k - n) \bmod N) + 1], \; n, k = 1, \ldots, N,$$

which may reduce the SNR. No overlap of consecutive symbols occurs ($L_w = 0$) and $Z = Z_{rem}$ given by (4). Intrasymbol windowing restricts the weighting to the samples of the current symbol only—thus the name.

3.2 Intersymbol Windowing

Intersymbol windowing [4, 5] introduces an additional cyclic extension of L_w samples at both the beginning and the end of a multicarrier symbol ($L = L_{cp} + L_w$ and $L' = L_w$) before applying the symmetric window function

$$u[k] = \begin{cases} u[N'-k+1], & k=1,\ldots,L_w \\ 1, & k=L_w+1,\ldots,N'-L_w \end{cases}, \quad (5)$$

which weights only these extended parts. Consecutive symbols are transmitted with an overlap of L_w samples. $Z = Z_{rem}$ given by (4) with $L = L_{cp}+L_w$ and $L' = L_w$, which maintains orthogonality among subcarriers. Note, however, that due to the overlap of L_w samples, the receiver purges only the leading $L_w + L_{cp}$ samples of each receive block and does not purge trailing samples.

No windowing-related receive processing is required ($R = I$). This kind of windowing appears in several communication standards, e.g., in wireline communications [9].

3.3 Nyquist Windowing

Nyquist windowing [6] is a variation of intrasymbol windowing applied together with zero padding. It allows intrasymbol windowing while preserving subcarrier orthogonality at the cost of very simple receive processing.

Z_{add} given by (2) introduces cyclic extensions of length $L = L_{nyq}$ and $L' = L_{nyq} + L_{cp}$, where L_{nyq} is a design parameter. The transmit window $u \in \mathbb{C}^{(N+2L_{nyq}+L_{cp}) \times 1}$ obeys

$$u[k] = \begin{cases} 1 - u[k+N], & k=1,\ldots,2L_{nyq} \\ 1, & k=2L_{nyq}+1,\ldots,N \\ 0, & k=N+2L_{nyq}+1,\ldots,N+2L_{nyq}+L_{cp} \end{cases}.$$

Note that the third condition corresponds to zero padding with L_{cp} samples.

Consecutive symbols are transmitted without overlap ($L_w = 0$). No dedicated receiver processing is required ($R = I$), except for the following: instead of purging samples, the receiver sums leading and trailing transients according to

$$Z = \begin{bmatrix} \begin{bmatrix} \mathbf{0}_{N-L_{\text{nyq}},L_{\text{nyq}}} \\ I_{L_{\text{nyq}}} \end{bmatrix} I_N \begin{bmatrix} I_{L_{\text{nyq}}+L_{\text{cp}}} \\ \mathbf{0}_{N-L_{\text{nyq}}-L_{\text{cp}},L_{\text{nyq}}+L_{\text{cp}}} \end{bmatrix} \end{bmatrix},$$

which maintains the orthogonality among subcarriers.

4 Performance Measure and Optimization

In order to allow a fair comparison of different system variants, we use the normalized information rate

$$R = \frac{N}{N''} \sum_k \log_2 \left(1 + \frac{(BHA(BHA)^{\text{H}})[k,k]}{(BC_n B^{\text{H}})[k,k]} \right) \qquad (6)$$

in bit per multicarrier symbol. All extensions and overlaps (cf. intersymbol windowing) are taken into account by the factor N/N''.

Note that R/N is a measure for the number of bits per subcarrier (of width F_s/N Hertz, where F_s is the sampling frequency) per multicarrier symbol (of length $1/F_s$ seconds), which corresponds to the number of bits per second per Hertz.

For each scheme, the datarate-maximizing power allocation is found by solving

$$\begin{aligned} p^{(\text{opt})} &= \arg\max_p R \\ &\text{subject to } q \leq m, \ |p|_1 \leq P_{\text{tot}} \end{aligned}, \qquad (7)$$

where P_{tot} denotes the total transmit power, q is the transmit power spectral density (PSD), and m is the transmit PSD mask. The computation of q is outlined in [8].

5 Comparison: Case Study

In order to illustrate the application of this framework as well as some interesting facts regarding the performance of different techniques, this section presents a case study. We consider an OFDM system with $N = 64$ subcarriers and a CP of length $L_{\text{cp}} \in \{1, 2, 4, 8, 16\}$. For simplicity, we assume a flat channel with white Gaussian noise, which yields the same subchannel SNR, denoted SNR, on all subcarriers ($C_n = SNR^{-1} I$). The PSD mask m, depicted in Fig. 1, defines an out-of-band region by dictating a transmit-PSD limit at the band edges that is 20 dB below the in-band level. The normalized in-band width is $B_{\text{in}} = 0.8477$, which corresponds to the band occupied by roughly 54 subcarriers.

The resulting transmit PSD of a standard transmitter performing optimal power loading in the sense of maximizing the achievable datarate according to (7) shows that the power on subcarriers close to the mask edges has to be reduced significantly.

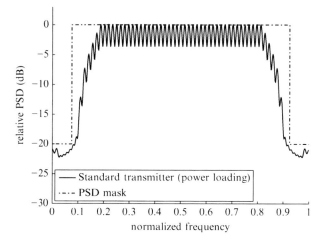

Fig. 1 Case study: the PSD mask (dashed dotted line) divides the band into an in-band region (of width $B_{\text{in}} = 0.8477$, which corresponds to roughly 54 times the subcarrier spacing of $1/64$) and an out-of-band region. The transmit PSD of a standard transmitter (solid line) using optimal power values (in the sense of maximizing the achievable datarate given by (6)) reveals that the power backoff on subcarriers close to the mask edges is significant ($N = 64$, $L_{\text{cp}} = 16$)

Windowing techniques yield transmit PSDs whose sidelobes decay faster, which allows a better exploitation of the available in-band region in the sense of transmitting with higher power on in-band subcarriers (in particular on subcarriers close to the mask edges).

In the following, we compare different windowing techniques in terms of achievable datarate and bit error rate for uncoded signalling. We use linear-slope window shapes for intersymbol windowing and Nyquist windowing. While the window shapes are fixed, the window lengths are chosen optimally for each setup in the sense of maximizing R. For intrasymbol windowing, we use the design proposed in [1], which maximizes the energy of the basis function's mainlobe.

Problem (7) is solved numerically using the semidefinite program solver SEDUMI [10] and the MATLAB interface YALMIP [11].

5.1 Achievable Datarate

The performance of different schemes in terms of R/N versus CP-length L_{cp} for a subchannel SNR of $SNR = 10$ dB is depicted in Fig. 2. Windowing techniques clearly outperform plain power loading. Nyquist windowing has an advantage over other schemes for channels with large delay spreads. Since no power is wasted during the zero-padded gaps, all in-band data subcarriers can be loaded with higher power values while obeying the PSD mask.

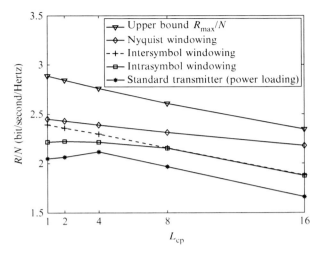

Fig. 2 Comparison: spectral efficiency R/N in bit/s/Hertz of different windowing techniques for the PSD mask depicted in Fig. 1. Upper bound (triangle) given by (8), Nyquist windowing (diamond), intersymbol windowing (plus), intrasymbol windowing (square), and power loading (star). Intersymbol windowing and Nyquist windowing employ linear-slope shapes of optimal length (in the sense of maximizing R/N). Intrasymbol windowing employs the maximum mainlobe-energy window suggested in [1].

As a reference, we include the simple upper bound

$$R_{\max}/N = \frac{NB_{\text{in}}}{N + L_{\text{cp}}} \log_2(1 + SNR) \qquad (8)$$

on R/N, which assumes that all in-band subcarriers can be loaded with unit power.

5.2 Bit Error Rate

While the achievable datarate is a valid and convenient measure from optimization point of view, it should be noted that uncoded signalling using the obtained power values needs to be applied with care. In the following, we assume uncoded transmission over a set $\mathcal{S}$ of in-band tones using QPSK signalling. The choice of $\mathcal{S}$ is key since the subcarrier(s) with the lowest transmit power value(s) dominate(s) the bit error rate (BER). Figure 3 depicts the BER results for $\mathcal{S} = \{10, 11, \ldots, 56\}$, which clearly reflect the benefit of transmit windowing.

Figure 4 illustrates the importance of carefully selecting $\mathcal{S}$: extending the set by three subcarriers on each side (yielding $\mathcal{S} = \{7, 11, \ldots, 59\}$) annihilates the benefit of intersymbol windowing and Nyquist windowing since low transmit power

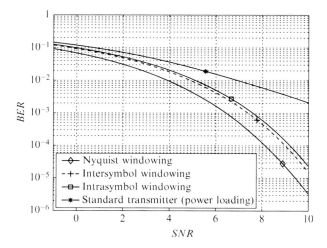

Fig. 3 Bit-error-rate performance for QPSK signalling over all subcarriers in the set $\mathcal{S} = \{10, 11, \ldots, 56\}$ ($L_{cp} = 4$)

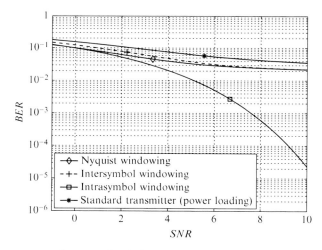

Fig. 4 Bit-error-rate performance for QPSK signalling over all subcarriers in the set $\mathcal{S} = \{7, 11, \ldots, 59\}$ ($L_{cp} = 4$)

values on these edge subcarriers destroy the overall BER performance. Intrasymbol windowing, on the other hand, maintains a quasi-constant power loading over all tones, which results in an acceptable BER performance.

6 Conclusions

Maintaining spectral compatibility with coexisting systems while fulfilling the datarate requirements is a key requirement for every multicarrier system. Transmit windowing is an efficient and low-complexity technique to keep out-of-band egress, which is typically tackled via frequency-domain guard bands, low enough while allowing higher power values on data subcarriers than with plain power loading.

Optimal power loading in the sense of maximizing the achievable datarate under a PSD constraint is a sensible approach to cast different windowing techniques into a framework and allows a fair comparison. Nyquist windowing has an advantage over other techniques since the zero-padding yields inactive periods in the transmit sequence that allow for higher power values on data subcarriers while obeying the PSD mask. The longer the CP, the more pronounced this effect becomes.

Care has to be taken when applying the resulting power values directly to uncoded transmission since the subcarrier(s) with the lowest transmit power dominate(s) the overall BER. The choice of the data-subcarrier set is thus crucial.

References

1. Y.-P. Lin and S.-M. Phoong. Window designs for DFT-based multicarrier systems. *IEEE Transactions on Signal Processing*, 53:1015–1024, March 2005.
2. G. Cuypers, K. Vanbleu, G. Ysebaert, and M. Moonen. Intra-symbol windowing for egress reduction in DMT transmitters. *EURASIP J. Appl. Signal Process.*, 2006(1):87–87, 2006.
3. T. Magesacher. Optimal intra-symbol transmit windowing for multicarrier modulation. In *Proc. Intl. Symp. on Communications, Control and Signal Processing ISCCSP 2006*, Marrakech, Morocco, March 2006.
4. J. M. Cioffi. *Advanced Digital Communication*. class reader EE379C, Stanford University, 2005. Class web page http://www.stanford.edu/class/ee379c/.
5. T. Magesacher, P. Ödling, and P. O. Börjesson. Optimal intersymbol transmit windowing for multicarrier modulation. In *Proc. Nordic Signal Processing Symp. NORSIG 2006*, Reykjavik, Iceland, June 2006.
6. M. Sebeck and G. Bumiller. Effective configurable suppression of narrow frequency bands in multi-carrier modulation transmission. In *Proc. 2006 IEEE International Symposium on Power Line Communications and Its Applications*, pages 128–133, March 2006.
7. T. Magesacher, P. Ödling, and P. O. Börjesson. Fair comparison of transmitter-based spectral shaping techniques. In *Proc. 11th International OFDM Workshop InOWo'06*, pages 323–327, Hamburg, Germany, August 2006.
8. T. Magesacher. Spectral compensation for multicarrier communication. *IEEE Transactions on Signal Processing*, 55(7):3366–3379, July 2007.
9. ETSI TM6. Transmission and multiplexing (TM); access transmission systems on metallic access cables; Very high speed Digital Subscriber Line (VDSL); Part 2: Transceiver specification. *TS 101 270-2, Version 1.1.5*, December 2000.
10. J. F. Sturm, "Using SeDuMi 1.02, a Matlab toolbox for optimization over symmetric cones,," vol. 11–12, 1999, pp. 625–653, available from http://sedumi.mcmaster.ca.
11. J. Löfberg, "YALMIP: A toolbox for modeling and optimization in MATLAB," in *Proceedings of the CACSD Conference*, Taipei, Taiwan, 2004, available from http://control.ee.ethz.ch/~joloef/yalmip.php.

Interference Mitigation for the Future Aeronautical Communication System in the L-Band

Sinja Brandes and Michael Schnell

Abstract In this paper, an OFDM based inlay system is considered, which is operated in the spectral gap between two adjacent channels of an already existing system in the L-band. The strong impact of pulsed interference from the existing L-band systems is mitigated by means of pulse blanking and clipping. When applying these well-known approaches to OFDM systems, the two main issues to be solved are the optimisation of the threshold for pulse blanking or clipping and the detection of the interference pulses in the received signal. Simulation results with the optimal threshold show that interference can be mitigated considerably. However, the performance of the interference-free case is not reached since the desired OFDM signal is impaired by pulse blanking and clipping.

1 Introduction

OFDM based overlay systems are a promising approach for increasing spectral efficiency and for overcoming the problem of spectral scarcity since they enable the operation of an OFDM system in a frequency band that is already in use by licensed systems. In this paper, an OFDM system operated in the aeronautical L-band (960–1164 MHz) is regarded. This Broadband Aeronautical Multi-carrier Communications (B-AMC) system [1] is a promising candidate for the future L-band Digital Aeronautical Communications System (L-DACS).

Large parts of the L-band are subdivided into 1 MHz channels and used by the distance measuring equipment (DME), an aeronautical navigation system based on radar technology. As depicted in Fig. 1, the OFDM system is intended to be operated in the gap between two adjacent DME channels that has a bandwidth of approximately 500 kHz. For realising a successful coexistence, mutual interference

---

S. Brandes (✉) and M. Schnell
German Aerospace Center (DLR), Institute of Communications and Navigation,
Oberpfaffenhofen, 82234 Wessling, Germany
e-mail: {sinja.brandes, michael.schnell}@dlr.de

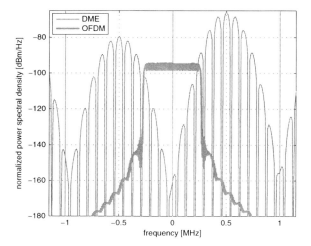

Fig. 1 Illustration of inlay concept: OFDM spectrum in gap between two adjacent DME channels

between the two systems has to be kept at an acceptable level. For reducing out-of-band radiation of the OFDM signal causing interference towards the licensed system, powerful techniques have been proposed e.g. in [2]. The mitigation of the impact of DME interference onto the OFDM system is topic of this paper.

The remainder of this paper is organised as follows. After introducing an appropriate interference model in Section 2, in Section 3, interference mitigation by means of clipping and pulse blanking is addressed. Special emphasis is put on the optimisation of the thresholds as well as on the detection of interference pulses. Simulation results in Section 4 show the performance of the interference mitigation techniques for different parameters. Finally, in Section 5 conclusions are drawn and an outlook on future work is given.

2 Modelling of Interference

The DME signal consists of pairs of Gaussian shaped pulses each having a duration of 3.5 μs with a separation of $\Delta t = 12$ μs or 36 μs. One pulse pair in the base band writes

$$p(t) = e^{-\alpha/2 t^2} + e^{-\alpha/2(t-\Delta t)^2} \qquad (1)$$

with $\alpha = 4.5 \cdot 10^{11} 1/s^2$. For generating the interference signal relevant for the OFDM system, these pulse pairs are modulated to relative carrier frequencies of the channels to the left and to the right of the OFDM bandwidth, i.e. $\Delta f_{c,i} = \pm 500$ kHz with $\Delta f_{c,i}$ denoting the relative carrier frequency of the ith interferer. In addition, for each interferer, $N_{\text{L},i}$ subsequent pulse pairs with different starting times $t_{i,l}$ randomly distributed in the considered time interval have to be considered, where each pulse pair has different power $P_{i,l}$ and phase $\varphi_{i,l}$. In total, the interference signal $i(t)$ in a certain time interval is composed of the contributions from N_{I} interferers resulting in

$$i(t) = \sum_{i=0}^{N_\text{I}-1} \sum_{l=0}^{N_{\text{L},i}-1} \sqrt{P_{i,l}}\, p(t - t_{i,l}) e^{j 2\pi f_{c,i} t + j \varphi_{i,l}}. \tag{2}$$

At the OFDM receiver (Rx), the Rx signal is composed of the desired OFDM signal and the interference signal. In order to model the impact of interference onto the OFDM signal as realistically as possible signal processing at the OFDM Rx has to be considered. To diminish influences of DME interference signals in channels at larger offsets and to avoid aliasing, a band-pass filter followed by a raised-cosine anti-aliasing filter is applied as proposed in [3]. When modelling the desired OFDM signal, the filters can be neglected as in the ideal case, they do not affect the OFDM signal lying in the pass-band of the filters. Furthermore, four-times over-sampling is introduced and has to be considered for the OFDM signal as well. Hence, the over-sampled Rx signal $r^{\text{ov}}[k]$ in the time domain writes

$$r^{\text{ov}}[k] = y^{\text{ov}}[k] + n^{\text{ov}}[k] + i^{\text{ov}}[k],\ k = 0, \ldots, V(N + N_{\text{GI}}) - 1, \tag{3}$$

with $y^{\text{ov}}[k]$, $n^{\text{ov}}[k]$, and $i^{\text{ov}}[k]$ denoting samples of the desired OFDM signal which is the transmitted signal affected by the radio channel, the additive white Gaussian noise (AWGN), and the interference signal, respectively. The Rx signal of one OFDM symbol including guard interval (GI) contains $V(N + N_{\text{GI}})$ samples, where V represents the over-sampling factor, N the size of the discrete Fourier transform (DFT), and N_{GI} the length of the GI. The resulting signal is transformed to the frequency domain by an $N \cdot V$-point DFT. Afterwards, the relevant subcarriers $R[n], n = 0, \ldots, N - 1$, lying in the OFDM bandwidth are extracted by an ideal low-pass filter. The resulting block diagram of an OFDM Rx including interference mitigation in the time domain is shown in Fig. 2.

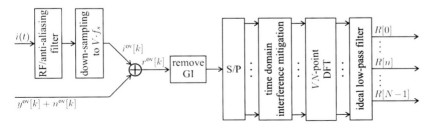

Fig. 2 Block diagram of OFDM Rx performing mitigation of pulsed interference

3 Interference Mitigation

Clipping and pulse blanking are well known approaches to combat pulsed interference. Both methods have been applied to DME interference in the E5- and L5-bands used by satellite navigation systems [4] and to impulsive noise in OFDM systems [5]. The combination of both approaches for mitigating impulsive noise has been proposed in [5]. Although very simple, both methods suffer from the drawback that the desired signal is clipped or blanked as well.

3.1 Clipping and Pulse Blanking

The amplitude of samples of the Rx signal affected by interference is reduced to a certain threshold T^{clip} when their amplitude exceeds T^{clip}. The clipping operation yields the clipped Rx signal $r'^{\text{ov}}[k]$ and writes

$$r'^{\text{ov}}[k] = \begin{cases} r^{\text{ov}}[k] & |r^{\text{ov}}[k]| < T^{\text{clip}} \\ T^{\text{clip}} \exp\{j \arg\{r^{\text{ov}}[k]\}\} & |r^{\text{ov}}[k]| \geq T^{\text{clip}} \end{cases} \quad (4)$$

$$k = 0, \ldots, V(N + N_{\text{GI}}) - 1.$$

Analogously, the pulse blanking operation using threshold T^{PB} is written as

$$r'^{\text{ov}}[k] = \begin{cases} r^{\text{ov}}[k] & |r^{\text{ov}}[k]| \leq T^{\text{PB}} \\ 0 & |r^{\text{ov}}[k]| > T^{\text{PB}} \end{cases}, k = 0, \ldots, V(N + N_{\text{GI}}) - 1. \quad (5)$$

3.2 Optimisation of Threshold

Clipping and pulse blanking reduce interference power, hence diminishing the impact on the OFDM signal and improving performance. This suggests setting the thresholds T^{clip} and T^{PB} as small as possible in order to keep the remaining interference power at a minimum. At the same time, clipping and erasing parts of the useful OFDM signal affects all OFDM subcarriers and degrades performance which demands for setting the thresholds as high as possible in order to reduce the number of affected samples. To determine the optimal threshold as a trade-off between these two contradictory effects, signal-to-noise and interference ratio (SINR) is employed as a figure of merit for the expected performance [5]. Before clipping or pulse blanking, useful signal, interference, and noise are uncorrelated. Hence, SINR is given by

$$\text{SINR} = \frac{E\{|y^{\text{ov}}[k]|^2\}}{E\{|n^{\text{ov}}[k]|^2\} + E\{|i^{\text{ov}}[k]|^2\}}. \quad (6)$$

After clipping has been applied, the power of the desired signal is attenuated and interference power is reduced. In addition, a distortion representing the impact on the desired OFDM signal is induced. Since this distortion is correlated with the useful OFDM signal $y^{\mathrm{ov}}[k]$, the SINR definition from (6) cannot be used. In [5,6], the Rx signal after clipping or pulse blanking $r'^{\mathrm{ov}}[k]$ has been split up into a useful and a distorted part yielding

$$r'^{\mathrm{ov}}[k] = \underbrace{\xi \cdot y^{\mathrm{ov}}[k]}_{\text{useful part}} + \underbrace{v^{\mathrm{ov}}[k]}_{\text{distorted part}}. \tag{7}$$

The distorted part of the clipped or blanked Rx signal contains AWGN, the remaining interference signal, as well as the distortion induced by clipping or blanking. The scaling factor ξ represents the attenuation of the desired OFDM signal due to clipping or pulse blanking. In [5,6], ξ is defined such as to achieve an uncorrelated useful and distorted part resulting in

$$\xi = \frac{\mathrm{E}\{r'^{\mathrm{ov}}[k] y^{*\mathrm{ov}}[k]\}}{\mathrm{E}\{|y^{\mathrm{ov}}[k]|^2\}}. \tag{8}$$

Since the distortion and the useful part of the clipped Rx signal are now uncorrelated, the SINR after clipping or pulse blanking SINR$'$ can easily be defined as

$$\mathrm{SINR}' = \frac{E\{|\xi y^{\mathrm{ov}}[k]|^2\}}{E\{|v^{\mathrm{ov}}[k]|^2\}} = \frac{E\{|\xi y^{\mathrm{ov}}[k]|^2\}}{E\{|r'^{\mathrm{ov}}[k] - \xi y^{\mathrm{ov}}[k]|^2\}}. \tag{9}$$

The optimal threshold is determined in simulations with simplified DME interference conditions. One DME interferer at $f_{c,0} = -0.5$ MHz with different representative power values is considered. The duty cycle is assumed to be 10,800 pulse pairs per second (ppps) such as to represent interference from three DME stations transmitting with maximum pulse rate. In the OFDM system, SNR is set to 5 dB which results in average SINR values after filtering ranging from 0.35 to 1.72 dB for the considered interference power values. The SINR vs. threshold curves in Fig. 3 show that clipping increases SINR and a clear maximum is observed at $T^{\mathrm{clip}} = 2.25$ which is the optimal threshold. SINR is not improved for interferers with $P_0 < -73$ dBm. However, when taking into account coding, some errors induced by clipping are corrected at the decoder such that all in all, SINR is improved for all interferers independent of their power. The optimal threshold is independent of the pulse rate since the trade-off between the two effects is the same despite different absolute SINR obtained for different pulse rates.

For pulse blanking, slightly higher SINR values are observed, but the optimal threshold is the same as for clipping, i.e. $T^{\mathrm{PB}} = 2.25$. After pulse blanking, the remaining interference power is smaller as the interference signal is erased rather than just reduced. This is counterbalanced by the higher impact on the useful OFDM signal, resulting in a similar trade-off as for clipping.

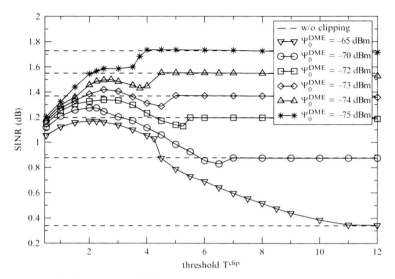

Fig. 3 Optimal clipping threshold T^{clip}; SNR = 5 dB, $f_{c,0} = -0.5$ MHz, 10,800 ppps

3.3 Detection of Pulses

The interference signal and the corresponding samples to be clipped or blanked have to be detected from the Rx signal. As the power level of the DME pulses usually exceeds the level of the desired signal, DME pulses can be detected in the time domain Rx signal. However, as illustrated in the upper part of Fig. 4, the detection is aggravated by a non-constant signal level which is an inherent property of the OFDM signal resulting in a high peak-to-average power ratio. Detecting interference pulses based on a simple threshold decision, peaks in the OFDM signal would falsely be identified as pulses. Clipping the corresponding samples would lead to additional bit errors while not reducing the impact of interference at all. An analysis of the peaks in the OFDM signal including noise has shown that this effect is not negligible. In particular, for low SNR, peaks with a width of up to eight samples are observed leading to more than 5% of the samples of an OFDM symbol being falsely clipped.

To circumvent this problem, the amplitude of the interference signal is reconstructed based on the correlation of the Rx signal and a DME pulse with known peak amplitude. The correlation of the Rx signal and a known signal is a well-known procedure for signal detection at Rx.

Prior to clipping or pulse blanking, the cross-correlation function $r_{\text{rp}}[u-v]$ between the time domain Rx signal $r[u]$ and a generic DME pulse $p[v]$ is generated as

$$r_{\text{rp}}[u-v] = \mathrm{E}\{r[u]p^*[v]\}, \quad u,v = 0,\ldots,VN-1. \quad (10)$$

The generic pulse $p[v]$ is a DME pulse which is not modulated to a carrier frequency to be able to detect both interferers at +0.5 and −0.5 MHz offset to the

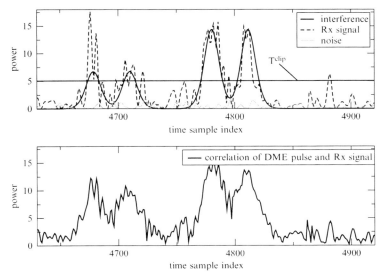

Fig. 4 Cross-correlation of Rx signal and generic DME pulse for detection of interference pulses, $V = 4$, SNR $= 5$ dB

center frequency of the OFDM system. The phase is set to 0 and the power is set to a fixed value, e.g. $P_0 = 1$.

In Fig. 4, the Rx signal, the interference signal, as well as the resulting cross-correlation function of the Rx signal and the generic DME pulse are given. In the cross-correlation function, maxima are observed at the same positions as peaks occur in the actual interference signal. For deriving the amplitude of the interference signal, the amplitude of the peaks in $r_{\text{rp}}[u - v]$ is adapted to the amplitude of the Rx signal averaged over four samples around the maximum to remove noise. Given the relatively low resolution of the interference signal induced by only four-times over-sampling, the accuracy of this estimation is sufficient for applying the threshold T^{clip} or T^{PB} to this estimation and clipping or blanking the samples in the Rx signal accordingly. Compared to clipping or pulse blanking based on the perfectly known interference signal, in most cases the same number of samples is clipped in the Rx signal. Hence, in the following, perfect pulse detection based on the actual interference signal is assumed.

4 Simulation Results

The performance of clipping and pulse blanking in combination with the optimal threshold determined in the previous section is evaluated in a realistic interference scenario retrieved from real DME channel assignments in a worst case area in Europe.

The victim OFDM Rx is positioned at Paris, Charles-de-Gaulles airport, at 15 km altitude to reproduce interference conditions at an en-route flight. The peak interference power originating from all surrounding DME/TACAN stations on the ground is determined via simple link budget calculations, taking into account free space loss and antenna patterns dependent on elevation angles. Typical interference conditions are observed when the OFDM system is operated at 995.5 MHz, for example. In the channels at ±500kHz offset, three TACAN stations with power and duty cycle as listed in Table 1 are observed.

In the simulations, the parameters of L-DACS 1 have been used [3]. For coding and modulation, a (133,171) convolutional code with rate 1/2 and QPSK modulation are applied, respectively. Propagation through the radio channel is modelled by an en-route channel model taking into account a strong line-of-sight path, Doppler frequencies of up to 1.25 MHz, and two delayed paths.

The performance of the OFDM system under the interference conditions from Table 1 is shown in Fig. 5 in terms of bit error rate (BER) versus SNR. The performance without interference is given as reference. With interference, but without applying any interference mitigation techniques, performance degrades signifi-

Table 1 FL en-route interference scenario

Station	Frequency	Interference power at victim Rx input	Pulse rate
TACAN	995 MHz	−67.9 dBm	3,600 ppps
OFDM	995.5 MHz		
TACAN	996 MHz	−74.0 dBm	3,600 ppps
TACAN	996 MHz	−90.3 dBm	3,600 ppps

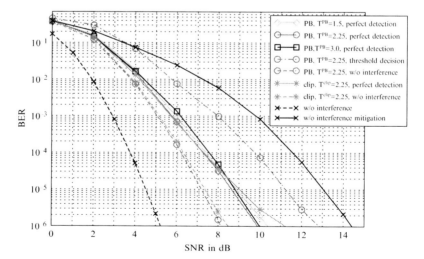

Fig. 5 BER performance with pulse blanking and clipping, ideal pulse detection, en-route interference scenario

cantly. When clipping or pulse blanking with the optimal threshold are applied, the impact of interference is reduced by about 4.3 dB at BER = 10^{-4}. Pulse blanking performs slightly better due to the slightly lower remaining interference power. For both mitigation techniques, a significant gap to the interference-free case remains. Additional simulations with applying pulse blanking but without adding the actual interference signal show that the minor part of this gap is explained by remaining interference power, whereas the impact on the useful OFDM signal preponderates. This gap cannot be reduced by improving the pulse blanking algorithm itself, but by compensating the impact of pulse blanking on the desired OFDM signal.

In addition, the influence of pulse detection has been investigated. When a simple threshold decision is employed, performance is very poor as too many samples of the OFDM signal and too few samples of the interference are clipped or blanked. With perfect pulse detection, an improvement of 2.5 dB at BER = 10^{-4} is achieved. The performance of the proposed correlation algorithm is close to the case of perfect detection.

5 Conclusions and Outlook

In this paper, the application of pulse blanking or clipping for the mitigation of pulsed interference in L-DACS 1 is addressed. For optimal performance, the two main issues are investigated, namely the optimisation of the threshold as well as the distinction of pulses from peaks in the desired signal. Simulations with a realistic interference scenario show that pulse blanking and clipping are capable of reducing the impact of interference by several dB. However, a significant gap to the interference-free case remains that is mainly explained by the impact of clipping or pulse blanking on the useful signal. To further improve performance, the combination with other approaches [3] and/or the compensation of the impact on the OFDM signal by exploiting the structure of the OFDM signal is proposed for future work.

References

1. M. Schnell, S. Brandes, S. Gligorevic, C.-H. Rokitansky, M. Ehammer, Th. Gräupl, C. Rihacek, and M. Sajatovic. B-AMC - Broadband Aeronautical Multi-carrier Communications. In *8th Integrated Communications, Navigation, and Surveillance (ICNS) Conference*, Bethesda, MD, USA, May 2008.
2. S. Brandes, I. Cosovic, and M. Schnell. Reduction of Out-of-Band Radiation in OFDM Systems by Insertion of Cancellation Carriers. *IEEE Communications Letters*, 10(6):420–422, June 2006.
3. M. Schnell, S. Brandes, S. Gligorevic, M. Walter, C. Rihacek, M. Sajatovic, and B. Haindl. Interference Mitigation for Broadband L-DACS. In *27th Digital Avionics Systems Conference (DASC)*, pages 2.B.2–1 – 2.B.2–12, St. Paul, MN, USA, October 2008.
4. Grace Xingxin Gao. DME/TACAN Interference and its Mitigation in L5/E5 Bands. In *ION Institute of Navigation Global Navigation Satellite Systems Conference*, Fort Worth, TX, USA, September 2007.

5. Sergey V. Zhidkov. Analysis and Comparison of Several Simple Impulsive Noise Mitigation Schemes for OFDM Receivers. *IEEE Transactions on Communications*, 56(1):5–9, January 2008.
6. P. Banelli and S. Cacopardi. Theoretical analysis and performance of OFDM signals in nonlinear AWGN channels. *IEEE Transactions on Communications*, 48(3):430–441, March 2000.

Part X
Demonstration

Ranging and Communications with Impulse Radio Ultrawideband

Álvaro Álvarez, David Blanco, Lorena de Celis, and Amparo Herrera

Abstract This paper describes the design and development of an Impulse Radio Ultrawideband demonstration platform, in order to evaluate and test the performance of UWB technologies for future advanced smart networking. The whole architecture is described, covering both hardware and software implementation, together with the Time of Arrival (ToA) ranging technique applied, and the achieved results. As a main output of the work performed, the first conclusion is that UWB technologies can be employed in order to perform accurate ranging estimations below 1 m, without heavy and costly signal processing on the nodes. The work described in this paper has been carried out in the scope of the FP7 Where Project.

1 Introduction

Ultrawideband (UWB) technologies have been deeply studied and analyzed during the last few years, as a candidate for new communication standards, for both high and low data rate systems. One of its main advantages it is that can be used at very low energy levels for either short-range high throughput or medium range low data rate communications by using a large portion of the radio spectrum. Focusing our attention in one of the UWB alternatives, Impulse Radio (IR) UWB, it is a technology that makes use of very short radio pulses, where the information can be coded following several approaches (amplitude, time or phase modulation). UWB has traditional applications in non cooperative radar imaging, while most recent applications target sensor data collection, precision locating and tracking applications, topics covered by the work explained in this paper.

Á. Álvarez (✉), D. Blanco, and L. de Celis
ACORDE TECHNOLOGIES S.A, Research and Development Department, Parque Científico y Tecnológico de Cantabria (PCTCAN), 39011, Santander, Cantabria, Spain
e-mail: {alvaro.alvarez, david.blanco, lorena.decelis}@acorde.com

A. Herrera
DICOM, University of Cantabria, Avd. Los Castros s/n, 39005, Santander, Cantabria, Spain
e-mail: amparo.herrera@unican.es

It is a common agreement between researchers and industry, to make use of the Ultra-Wideband (UWB) definition released by the Federal Communications Commission (FCC), which stands that UWB may be used to refer to any radio technology having bandwidth exceeding the lesser of 500 MHz or 20% of the arithmetic centre frequency.

Within the following chapters, an IR-UWB architecture proposal and a demonstration platform implementation for a Low Data Rate (LDR), location and tracking (LT) system, as well as a software application for its control and user interface is described. The demonstrator developed also shows that one of the main advantages of UWB communications is that the transmission is done without large interference with any other traditional 'narrow band' or wideband – continuous carrier wave signals using the same frequency band.

2 UWB System Description

The developed system is based on Impulse Radio Ultrawideband system (like radar systems), in order to allow the implementation of both localization and communication features on the same hardware platform. While the communication is done making use of an OOK modulation, it is possible to combine the modules employed for communication, with specifically developed modules to get the Time Of Arrival (ToA) information from the Impulse Radio signal, and estimate the distance (ranging) between a variety of UWB devices.

In order to test both features, a simple MAC with three commands has been developed (it was not on the scope of this research work to develop a strong access layer, but it was required in order to operate the platform). This MAC integrates a Ranging Request, Ranging Response and Data Indication protocol commands.

The hardware platform is on an ATMega microcontroller, which handles and interfaces with a FIFO, located between the RF front-end and the processing block, employed in order to unsynchronize the baseband and radio sub-systems.

Table 1 Main IR-UWB platform characteristics

	Communication + ranging UWB
UWB signal	Impulse radio
Center frequency	4 GHz
Bandwidth	2 GHz
Binary rate	13 Mbps(on air)/1 Mbps(interface)
Data modulation	OOK
Ranging precision	60 cm (5 cm with post-processing)
Coverage range	NLOS < 12 m, LOS < 30 m
Power spectral density	−25 dBm/MHz
Antenna connection	Single ended
Power consumption	1.25 W

3 UWB Transceiver System Description

The UWB platform is a full transceiver, this means that comprises both transmitter and receiver blocks on the same board. Both sub-systems are located in parallel and make use of the same processing and baseband resources (double bus and FIFO memory).

On the transmitter side, the following requirements were taken into account:
While the main requirements on the receiver side were:

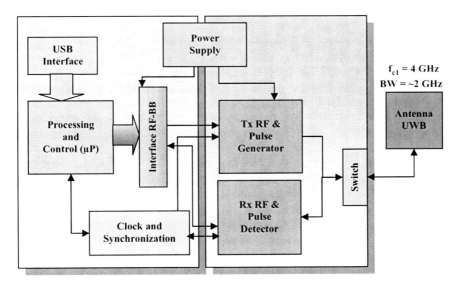

Fig. 1 UWB transceiver architecture

Table 2 Transmitter parameters

Host interface	USB
Modulation type	OOK
Packet size	1 Kbyte
Binary rate	1 Mbps (due to USB limitation)
PSD (approx)	−25 dBm/MHz
Operation bandwidth	3–5 GHz

Table 3 Receiver parameters

Host interface	USB
Rx chain gain	>50 dB
Voltage threshold for pulse detection	300 Mv
Synchronization time	<100 µs
Operation bandwidth	3–5 GHz

3.1 Hardware Architecture

The UWB platform system is divided into two main sub-systems, the radio module, in charge of the UWB radio signal (between 3 and 5 GHz), and the post-processing unit, in charge of the baseband signal processing and communication/ranging features. The whole platform is based on a COTS (Commercial-off-the-shelf) solution, and no proprietary ASIC solution has been implemented.

The radio frequency system is separated into two different branches, the first one for the transmitter part, including a baseband amplifier, a pulse generator, and filtering and amplification stages, and the second for the receiver chain, with several amplification and filtering stages, a pulse detector and a pulse inverter circuit. Both chains are combined and multiplexed in time (no Tx/Rx is being done at the same time) with a RF switch.

Interfacing with this radio module, is the antenna, for the wireless link, and the baseband block. This block contains a synchronization module, a FIFO memory for data transmission buffering and data reception processing, and the core of the system is based on a ATMega microcontroller, able to monitor the status of the whole system, transmit and receive data, translate the commands to/from the control system.

The platform is equipped with a USB Interface, able to provide up to 1 Mbps data throughput, while on the wireless side, it transmits a UWB signal from 3 to 5 GHz, making use of filtered Gaussian monopulses, at 13 Mpps (Mpulsespersecond).

One of the most important topics took into account during the design and development for this prototype, was that it was not a specific regulation for the UWB devices in Europe, therefore, these devices are intended just for prototype test and technology research, and currently, they cannot be used for commercial purposes.

The main characteristics of the wireless link are:

The whole system needs one single power supply of 7.5 V and takes up to 260 mA. Internally there are a series of regulators which provides 5, 3.6 and 3 V where required. The total DC power consumption is 1.95 W, while the effective power consumption is 1.3 W.

Table 4 Signal properties

Standard	None (proprietary solution from PHY, MAC and application topology)
Center frequency	4 GHz
Bandwidth	2 GHz @ −6 dB PSD
Data rate	Nominal peak value of 13 Mbps
Modulation	OOK
Mean transmit power	2 dBm (peak value)
Rx sensitivity	Theoretically −68 dBm

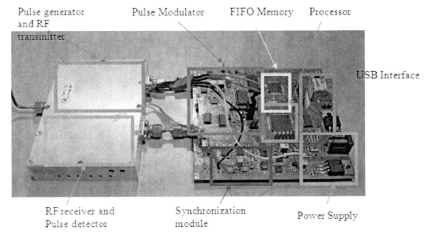

Fig. 2 Subsystems and modules of the transceiver prototype

3.2 Baseband and Control Software

The software employed for this application is split in two chapters:

3.2.1 Embedded Microprocessor Software

This software is written in C language, and its purpose is to perform the synchronization of the received data, as well as the data packet transfer between the USB and the radio module.

The code is complied making use of the AVR studio application, and it follows a proprietary solution for this UWB platform version.

3.2.2 Application and GUI Software

The Graphical User Interface developed for this platform is focused in shows its main capability, range acquisition. Taking this issue into account the application is designed to connect two devices, base stations, and show the distance with a mobile device.

The distance displayed is obtained with several range measurements. The number of samples involved in the calculation can be decided by the user just like the time of each request. The user can request only one measurement (with several samples) or he can start a periodic process to request several range measurements.

Other important features offer by the application is the option to calibrate each system (base station-mobile device). The platform can work with only one base

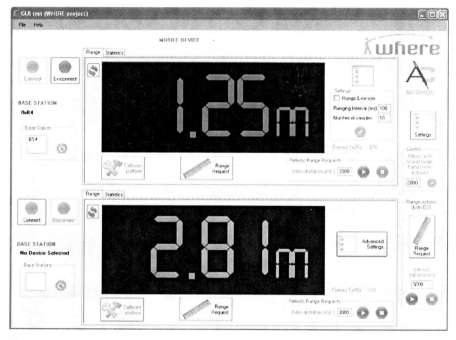

Fig. 3 GUI interface

station connected or with two of them. In case two base stations are connected the user can request the distance to the mobile device from both of the base stations. This process can be request once or several times using a periodic way.

In order to provide an easier way to process the data in a future or to obtain some graphics to show the results, a tab has been designed to show and save all the data collected (graphic, excel files, . . .).

4 Ranging and Positioning Technique

Due to a enhanced synchronization module, it is possible to get from the hardware, the Time Of Arrival (ToA) of the Impulse Radio signal, therefore to do it, it is required to synchronize at least two UWB devices (multiple ranging can be done by peer to peer connections among the different units).

On each peer to peer connection, those two different devices take different roles. First one acts as a base station, which is the device that starts the range procedure and estimated the distance, and the other works as a mobile unit or blind node. The mobile unit only has to response as fast as possible to the base station, and then, via the synchronization between the transmitted and received signal, it is possible to compute the distance between both devices.

Ranging and Communications with Impulse Radio Ultrawideband 393

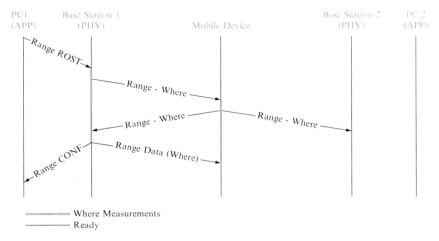

Fig. 4 Ranging protocol

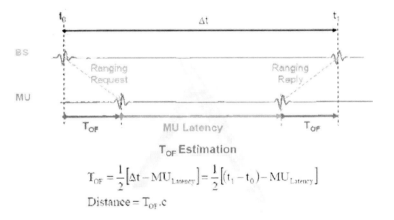

Fig. 5 Distance estimation through ToA

This platform uses the time between primitives to estimate the distances between to devices. A high level description of the process is depicted in Fig. 4.

When a range request is required from the Base Station, a range request primitive is sent. The mobile device received this packet (Range-Where) and response immediately. The base station is in charge to send the distance estimated to the application layer to show the data and generate a response frame for the mobile device to give it the distance estimated.

The distance estimation procedure is based on Two Way Ranging (TWR) combined with Time of Arrival (ToA), which leads to an average precision of 2 ns (60 cm). Moreover, basic signal processing allows up to 20 cm precision, advanced techniques (not yet implemented) shows up to 5 cm.

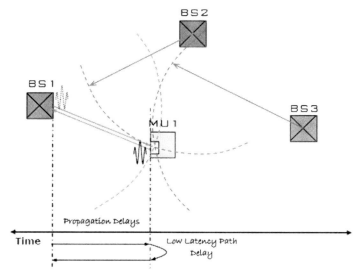

Fig. 6 Trilateration procedure

The implementation of the algorithm makes use of the information got by each individual base station, and then collaborative processing via trilateration:

5 Results

The test and evaluation of the platform for both communications and ranging features have been carried out successfully. These results show:

- Data Rates up to 13 Mbps (on air) with simple OOK modulation
- Range up to 30 m for indoor environments with LOS
- 60 cm precision on single shot measurements
- 10 cm precision with minor signal processing
- Communication and ranging capabilities with the same hardware platform

6 Conclusions

An IR-UWB demonstration platform, combining both communications and ranging capabilities has been designed, developed and tested, showing successful results, with distance estimations in indoor environments below 60 cm. The paper describes the complete specifications of the system, taking into account both hardware and software developments, the ranging or position technique implemented, as well as other important parameters like size, power consumption or regulatory and standardization status of the developed technology.

The work described in this paper was partially funded by the FP7 Where project.

References

1. http://hraunfoss.fcc.gov/edocs_public/attachmatch/FCC-02-48A1.pdf
2. Where Project Deliverable, D5.1 "Where adapted Platforms Description", http://www.ict-where.eu, November 2008
3. Maria-Gabriella Di Benedetto and Guerino Giancola, Understanding Ultra Wide Band Radio Fundamentals, Prentice-Hall PTR, June 2004
4. Where Project Deliverable "Survey on localisation in communication networks" 2008
5. PULSERS II deliverable D3a5.3 "LDR-LT system verification platform report – PHY to Application layers", August 2008
6. Huseyin Arslan, Zhi Ning Chen, Maria-Gabriella Di Benedetto, Ultra Wideband Wireless Communication, John Wiley & Sons, Inc., Hoboken, New Jersey, September 2006
7. Aaron Michael Orndorff, Transceiver Design for Ultra-Wideband Communications, thesis of Virginia Polytechnic Institute and State University, 2005

Generic SDR Platform Used for Multi-Carrier Aided Localization

Igor Arambasic, Javier Casajus, and Ivana Raos

Abstract This paper describes the Software Defined Radio (SDR) platform that is being developed and implemented in the scope of the EU seventh Framework WHERE Project. The platform is based on one full-length PCI carrier board populated with one DSP module, one dual high-speed ADC/DAC module and one module corresponding to IF/RF front-end. It is equipped with two omni-directional antennas working at 2.4 GHz with 100 MHz bandwidth. The platform is not restrained only to the PC tower box since its mobility is guaranteed with small portable PCI bus extension chassis supporting PCMCIA laptop connections. It is intended for exploring the localization possibilities of different multi-carrier interfaces.

1 Introduction

The presented SDR platform is developed using hardware produced by Sundance [1]. It is based on one full-length PCI carrier board (SMT310Q) which provides access to four, industry standards, Texas Instruments Modules (TIMs) format. The benefits of system based on this PCI carrier board approach include low-cost, scalable system configuration, flexible task allocation and above all expandable system which allows future expansions. In fact, the SDR platform used in WHERE project [2] is an expansion of SMT8036 SDR development kit which has been used for communication research and system demonstration purposes at the Universidad Politecnica de Madrid.

The SMT8036 kit includes C64xx-based module (SMT365) combined with a dual high-speed ADC/DAC module (SMT370), both plugged on a SMT310Q carrier board. The SDR platform implemented in WHERE includes an additional module which is the IF/RF front-end (SMT349). This module is delivered together with two omni-directional antennas working at 2.4 GHz with 100 MHz bandwidth.

I. Arambasic (✉), J. Casajus, and I. Raos
Universidad Politecnica de Madrid (UPM), Ciudad Univeristaria s/n, 28040 Madrid, Spain
e-mail: {igor, javier, ivana}@gaps.ssr.upm.es

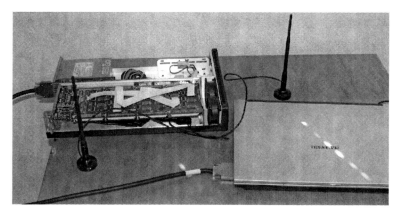

Fig. 1 The carrier board is connected to laptop

Since WHERE SDR platform is imagined as dynamic and mobile system which is later to be used for measurements campaigns, the carrier board should not be restrained to large PC tower boxes. For this reason a PCI extension bus chassis is acquired. This chassis includes its own cooling fans and power supply. It can be connected to the PCMCIA slot of the laptop (as seen in Fig. 1) and as such offers flexibility and mobility to the system.

2 Platform Overview

The block diagram of WHERE SDR platform is depicted in Fig. 2. The SMT310Q carrier board is connected to the host over PCI bus. The communication with the host computer is done only via SMT365 located in the root TIM slot. In fact the entire platform is controlled through this DSP as ADC/DAC and RF module are both programmed by DSP using comport connections on the carrier board. The connection between the SMT370, SMT349 and corresponding antennas is done via coaxial cables and MMBX connectors while data transmission between the DSP and ADC/DAC converters is achieved with Sundance High Speed Bus (SHB) protocol. The SHB connectors give 32-bit communication interfaces which operate at a fixed clock rate of 100 MHz and can thus support traffic between the modules of up to 400 MBytes/s.

2.1 SMT310Q – PCI Carrier Board

SMT310Q has an 'on-board' XDS-510 compatible JTAG controller which allows Code Composer Studio and 3L Diamond applications to be used for debugging or

Generic SDR Platform Used for Multi-Carrier Aided Localization

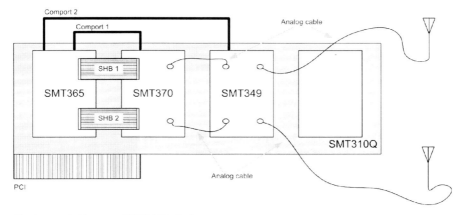

Fig. 2 Block diagram of WHERE platform

uploading the software to the modules. The interconnection of modules is done using comports which support communications at 20 MBytes/s. The external comports cables are not necessary since the board supports software-configurable routing matrix between the TIM modules. The board also includes on-board V3 PCI bridge chip which provides DMA access, mailbox events, and interrupts control. The SMT310Q is connected to the host PC through standard PCI bus which runs at 33 MHz. Hence, the communication between the host and root TIM site is implemented at burst speed in the range of 60–130 MBytes/s. This fact labels PCI access as a possible bottle neck for real time presentations of measurements on the host screen.

2.2 SMT365 – DSP Module

The presented SDR platform is based on Sundance SMT365 module at the root TIM space. The main components of this single width TIM module are one Texas Instruments DSP and one Xilinx FPGA Virtex-II device with the following characteristics:

- TMS320C6416 fixed-point processor running at 600 MHz
- Xilinx Virtex-II device XCV2000
- Six 20 MBytes/s Comports
- 8 MBytes of ZBTRAM (133 MHz)
- 8 MByte Flash ROM for boot code and FPGA programming
- High bandwidth data input/output via 2 SHBs

2.3 SMT370 – ADC/DAC Module

The second TIM slot is filled with SMT370 which is a single width TIM module that offers dual high-speed ADC/DAC processing characterized with the following features:

- Two 14-bit ADCs (AD6645-105) sampling at up to 105 MHz, DC coupled
- Dual 16-bit TxDAC (AD9777) sampling at up to 160 MHz without interpolation and up to 400 MHz with interpolation
- Xilinx Virtex-II FPGA (XC2V1000-6)
- Two SHB connectors
- Two 20 MBytes/s communication ports
- User defined pins for external connections
- 50-Ohm analogue inputs and outputs, external triggers and clocks via MMBX connectors

The 14-bit ADC converter is able to sample from 30 MHz up to 105 MHz which makes it suitable for multi-carrier 3G applications. It also includes the option to reduce sampling rate by performing decimation on the data flow. The 16-bit dual interpolating DAC consists of two data channels that can be operated independently or coupled to form a complex. The chip features include a selectable 2x/4x/8x interpolation filter, an Fs/2, Fs/4 or Fs/8 digital quadrature modulation with image rejection, a direct IF mode and a programmable channel gain and offset control.

This module is also populated with 32 MBytes NtSRAM memory which allows a 'pattern generator' function where a pattern (or periodic frame) is stored into the memory, read back continuously and send to the DAC. This configuration allows to board to work as a periodic generator in stand-alone mode. In this way, SMT370 works as a loadable Arbitrary Waveform Generator (AWG) and this mode is used for demonstration purposes. The board offers the possibility to output data in either two's complement or binary format. It is also possible to output a 16-bit counter on each SHB half for system testing purpose as it then becomes easier to detect any missing data.

The communication between SMT370 and SMT365 is done at two levels. The first level corresponds to the programming and control of SMT370 done via comports while second level refers to transmission of data between the modules which is done at higher speed via SHB interface.

2.4 SMT49 – IF/RF Front-end Module

SDR platform of WHERE is completed with the introduction of SMT349, corresponding to IF/RF front-end module located in third TIM slot. This single width TIM completes the transmission and reception paths of digital radio system and its main features include:

- Xilinx Virtex-II FPGA (XC2V1000-4)
- Two SHB connectors
- Two comports
- 50-Ohm analogue RF/IF inputs and outputs and external clock via MMBX connectors

The analogue section of SMT349 includes two RF transceiver modules, operating in the 2.4 GHz ISM band. The up conversion from the 70 MHz IF to the 2.4 GHz is done in two stages. The first stage converts the first IF to 374 MHz, while the second converts it to the RF frequency. Down-conversion approach follows the opposite procedure converting RF frequency to 374 MHz and than in the second stage to 70 MHz. The RF centre frequency of both transceivers is controlled by one synthesizer which is set by the FPGA. The FPGA also controls the AGC, TX power and TX/RX switch for each of the two IF/RF sections.

Technical specifications for the IF/RF part are:

- Input signal filtered by first IF tuned to 70 ± 8 MHz
- 70 MHz BPF characteristics
- -1 dB $\rightarrow +/- 8$ MHz
- -20 dB $\rightarrow +/- 12$ MHz
- -50 dB $\rightarrow +/- 15$ MHz
- 70 MHz IF is converted to a second IF of 374 MHz
- RF output signal is in the 2.4–2.5 GHz ISM band
- Transmitter gain control done using 31 dB at 1 dB steps
- RF input signal is in the 2.4–2.5 GHz ISM band
- Receiver gain control done using 2×31 dB at 1 dB steps
- IF output signal is at 70 ± 8 MHz

3 Software Features

Due to complex nature of software controlling the described SDR platform, the logical way would be to classify it into various sections depending on the interface and required debugging information. For example, the GUI is done in Labview since it offers flexible interface and rapid memory access favorable for data graphic presentation. More on, four wireless interfaces, whose localization skills are to be explored, are developed and tested in Matlab before passing them on to C and finally the last step is to implement the interfaces inside SMT365 module which is achieved by means of 3L Diamond IDE and Code Composer Studio.

The execution of entire software package begins with program loader developed in Visual C, that establishes a contact with the carrier board, and loads the application into the root DSP. The application is previously developed in C and debugged under 3L Diamond environment directly on SMT365 root module. The purpose of this application is to start and control the entire hardware initialization process and consequently prepare the platform for corresponding signal processing.

The platform initialization includes DSP boot-up, configuration of ADC/DAC of the SMT370, and of the IF/RF parameters of SMT349. The configuration is done from the root DSP through the corresponding comport connections available on the SMT310Q board. After the initial parameters are set, a signal pattern is generated at DSP and transmitted to SMT370 via SHB connection. It is than stored in the local memory of SMT370 module. Since the module is configured as a periodic generator, this pattern is read back continuously and sent to the DAC.

At this point the loader execution on the host side is terminated, and the GUI in Labview takes over the control. The communication between Labview and DSP on the carrier board is done by means of dll library, also developed in Visual C. This library is in fact a two-way interface as DSP receives the expected number of samples, sampling frequency etc. from the Labview, while it transmits the received AD samples, calculated location parameters, control status etc. in the opposite direction.

This analog output of DAC converter is connected via MMBX-MMBX coaxial cable with the SMT349 module where FPGA than routes the data through the transmission RF chain. On the receiver side (which can be implemented on the same carrier board) the RF signal received by the antenna is guided through RF chain via on-board FPGA. The base-band signal is than passed with via another MMBX-MMBX coaxial cable to ADC of SMT370 and is afterwards routed to DSP via second SHB connection.

4 Parameter Measurement Procedure

The main objective of the measurement scenario is to provide real data which is to be used in WHERE project for different signal exploration, namely hybrid data fusion, tracking, cooperative positioning and estimation of location dependent channel information. The data will be available under the form of empirical likelihood functions for location dependent magnitudes, namely:

- Received signal power
- Time of arrival
- Angle of arrival

Thus, the empirical likelihoods $p(\theta|x, y)$ will be measured, where θ is any of the above magnitudes and (x, y) is the position on a rectangular grid within the measurement space. Once the experimental setup is defined the likelihoods associated to several specific air interfaces will be estimated. In this case the kind of modulation and the bandwidth impose restriction on the accuracy of the measurement that combine with the environment and variability to yield the final likelihoods. Two multicarrier interfaces that are implemented are: 802.11 (OFDM in general), 802.11g (WiFi). In addition two more wireless interfaces, namely: 802.15.4 (ZigBee) and 802.15.4a (CSS – Chirp Spread Spectrum) are also supported.

Because the resulting data are to be used for the development and analysis of cooperative positioning techniques, the likelihoods are measured pairwise. That is

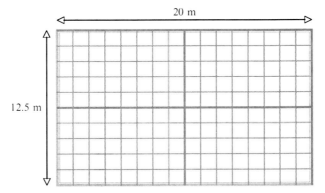

Fig. 3 Measurement grid where transmitter positions are shown in blue and receiver positions are in red

to say, given a grid position (x, y) there is a different empirical likelihood for every possible position of the transmitter on the same grid. In fact the measured likelihoods are $p(\theta|\mathbf{r}, \mathbf{r}_i)$, where $\mathbf{r}$ and $\mathbf{r}_i$ are the position vector of a generic grid point and the position vector of the transmitter respectively.

Assuming that the measurement space is a rectangular area, as shown in Fig. 3, with sides of 12.5 and 20 m respectively, and assuming a grid of step 1 m, the number of grid points in the space is very large (more than 31,000). For this reason the number of measured pairs is reduced. However, in order to be able to use the obtained measurements for the cooperative positioning we will restrict the position of the transmitters to be on grid point situated on two sides of the perimeter (major and minor sides) and along the two middle lines of the area. This way, for the example shown on Fig. 3, the resulting number of pairs is reduced to 5,625.

At the time the measurements are taken, the plan is to use the optimal measurement procedures and ad-hoc methods specifically developed for signals supported by the analyzed wireless interface standard. For optimal procedures the synchronization is assumed to be perfect. In this case the two platforms are used, one as a transmitter and another as receiver, and the synchronization between the platforms is achieved via cable.

5 Conclusions

The described SDR platform used in the scope of WHERE project is based on the commercial hardware developed by Sundance. As such it is not intended for ranging purposes and the expected precision of localization results is limited. However, the advantage is the software flexibility the platform offers as well as the scalable hardware system configuration which can be easily expanded if necessary. The platform is implemented for exploring the localization possibilities of two multicarrier wireless interfaces, namely: 802.11 (OFDM in general), 802.11g (WiFi). Two additional wireless interfaces are also explored: 802.15.4 (ZigBee) and 802.15.4a (CSS – Chirp Spread Spectrum). The measurements will take place in short range LOS situation in the typical indoor office of around 250 m^2.

Acknowledgement This work has been performed in the framework of the ICT project ICT-217033 WHERE, which is partly funded by the European Union.

References

1. Sundance Multiprocessor Technology, www.sundace.com
2. WHERE project, http://www.ict-where.eu/

Printed in the United Kingdom by
Lightning Source UK Ltd., Milton Keynes
138723UK00001BA/53/P